CHARLES BABBAGE

ON THE PRINCIPLES AND
DEVELOPMENT
OF THE CALCULATOR

This machine is held in the Science Museum, London S.W. 7. It was made from the drawings of the Analytical Engine left by Charles Babbage and was constructed partly by his son and partly by the firm of R. W. Munro. It is the "mill" and printing mechanism, capable of performing the four arithmetical operations and printing the result to 29 places. Illustration British Crown Copyright, Science Museum, London.

CHARLES BABBAGE
ON THE PRINCIPLES AND DEVELOPMENT OF THE CALCULATOR

and Other Seminal Writings
by Charles Babbage and Others

Edited and with an Introduction by
PHILIP MORRISON
and
EMILY MORRISON

DOVER PUBLICATIONS, INC.
New York

Published simultaneously in Canada by McClelland and Stewart, Ltd.

Published in the United Kingdom by Constable and Company Limited, 10 Orange Street, London, W.C.2

BIBLIOGRAPHICAL NOTE

This new Dover edition, first published in 1961, contains the following:

Selected unabridged chapters from *Passages from the Life of a Philosopher* by Charles Babbage, originally published by Longman, Green, Longman, Roberts, & Green in 1864.

Selected unabridged essays from *Babbage's Calculating Engines,* edited by Henry Prevost Babbage and originally published by E. and F. N. Spon in 1889.

"On the Age of Strata, as Inferred from the Rings of Trees embedded in Them" by Charles Babbage, in unabridged republication as originally published in *The Ninth Bridgewater Treatise* in 1837.

A new Introduction, especially written for this edition, by Philip Morrison and Emily Morrison.

The editors and publisher are grateful to Mr. Stephen A. McCarthy, Director of the Cornell University Library, who kindly made available a copy of the original edition of *Babbage's Calculating Engines* for editorial purposes.

International Standard Book Number: 0-486-24691-4

Manufactured in the United States of America

Dover Publications, Inc.
31 East 2nd Street
Mineola, N.Y. 11501.

ACKNOWLEDGMENTS

We should like to thank Director Stephen S. McCarthy of the Cornell University Library for lending *Babbage's Calculating Engines* to Dover Publications, Inc.; Mr. J. C. Cain and Dr. H. R. Calvert of the Science Museum, London S.W.7, for information about the Museum's collection; Professor F. E. Mineka of Cornell University for supplying an interesting anecdote; Professor W. F. Willcox of Cornell University for reminiscences about Hollerith; and Jane S. Wilson of Ithaca, New York, for bibliographical assistance.

CONTENTS

PART III

APPENDIX OF MISCELLANEOUS PAPERS

ILLUSTRATIONS

INTRODUCTION

I

CHARLES BABBAGE is a name known fairly widely today; in his own time the value of his work was recognized by few of his contemporaries, and he was held a crackpot by his London neighbors. His name has emerged from obscurity in the past generation because it has become increasingly clear that he was a man far ahead of his time. That knowing and critical figure, J. M. Keynes, his students still recall, admired Babbage highly. Today there are applied mathematicians everywhere who share his passion for developing calculating machines; their technical resources have vastly improved, but the fundamental principles of design remain very similar. The British magazine *Nature* entitled a discussion of one of the first large American calculators "Babbage's Dream Comes True," and described that Harvard relay computer, Mark I, as a realization of Babbage's project in principle, but with the benefit of twentieth-century mechanical engineering and mass production methods for its physical form. A few years ago B. V. Bowden in his excellent book on calculating machines, *Faster than Thought*, said that Babbage enunciated the principles on which all modern computing machines are based. Unquestioned pioneer in the field of large-scale mathematical machines, Babbage was, in a sense, the unheralded prophet of the even newer field now known as operational research, foreshadowed in his book *Economy of Manufactures and Machinery*. This study of scientific manufacturing processes of all kinds, written as a by-product of his interest in mathematical machines, was in fact the only major undertaking he actually completed. He was ahead of his contemporaries in still a third way: he made a determined campaign for Government subsidy of scientific research and education at a time when research was still, to a large extent, a gentleman's hobby.

II. LIFE

BORN IN Devonshire in 1792, Charles Babbage was the son of a banker who later left him a considerable fortune. Because of poor health, he was privately educated until he entered Trinity College at Cambridge in 1810. He was already passionately fond of mathematics before coming to college and was discouraged to find then that he knew more than his tutor.   He soon had a great circle of friends, or rather, a number of circles—chess- and whist-playing groups, fellow members of the Ghost Club, and boating companions, all in addition to mathematical colleagues.   Of the latter, his most intimate friends were the younger Herschel (later Sir John) and George Peacock (later Dean of Ely).   The three undergraduates entered into a compact that they would "do their best to leave the world wiser than they found it."   In 1812, as their first step toward the achievement of this goal, the three, together with several others, founded the Analytical Society, hired rooms for it, read mathematical papers, even published transactions.   Babbage, Herschel, and Peacock translated Lacroix's *Differential and Integral Calculus* and published two volumes of examples (see page 24).   In spite of considerable opposition, the Society fought valiantly to put "English mathematicians on an equal basis with their Continental rivals," and actually had a profound effect on the future development of English mathematics.   Babbage believed that he was certain to be beaten in the tripos examinations by both Herschel and Peacock, and preferring to be first at Peterhouse rather than third at Trinity, he transferred in his third year.   In fact, he stood first in Peterhouse in 1814, and received his M.A. in 1817.   For about ten years after his graduation he published a variety of mathematical and physical papers, mostly on the calculus of functions, but also including one on Euler's study of the knight's move in chess and one on barometric altitude measurements.

Babbage, Herschel, and Peacock continued to be friends after they left school.   Each in his own way lived up to their joint compact, though their careers were very different.   Peacock devoted himself to mathematics and astronomy until finally he decided to join the ministry.   He took his D.D. in 1839, and shortly thereafter became Dean of Ely, a post he filled with great vigor and success.   Herschel, after a brief apprenticeship at law, decided to follow his great father into astronomy.   As a crowning achievement, to supplement his father's work on stars of the Northern Hemisphere, Herschel went to the Cape of Good Hope in 1833, and in four years completed his observations of the southern stars.   After his triumphant return to England, his main work was compiling his great catalogues of nebulae and stars.   He was knighted by the Crown, served as Master of the Mint, and avoided all scientific feuds; his biographers all report that his was a life full of serenity and innocence.   In contrast, Babbage published at thirty his "Observations on the Application of Machinery to the Computation of Mathematical Tables," and was received with general acclaim and presented the first award of the Gold Medal ever given by the Astronomical Society—but spent the rest of his life fruitlessly trying to bring his machines to completion. He was led by his broad interests into many byways, from a vigorous campaign against the policies of the Royal Society to the study of ciphers, and from speculative geology to the design of tools for lathes and shapers, but always his work centered on his beloved engines. His career was a long series of disappointments, and to friends who visited him in 1861 he said that he had never had a happy day in his life and spoke "as though he hated mankind in general, Englishmen in particular, and the English government and organ-grinders most of all . . . in truth Mr. Babbage was a mathematical Timon."

Actually Charles Babbage was a most social and gregarious fellow, with a considerable sense of humor, as one can see in his autobiography. Charles Darwin wrote, "I used to call pretty often on Babbage and regularly attended his famous evening parties." Babbage was an enthusiastic conference man, instrumental in founding the Astronomical Society (1820), the British Association for the Advancement of Science (1831), and the Statistical Society of London (1834). Darwin also recalled a brilliant dinner at his brother's house at which even Babbage who "liked to talk" was outdone by Thomas Carlyle; and an Edinburgh professor who was asked to dinner by Babbage reported that "it was with the greatest difficulty that I escaped from him at two in the morning after a most delightful evening." Encouraged to travel for his health, he made many trips to the continent of Europe, and was equally interested in meeting members of the aristocracy, fellow mathematicians, and skilled mechanics. He was a friend of Laplace and of Alexander Humboldt and knew Poisson, Fourier, and Biot. "It is always advantageous," he advised, "for a traveller to carry with him anything of use in science or in art if it is of a portable nature, and still more so if it has also the advantage of novelty." Among his most useful objects of this sort were some gold buttons stamped by steel dies with ruled parallel lines 4/10,000 of an inch apart, produced by a designer of machine tools named Sir John Barton, who was Comptroller of the Mint. The rainbow patterns playing on these small diffraction gratings, which indeed they were, provided a splendid opening gambit in conversations with strangers.

III. MACHINES

BABBAGE's transformation from a cheerful young man into a bitter old one was to a large extent the result of his devotion to his mathematical machines.   He has provided us with two versions of the origin of his ideas about machines, but the one written in 1822 seems more plausible than the other, which appeared in his autobiography some forty years later.   According to the first story, Herschel brought in some calculations done by computers for the Astronomical Society. In the course of their tedious checking, Herschel and Babbage found a number of errors, and at one point Babbage said "I wish to God these calculations had been executed by steam." "It is quite possible," remarked Herschel.   From this chance conversation came the obsession that was to rule Babbage for the rest of his life.   The more he thought about it, the more convinced he became that it was possible to make machinery to compute by successive differences and set type for mathematical tables.   He set down a rough outline of his first idea, and made a small model consisting of 96 wheels and 24 axes, which he later reduced to 18 wheels and 3 axes.   In 1822, in addition to his above-mentioned original note published by the Astronomical Society, he wrote an article "On the Theoretical Principles of the Machinery for Calculating Tables" for *Brewster's Journal of Science*, and a letter on the general subject to the President of the Royal Society, Sir Humphry Davy.   In this letter, Babbage pointed out the advantages such a machine would have for the Government in producing the lengthy tables for navigation and astronomy, and proposed to construct a machine on an enlarged scale for the Government's use. There had been machines since the time of Pascal for carrying out single arithmetical operations, but they afforded little saving of time or security against mistakes.   The Astronomical Society received Babbage's proposal with the highest enthusiasm, and the Royal Society reported favorably on his project for building what he called a Difference Engine.   In an interview held in 1823 between Babbage and the Chancellor of the Exchequer, a rather vague verbal agreement was made whereby the Government would grant funds for the enterprise which was expected to take three years.   Work proceeded actively for four years although Babbage was constantly having new ideas about the machine and scrapping all that had been done before.

In 1827 Babbage went abroad for a year on the advice of his physician, and during this period made a study of foreign workshops and factories to supplement his considerable familiarity with British manufacturing processes.   He later used this information in his book *Economy of Manufactures and Machinery*, published in 1832, about which we shall have more to say later.   While still abroad, he learned that he was to be appointed Lucasian Professor at Cambridge, the chair once held by Isaac Newton. Although he hesitated because of his work on the Difference Engine, he decided to accept and hold the position for a few years. He remarks in his autobiography that this was the only honor he received in his own country. He resigned in 1839 to devote himself completely to his machines, even though during his entire term of office he neither resided at the college nor taught there. His income as Lucasian Professor was between eighty and ninety pounds a year!

Upon returning to London in 1828, Babbage made a new application for funds to the Treasury. The Royal Society again reported favorably, the Duke of Wellington inspected the model, and once more the Government granted liberal funds for the work, and decided to build a fireproof building and workshop on land leased next to his home. In 1833, when arrangements were made for moving the engine and the work to the new shops, Babbage and his excellent engineer Clement reached a crisis. There had for years been differences about the various delays in salary payments and Clement refused to continue work in the new buildings without new and expensive arrangements. At this point Clement abruptly stopped work and dismissed those of his men who were working on the Babbage job. After months of dispute he allowed the drawings and the parts of the engine to be moved to the new building, but he was legal owner of his tools and retained all those tools which had been so laboriously built in his shop at Babbage's and the Government's expense over six or eight years of effort.

Twelve months after work on the Difference Engine had stopped, Babbage thought of an entirely new principle for a machine which would wholly supersede and transcend the Difference Engine. The Analytical Engine, as he called it, would have far more extensive powers, more rapid operation, and yet a simpler mode of construction than his original design. In 1834 Babbage requested an interview with the First Lord of the Treasury to explain his new idea and get an official decision on whether to continue and complete the original Difference Engine, or to suspend work on it until the new idea could be further developed. For eight years he pressed for an answer either from the First Lord of the Treasury or from the Chancellor of the Exchequer, with whom he corresponded at some length. At last he was advised that the Chancellor and the Prime Minister, Sir Robert Peel, concluded that the Government must abandon the project because of the expense involved. The Government had already spent £17,000 and Babbage had contributed a comparable amount from his private fortune. The parts of the machine already completed and the drawings for the whole machine were deposited in the Museum of King's College, London, were shown at the International Exhibition of 1862, and were eventually delivered to the South Kensington Museum, where they are now.   The part on exhibit is in working order and has recently been taken apart, thoroughly cleaned, and reassembled so that an exact copy could be made for the International Business Machine Corporation's museum.   The copy was built by the firm of R. W. Munro, who built the "mill" for Babbage's son, see frontispiece.

Much embittered by the Government's withdrawal, Babbage turned his attention to the development of the Analytical Engine, and maintained a staff of draftsmen and workmen to work on drawings and experimental machinery for future construction.   As always, Babbage would start work on a model and then abandon it in an unfinished state to start work on a new one.   In 1848, after working for several years on the Analytical Engine, he decided to make a complete set of drawings for a second Difference Engine, which would include all the improvements and simplifications suggested by his work on the Analytical Engine.   He again offered to give the completed drawings and notations to the Government provided they would build it, and again his offer was turned down by the Chancellor of the Exchequer, whom Babbage termed "the Herostratus of Science, [who] if he escape oblivion, will be linked with the destroyer of the Ephesian Temple."

IV. DIFFERENCE ENGINE

In 1834 the *Edinburgh Review* had published an account of the principles of Babbage's Difference Engine.   Inspired by this article, a well-to-do Stockholm printer, George Scheutz, undertook to make a machine of his own.   With a little belated financial assistance from his own Government and from members of the Swedish Academy, Scheutz and his son Edward completed, after many years of work, a difference engine of their own, which Scheutz brought to England for exhibition in 1854.   Somewhat to Scheutz' surprise, Babbage did everything in his power to help, and in a speech before the Royal Society recommended Scheutz and his son for one of the Society's medals.   Babbage's own son Henry used the Scheutz machine to demonstrate his father's pet system of "mechanical notation."   The Swedish machine won a Gold Medal in Paris in 1855.   Babbage and his son had prepared a series of drawings to accompany the machine and explain its operation. The Scheutz machine was bought for $5,000 in 1856 by an American businessman for the Dudley Observatory in Albany, New York, whose first building was to be dedicated that same year. G. W. Hough, first Director of the Dudley Observatory, was certainly a man to appreciate the machine, for he himself developed, with much effort, both a printing barometer and a very early form of recording chronograph for daily use at the Observatory. The Scheutz machine computed four orders of differences and displayed its results by setting type to eight decimal places. An engraving of the machine standing on its four fluted hardwood legs appears as plate 4 of Volume I of the *Annals of the Dudley Observatory*. The machine was used in Albany for many years in computing Ephemerides and various correction tables, and exists today in a private collection in Chicago. In 1863 an exact copy was made for the British Government and used by W. Farr for the computations for the *English Tables of Lifetimes, Annuities, and Premiums* published by the Registrar-General.

Calculating machines were by no means new; they had been devised by such luminaries as Napier, Pascal, and Leibnitz. Their devices were meant to serve as "desk calculators," like the wonderful and ubiquitous machines whirring on so many desks today, or like the humble slide rule. But detailed mechanical realization of the designer's ideas had not yet been skillful enough to make the machines more than mere curiosities in 1830. Babbage had higher ambitions; he planned to make a machine suited for the computation and direct setting-up in type of lengthy mathematical tables. He was greatly concerned about the errors introduced in the processes of printing and publishing tables, and listed and analyzed many repeated errors. He remarked in the first enthusiastic year of his public campaign to see his machine initiated with government aid: "Machinery which will perform . . . common arithmetic . . . will never be of that utility which must arise from an engine which calculated tables."

The Difference Engine was an embodiment in wheels and cranks of the principle of constant differences. The non-mathematical reader may find Babbage's own account in the *Life of a Philosopher* (see pp. 38–51 in accompanying text) entirely clear. We summarize it here more compactly.

Let us actually construct a table of the squares of the successive integers: 1^2, 2^2, 3^2, 4^2, etc. . . ., using exactly the method of the machine. We set up three columns: the first two columns are work columns, and the answers will appear in column C. To begin we have only to specify their three initial entries and a fixed pattern of

$$
\begin{array}{ccc}
\text{A} & \text{B} & \text{C} \\
 & & 1 \\
 & 1 & \\
2 \to & & \\
 & \overline{3} \to & \\
2 \to & & \overline{4} \\
 & \overline{5} \to & \\
2 \to & & 9 \\
 & \overline{7} \to & \\
 & & \overline{16}
\end{array}
$$

procedure.   From this we can generate the table as far up as we have the patience to go, using the single mathematical operation of addition. The three staggered columns are shown in the above with the pattern of procedure.   The three given numbers are a 2 for column A, 1 for B, and 1 for C.   The function of the digit 2 in A is to show that we are constructing a table of the second powers, while the first entries for columns B and C simply tell where the table begins.   Complete column A by filling in 2's as far down as you like.   Now complete column B by adding in the entries from column A again and again as shown by the arrows.   Once column B is constructed its values are fed in turn into column C by addition in exactly the same way.   The result in column C is automatic and almost painless construction of a table of the squares of integers, made wholly by this very repetitive pattern of simple addition.

More mathematical readers will recognize the procedure in an obvious notation of the difference calculus: $\Delta_n^2 y = 2$; $\Delta_1 y = 1$; $y_0 = 1$.   This gives the general rule and the specific initial conditions.

All the work here performed consisted of additions (including the carrying of units to the next higher place), storage (or memory) of previous results, and the repeated addition of the column entries in a certain simple and invariable order.   In 1822 Babbage made himself a small machine which could do exactly the operation given above, up to five-place numbers.   This was the harbinger of Difference Engine No. 1.   The frontispiece to the *Life of a Philosopher* (page 2) is a facsimile of a woodcut of a very similar fragment of Difference Engine No. 1 itself, which has been preserved in the Science Museum in London.   By the use of toothed wheels on shafts, not much different in principle from the familiar figure wheels of the mileage indicator on an auto speedometer, the Difference Engine can carry out the operations exemplified above. In the present frontispiece, the first table entry is set at the bottom of the column of figure wheels at the right; column B is the central vertical set of wheels, column A the left-hand vertical shaft. Turning the crank once presents a new figure in the table column, the other columns taking their proper values. In the machine as designed, but never constructed, the number indicated in the table column would each time be transmitted through a set of levers and cams to a collection of steel punches, which would then be in position for stamping the number on a copper engraver's plate. The plate was moved with each turn of the mechanism so that the punched number would appear in the proper place on the printed page. Mechanically all this was far from simple. Recall that standardized machine parts, without hand-fitting, were yet a rarity. Clocks, which most closely resembled this type of mechanism, were still principally hand-fitted. Babbage's plans were on a grand scale—one of his most conspicuous failings—and called for no less than twenty-place capacity, up to differences of the sixth order. The variety and number of bolts and nuts, claws, ratchets, cams, links, shafts, and wheels may be imagined! All of these parts were designed with skill and care, with supplementary mechanism intended to minimize wear, prevent improper registration, and so on. Some of the modern practices of instrument design were foreshadowed, and there is no doubt that the technical devices used were superior for their time. The presence of gauges, of a shaper, of a kind of embryonic turret lathe, of die-cast pewter gear wheels and the pressure molds in which they were made, is evidence enough of that. Babbage even studied the action of cutting tools, and rationalized tool-grinding. But perhaps the very care and thoroughness of the design was its greatest weakness, for it was far from completion when the controversy which ended its financial support became the cause of its postponement after years of work. And the rise of Babbage's own interest in a far grander (and still more unrealizable) project at last killed the Difference Engine. Its state at death was most incomplete, but all the drawings, and a considerable number of the tools, gauges, jigs, and a quite respectable amount of development of methods and machinery had been completed. Precise information cannot be found, but a reasonable estimate would seem to show that Babbage's engine would have cost about fifty times what the similar though more modest Swedish version sold for in the fifties. It would have been some two tons of novel brass, steel, and pewter clockwork, made, as nothing before it, to gauged standards.

V. ANALYTICAL ENGINE

WHAT BABBAGE saw after his work with the Difference Engine was a really grand vision. He had early conceived the notion he picturesquely called "the Engine eating its own tail" by which the results of the calculation appearing in the table column might be made to affect the other columns, and thus change the instructions set into the machine. On this insight, and after a striking mathemetical digression into difference functions new to mathematics, and suggested only by the operation of the engine, he built a great program. It was nothing less than a machine capable of carrying out *any* mathematical operation instead of only the simple routine of differences we have inspected. Such a machine would need instructions both by setting in initial numbers, as in the Difference Engine, and also far more generally by literally telling it what operations to carry out, and in what order. Capable of repeated additions, of multiplication which is hardly more than that, and of reversing the procedure for subtraction and division, the arithmetical unit would do these operations upon command. It would work on previously obtained intermediate results, stored in the memory section of the Engine, or upon freshly found numbers. It could use auxiliary functions, logarithms, or similar tabular numbers, of which it would possess its own library. It could make judgments by comparing numbers and then act upon the result of its comparisons —thus proceeding upon lines *not* uniquely specified in advance by the machine's instructions. All this, which forms the backbone of modern computing development, was to be carried out wholly mechanically with not even a simple electrical contact anywhere in the machine, nor, of course, a tube or a relay. The scale, as usual, was grand. The memory was to have a capacity of a thousand numbers of fifty digits—respectable even by today's standards. Of course the speed of today was wanting. The multiplication which takes not a millisecond in the fast electronic giants of today, and some seconds in a punch-card business machine installation, would have taken the Analytical Engine two or three minutes.

This operation depended upon punched cards (see Note on page xxxiii). They were not the fast-shuffled Hollerith cards moving over handy electrical-switch feelers, but cards modeled on the already well-worked-out scheme of the Jacquard loom. Punched holes in these cards would supply the machinery with numerical constants and directions for operation. The cards would be interposed into long lines of linkages within the engine. Whether or not holes came in the right places would determine the passage of feeler wires capable of linking together the notion of "chains" of columns and whole sub-assemblies. Thus, numbers or even arithmetical processes, transfer from column to column, storage, inspection of given columns already in the store, intercomparison of results, and so on—could be told to the machine. All this was done purely mechanically, and the process was elaborately safeguarded against the perils of friction, wear, jamming, and even errors by human attendants who, at the signal of the machine, were to set in cards at programmed points in the process. This is the barest sketch of the machine. Only looking at the visible complications of a modern machine, and translating them into the still self-conscious machinery of more than a century ago, can do justice to the plan. Charles Babbage would be proud to see how completely the logical structure of his Analytical Engine remains visible in today's big electronic computers.

Babbage was too much concerned with the development of his engines to publish any description of them, but in 1840 he was invited to Turin to discuss his Analytical Engine. In the audience was L. F. Menabrea (later a general in Garibaldi's army) who summarized Babbage's ideas in a paper published in 1842 in the *Bibliothèque Universelle de Genève*. This paper was translated into English and extensively annotated by the Countess of Lovelace (daughter of Lord Byron) and published in *Taylor's Scientific Memoirs*. It is reprinted in this volume (page 225). The Countess thoroughly understood and appreciated Babbage's machine, and has provided us with the best contemporary account—an account which even Babbage recognized to be clearer than his own. Miss Byron studied mathematics and, with the encouragement of her mother's various intellectual friends, her interest continued after her marriage. The Countess often visited Babbage's workshop, and listened to his explanations of the structure and use of his Engines. She shared with her husband an interest in horse racing, and with Babbage she tried to develop a system for backing horses; Babbage and the Count apparently stopped in time, but the Countess lost so heavily that she had to pawn her family jewels. Apparently Babbage was willing to try anything once in an effort to raise funds for his Engine. He once designed a tit-tat-toe machine, which he intended to send round the countryside as a travelling exhibit to raise money for his serious machines. The tit-tat-toe machine was designed to recall a splendid eighteenth-century automaton, with the figures of two children, a lamb, and a cock, alternately clapping, crying, bleating, and crowing. Underneath the bric-à-brac was to be the mechanism for a genuinely automatic machine, slow-moving but unbeatable at tit-tat-toe.

Babbage was persuaded to abandon the exhibition of this machine as an unprofitable venture by someone wise in the ways of the theater, who advised him that it was impossible to compete with General Tom Thumb, the reigning favorite of the day.

VI. MATHEMATICAL INTERESTS

ALTHOUGH BABBAGE never strayed very long from his calculating Engines, his tremendous scientific curiosity led him into many byways —some stemming directly from the main line of his machines, and some that were far afield. The machines themselves were, of course, a direct result of Babbage's great interest in mathematical tables, and he was much impressed with the importance of having them easy to read as well as accurate. In 1826, after a vast amount of labor, he published a table of logarithms from 1 to 108,000 in which he paid great attention to the convenience of calculators who would be using the tables. His work was much appreciated by computers both in England and abroad, and several foreign editions were published from his stereotype plates, with translated preface. In the same year he published a short book called *A Comparative View of the Different Institutions for the Assurance of Life*, which was one of the first clear, popular accounts of the theory of life insurance. He was led into this field as a result of his interest in calculating tables of mortality, and his tables were adopted by several German companies from the German edition of his book. In England his life tables were used by life insurance companies until a new set of tables was compiled by the Government in about 1870 on a Difference Engine built especially for the purpose (see page xvii). In 1831, in an effort to determine which was easiest to read, Babbage printed a single copy of his tables of logarithms in 21 volumes on 151 variously colored papers with ten different colors of ink, and also in gold, silver, and copper on vellum and on various thicknesses of paper.

He was constantly calling to the attention of scientific societies and government offices the number and importance of errors in astronomical tables and other calculations. At one of the first meetings of the British Association for the Advancement of Science, Babbage recommended a calculation of tables of all those facts which could be expressed by numbers in the various sciences and arts, which he called "the Constants of Nature and Art." At another BAAS meeting, in remarking on the vital statistics of an Irish parish, he said "to discover those principles which will enable the greatest number of people by their combined exertions to exist in a state of physical comfort and of moral and intellectual happiness is the legitimate object of statistical science."

He even extended his demand for statistical accuracy to poetry; it is said that he sent the following letter to Alfred, Lord Tennyson about a couplet in "The Vision of Sin":

> "Every minute dies a man, / Every minute one is born": I need hardly point out to you that this calculation would tend to keep the sum total of the world's population in a state of perpetual equipoise, whereas it is a well-known fact that the said sum total is constantly on the increase. I would therefore take the liberty of suggesting that in the next edition of your excellent poem the erroneous calculation to which I refer should be corrected as follows: "Every moment dies a man / And one and a sixteenth is born." I may add that the exact figures are 1.167, but something must, of course, be conceded to the laws of metre.

It is a fact that the couplet in all editions up to and including that of 1850 read "Every minute dies a man, / Every minute one is born," while all later editions read "Every moment dies a man, / Every moment one is born."

Like many mathematicians he was fascinated by the art of deciphering, and believed firmly that every cipher could be deciphered with sufficient time, ingenuity, and patience. He began composing a series of dictionaries in which words were arranged according to the number of letters they contained, then alphabetically by the initial letter, then alphabetically by the second letter, etc. This work was never finished, nor were the grammar and dictionary he began to write when, as a young man, he first heard of the idea of a universal language. He also wrote a paper, never published, "On the Art of Opening all Locks," and then made a plan to defeat his own method. During all of his travels he never missed an opportunity to measure the pulse and breathing rate of any animals he happened to encounter, and prepared in skeleton form a "Table of Constants of the Class Mammalia."

Babbage made one excursion into the field of apologetics with an incomplete work entitled *The Ninth Bridgewater Treatise, A Fragment*, published in 1837. The regular Bridgewater series had been supported by a bequest which called for the preparation of eight treatises which would give evidence in favor of natural religion. Babbage decided to add a ninth, at his own expense, on the same general subject, but with particular arguments against the prejudice, which he felt was implied in the first volume of the series, that the pursuits of science, and of mathematics in particular, are unfavorable to religion.

He used his experiences with his calculating engine to bolster his arguments on the nature of miracles and in favor of design. As B. V. Bowden says, "he thought of God as a Programmer." He repeatedly alludes to the possibility of defining a series by such a complicated rule that the first hundred million terms might proceed according to an obvious scheme, and the next number violate it, while the rest of the sequence continued according to the first plan. He described the programming of a calculator for generating such a series. In this argument he felt he had shown a possible origin of miracles in a world otherwise controlled under God by orderly natural law.

In his autobiography Babbage jokingly traces his ancestry to the prehistoric flint workers because of his "inveterate habit of contriving tools." He reports that as a child he had a great desire to inquire into the causes of all events, and that his invariable question on receiving any new toy was "Mamma, what is inside of it?" If the answer did not satisfy him, the toy was broken open. He remembers as a small boy being fascinated by an exhibition of clockwork automata, especially one of a small figure of a silver lady dancing. He ran across the silver lady many years later, and acquired her for his drawing room, where she was dressed in elaborate robes and displayed on a pedestal in a glass case. She held a place of honor next to the portion of his Difference Engine which he also had on exhibit at home. He would set either or both of them into operation for the entertainment of his guests: the lady to dance, and the Engine to print a small table, and noted ruefully that on one occasion his English friends were gathered about the silver lady while an American and a Hollander studied the Difference Engine. Babbage recounts a conversation he once had with the Countess of Wilton and the Duke of Wellington, who had called at his home to see the Difference Engine. The Countess asked what he considered was his greatest difficulty in designing the machine. He replied that his greatest difficulty was not that of "contriving mechanism to execute each individual movement . . . but it really arose from the almost innumerable *combinations* amongst all these contrivances," and compared his problems to those of a general commanding a vast army in battle. He was pleased to have Wellington confirm his analysis.

VII. MACHINERY

JUST AS the mathematical machine-designing team of today soon becomes involved in a welter of problems about the properties of vacuum tubes and electronic circuits, so Babbage became deeply involved in the problems of the machine shop and the drafting room. During the course of the work many ingenious mechanical devices were perfected and even some of those that were rejected for use on the calculator were not entirely wasted, but were introduced with success into other machinery as, for example, into a spinning factory at Manchester. To create the great variety of new and complex forms with the required precision for the Difference Engine, a number of new tools to use with a lathe were invented. To test the "steadiness and truth" of the tool-holders used in making some key gun-metal plates of the Difference Engine, Babbage reports that he "had some dozen of the plates turned with a diamond point," and he was delighted to observe the resulting grating spectra or "Frauenhofer images," as he called them. With his own sketches and with drawings prepared by a full-time draftsman, Babbage placed the construction of the machine in the hands of the engineer Joseph Clement. Clement was one of the great machine-tool builders of the century, who earlier had been a draftsman for Henry Maudslay, the introducer of the slide rest and the screw cutting lathe. Babbage tells of an order Clement once received from America to construct a large screw in "the best possible manner." This he proceeded to do, according to his standards, with a precision, and consequently a bill, far greater than his customer expected. The customer was required to pay some hundreds of pounds, although he had anticipated a bill of twenty pounds at most! The mainstay of Clement's custom shop was the first large planing machine, although from the sums expended by Babbage it appears that the Difference Engine was one of the largest single jobs in Clement's shop, somewhere between one fifth and one third of its whole effort. In Babbage's own writings it tends to appear as though Clement were in fact a full-time employee of Babbage, but this seems to be inaccurate. Babbage broke with Clement in 1833 and seems never to have carried any further projects beyond the stage of drawings and experimental parts.

Among the workmen in Clement's shop was one J. Whitworth, who became Sir Joseph Whitworth, Bart., leader in the machine-tool industry in the nineteenth century. It was Whitworth who first brought about the standardization of screw threads, and the Whitworth thread remained the British standard until 1948. He insisted upon the use of gauge blocks, recognizing end measurement as better than measurement between scratch lines. Even in the 1830's he could work to standards of a micro-inch. He is frequently credited with the familiar machinist's scheme of preparing plane standard surfaces three at a time by hand scraping, but this ascription seems to be doubtful. Whitworth's independent career began when he left Clement, probably because of the curtailment of the Babbage contract, and set up his own shop at Manchester. His was probably the first shop to build machine tools mainly for sale to other tool manufacturers.

Babbage wrote a paper "On the Principles of Tools for Turning and Planing Metals" for a three-volume reference work on the lathe, published in 1846 by Holtzapffel & Co. The publisher acknowledges that "The cultivation of Mechanics by Gentlemen . . . has given rise to many ideas and suggestions on their part, which have led to valuable practical improvements," and offers instruction to amateurs in "Turning or Mechanical Manipulation generally," either at Holtzapffel's shop or at the gentlemen's private residences. The parallel is complete: like today's mathematical-machine builder, today's scientific hobbyist is apt to use vacuum tubes; he is more likely to build amplifiers than do ornamental turning on a lathe. The drawings which Babbage had for his machines covered over 400 square feet of surface. They were described by experts at that time as perhaps the best specimens of mechanical drawings ever executed, done with extraordinary ability and precision. In the course of preparing them, Babbage invented a scheme of mechanical notation to make clear in a drawing the action of all the moving parts of a piece of machinery. Since Babbage's machinery was particularly complex to describe in motion, he was extremely proud of his notation. He prepared and had printed a short paper describing his principles of mechanical notation, which he gave away in considerable numbers during the Exhibition of 1851, requesting readers to send him any criticisms or suggestions. This paper is reprinted on page 357 of this volume.

Babbage's most successful book, and in fact the only work of any consequence which he ever *completed*, was the *Economy of Manufactures and Machinery*, published in 1832. Only one brief excerpt from this work is reproduced in the present volume (page 315). Although originally intended as a series of lectures at Cambridge, Babbage published the work in book form, with a condensed version prepared as a prefix to the appropriate volume of the *Encyclopedia Metropolitana*. It ran through several editions, was reprinted in the United States, and was translated into German, French, Italian, and Spanish, in spite of the trouble Babbage reports having with booksellers because of his chapter analyzing the book trade. As a result of supervising the construction of his own Engine, he became interested in the general problems of manufacturing and visited factories in England and on the Continent. He learned from a workman how to punch a hole in a sheet of glass without breaking it, and found a demonstration of this skill a useful method of winning the confidence of the various craftsmen with whom he spoke. (See page 129 for this interesting method.) The book includes a detailed description and classification of the tools and machinery used in various manufacturing operations which he observed, together with a discussion of the "economical principles of manufacturing." In the mood of an operational research man of today, Babbage takes to pieces the manufacture of pins—the operations involved, the kinds of skill required, the expense of each process, and the direction for improvements in the then current practices. He makes a number of suggestions about methods for analyzing factories and processes and finding the proper size and location of factories, and stresses the need for studying the work of contemporary inventors in other countries. He points out that the division of labor, so important for manufacturing, can be applied also to mental operations, and cites as an example the work of G. F. Prony, director of the École des Ponts et Chaussées, who successfully organized three groups of workers—skilled, semi-skilled, and unskilled—to prepare a great set of mathematical tables. Prony began work in 1784 under the same impetus which led to the establishment of the metric system in revolutionary France. He undertook to construct elaborate trigonometric tables based on the division of the quadrant into a hundred parts. A necessary auxiliary work was an unprecedented table of logarithms. This tabulated the logarithms of the natural numbers up to 200,000, carried out to fourteen decimal places. Prony realized that life was too short for such an effort (one sixth of that work had cost Briggs six or eight years). The story goes that, happening to read the new book of Adam Smith on the division of labor, he proceeded to organize the computations on this basis. His most skilled handful included mathematicians of the stature of Legendre. Their task was, of course, no mere computation but the choice of the best analytical expressions for numerical evaluation. Their formulae were transmitted to a group of about eight well-trained computers who put them into the appropriate numerical form. The computers of unskilled kind varied in number from 60 to 80. Their task was nothing more than addition and subtraction, according to the rules that were specified. It seems that nine tenths of them literally knew no more than addition and subtraction, and these turned out to be the best computers. Two teams worked independently and in duplicate and finished in about two years. The final "Tables du Cadastre" were never published, but remained in two copies, each of seventeen manuscript folio volumes. These were frequently consulted as checks by other computers, including Babbage himself who visited the Observatory in Paris on this errand.

Also included in the *Economy of Manufactures* is a panegyric for the "Science of Calculation . . . which must ultimately govern the whole of the application of Science to the Arts of Life." Babbage reports as one of the best compliments he ever received on the book a remark by an English workman he met, who said "that book made me think." One profoundly practical result of his operational research method was the introduction of the penny post in England; Sir Rowland Hill was encouraged to do this by Babbage's analysis of postal operations, which showed that the cost of handling the mail in the post office was greater than the cost of transportation. This pioneer work, *Economy of Manufactures*, is good reading even today.

VIII. OTHER INTERESTS

BABBAGE MADE a number of suggestions for practical inventions of various kinds. Much interested in railroads, he attended the opening of the Manchester and Liverpool Railway and made several suggestions for ways of preventing accidents, including a method for separating a derailed engine from a train. In 1838 he was consulted by Isambard Brunel and the directors of the Great Western Railroad and spent five months doing experiments, which consisted largely of tracing on paper the curves of motion made by the special car in which he worked. On the basis of these experiments, he recommended use of a broad gauge and proposed an automatic speed-recording device for every engine. He suggested a numerical system of occulting lighthouses, and sent a description of his scheme to the authorities of twelve maritime countries. The United States Congress appropriated $5,000 to try his scheme experimentally, and the results of these experiments were published in 1861 in an extremely favorable report recommending adoption by the U.S. Lighthouse Board. An experience in a diving bell in 1818 led Babbage to consider the question of submarine navigation, and he prepared drawings and a description of an open submarine vessel with air for four persons for two days. Such a vessel, he thought, could be screw-propelled, and might enter a harbor and destroy even iron ships! He also suggested the use of a rocket apparatus to boost projectiles and the use of mirrors for indirect fire for artillery. Once at an opera, much bored by the performance, Babbage had the notion of using colored lights in the theater. He did some experiments, using cells formed by pieces of parallel glass filled with solutions of various colored salts, and even devised a rainbow dance to demonstrate this new technique.

Babbage did some physics in Cambridge, and published a paper with Herschel in 1825 on magnetization arising during rotation, based on experiments of Arago. Babbage was also much interested in geology and astronomy. After the eclipse of 1851 he suggested the germ of the idea of the coronagraph for seeing the sun's prominences without an eclipse, but he had not at all analyzed the problems of scattered light. It seemed possible to him that one could get a record of the succession of hot and cold years in the past by examining and comparing tree rings in ancient forests (see page 367); this method was rediscovered early in this century and used to great advantage in southwestern United States. He wandered once again into the field of archaeology in his last scientific paper, entitled "On Remains of Human Art, mixed with the Bones of Extinct Races of Animals," published in 1859. Babbage once proposed to write a novel in order to help finance the completion of his Analytical Engine. He planned to devote a year to preparing a three-volume novel with illustrations, which was to earn 5,000 pounds. He was discouraged from pursuing this project by a poet friend wise in the pitfalls of literary fortune.

Unlike Darwin, who wrote his autobiography for his own children, with no thought of publication, Babbage says that he wrote his *Passages from the Life of a Philosopher* to "render . . . less unpalatable" the history of his calculating machines by an account of his own "experience amongst various classes of society." He was over seventy when he prepared this collection of anecdotes and ideas, and more obsessed than ever with his beloved engines. The book's characteristic combination of peevishness and humor is apparent as early as the title page. Babbage was a bitter man, and his autobiography is as much a record of his disappointments as of his achievements. The largest part of the book is, of course, devoted to his engines, with an account of their theory and principles of construction, and the sad tale of their neglect by the British Government. In 1832, and again in 1834, he ran unsuccessfully for Parliament on a Liberal (Whig) platform, and he includes in his autobiography parts of an amusing electioneering play used in connection with his campaign. In the play, entitled *Politics and Poetry* or *The Decline of Science*, Babbage is characterized as Turnstile, "a fellow of some spirit; and devilish proud," and again as "a sort of a philosopher—that wants to be a man of the world." In a chapter entitled "Street Nuisances," Babbage describes his one-man battle against street musicians, which brought him as much fame, in London at least, as all his scientific accomplishments combined. Babbage maintained that his ideas vanished when the organ-grinder began to play, and calculated that such interruptions destroyed one fourth of his working power. He waged a vigorous campaign of letters to newspapers and to members of Parliament, and personally hauled many individual offenders before a magistrate. One of the latter once asked Babbage whether a man's brain would be injured by listening to a hand organ, and Babbage replied "certainly not, for the obvious reason that no man having a brain ever listened to street musicians." That this particular battle of Babbage's was not taken seriously is made clear by his obituary notice in the sober London *Times* in 1871, which remarks somewhat cruelly in the first paragraph that Babbage lived to be almost 80 "in spite of organ-grinding persecutions."

IX. DEFENSE OF SCIENCE

Even before he had a personal grievance against the British Government for its failure to support his own machine, Babbage had sharply criticized the Government for its neglect of science and scientists. Never a man to avoid a fight for what he considered a good cause, Babbage had for years led an assault on the decline of science in England. He published two stinging tracts, *Reflections on the Decline of Science in England, and on some of its Causes* (1830), and *The Exposition of 1851; or Views of the Industry, the Science and the Government of England* (1851). Bitter because few Englishmen pursued science for its own sake, he attacked the neglect of science in the educational system and urged Government subsidies for scientists. He felt that scientists should hold many Government posts and that pure science should be encouraged. "It is of the very nature of knowledge that the recondite and apparently useless acquisition of today becomes part of the popular food of a succeeding generation," he wrote. Chief target of his diatribes on the neglect of science in England was the Royal Society, to which he had been elected while still at Cambridge. He submitted a plan for sweeping reforms to the Society, which rejected it without discussion. His plan included such items as requirements for publication of scientific articles as a test for membership, instituting democratic election procedures, and free discussion of policies at meetings. Infuriated by the Society's refusal to consider his plan, he continued to condemn what he termed the intrigues of the Society, pronounced its secretaries third-rate, and its president elected on the basis of rank rather than scientific interests. Babbage would perhaps be disappointed to find that within the last decade the BAAS which he helped to found was headed, nominally at least, by the Duke of Edinburgh! Babbage described the Council of the Royal Society as "a collection of men who elect each other to office and then dine together at the expense of the society to praise each other over wine and to give each other medals." Although some of Babbage's accusations against the Royal Society and the Government for their neglect of science were exaggerated, his position was fundamentally sound, and he found a good deal of support. Babbage has been aptly characterized as "a scientific gadfly" who "successfully needled his contemporaries into general agreement."

Babbage blamed the Royal Society for conditions at the Royal Observatory at Greenwich; on one occasion he had been refused a copy of some of the Greenwich observations, and later located five tons of the Greenwich tables in a shop which had bought them by the pound for making pasteboard. Babbage remarked that the Astronomer Royal was certainly the man best fitted to decide what should be done with his own publications, but did not think it possible to invent a more extravagant way of compensating a public servant than to establish an observatory and computing center for the production and printing of astronomical tables simply as a source of wastepaper! He had no great love for the then Astronomer Royal, Sir George Airy, in any case, for that official had recommended no further Government support for the Difference Engine, and had refused to consider the possibility of mechanizing his own computations.

X. CONCLUSION

BABBAGE ONCE said that he would gladly give up the remainder of his life if he could be allowed to live three days five hundred years hence and be provided with a scientific guide to explain the discoveries made since his death. He judged that the progress to be recorded would be immense, since science tends to go on with constantly increasing rapidity, and Babbage always took a confident view about human progress. The wide range of his practical and scientific interests and his clear commitment to the notion that careful analysis, mathematical procedures, and statistical calculations could be reliable guides in almost all facets of practical and productive life give him still a wonderful modernity.

More than one spiritual contemporary of Babbage is flying today from site to site on the missions of the Atomic Energy Commission and the Rand Corporation.   His whole story bears witness to the strong interaction between purely scientific innovation, on the one hand, and the social fabric of current technology, public understanding, and support on the other.   His great engines never cranked out answers, for ingenuity can transcend but it cannot ignore its context.   Yet Charles Babbage's monument is not the dusty controversy of the books, nor priority in a mushrooming branch of science, nor the few wheels in the museum.   His monument, not wholly beautiful, but very grand, is the kind of coupled research and development that is epitomized today, as it was foreshadowed in his time, by the big digital computers.

September, 1959   PHILIP AND EMILY MORRISON
Ithaca, New York

HISTORY OF PUNCH CARDS

" . . . the Analytical Engine *weaves Algebraical patterns*, just as the Jacquard-loom weaves flowers and leaves . . . "

ADA AUGUSTA, COUNTESS OF LOVELACE

THE USE of punched holes in paper cards for the digital storage of information can be traced step by step back to the early eighteenth century.

That history does *not* begin with writing; it is directly related rather to the weaving of elaborate figured silks. For the essence of the problem is the ability for easy storage of large amounts of information to be read, not visually, but mechanically. Writing and other graphic arts are intended essentially for visual examination. Even in these days it is plain that use of merely optical coupling is more complex than direct mechanical means for conveying information to machines.

The weaving of ornamentally figured silk textiles was well developed in China as early as 1000 B.C. The product of this art, though probably not the looms which made it possible, is known to have appeared in the West by late Roman times. A complex figure on the cloth implies that in the weaving process a specified set of the longitudinal threads (the warp) was lifted to allow the passage of the transverse weft thread to form each single line of the final pattern. The strong, fine, lustrous silk fiber has been the principal medium which made the elaborate labor of complex ornamental weaving worthwhile. Such weaving was always at least a two-man job. It employed a special loom, called the drawloom, which was certainly used in the Italian silk centers in the Middle Ages, but which represents a tradition going back, at least by analogy, to the earliest Chinese devices. In the drawloom a cord is attached to each warp thread. These cords are united in groups appropriate to the pattern to be woven. When these cords are pulled in the correct sequence, the weaver is able to throw the shuttle, with its weft thread attached, through the opening made by the lifted warp threads. An unskilled assistant to the weaver pulls the cords as indicated by a painted chart usually made on squared paper, one square to each thread, both warp and weft. Squared paper was used even before the invention of printing.

xxxiii

The preparation of this squared paper for the drawboy might be aided by the use of a simple stencil, in which holes are pricked to allow the quick copying of the coordinate mesh.   It must be remembered that information is needed in large amounts for the weaving of a complex ornamental pattern.   Even the most ancient Chinese examples required that about 1,500 different warp threads be lifted in various combinations as the weaving proceeded.   The design repeats after a number of weft threads, but the drawboy had to pull 40 or 50 distinct bundles made of these 1,500 lifting cords in the correct sequence for each transverse weft.   In the seventeenth century it took a skilled weaver two or three weeks to set up a drawloom for a particular pattern.   In 1725 Basile Bouchon, in the silk center of Lyons, designed a mechanism for automatically selecting the cords to be raised.   The cords were passed through eyes in a row of horizontal needles arranged to slide in a box.   The selection of the cord was made by pressing against these needles a roll of perforated paper.   On this roll a set of punched holes, spaced to fit the needles, was perforated line after line, as in a player piano roll or a monotype type-casting machine, both of the later nineteenth century.   It may be that M. Bouchon got his idea from the squared-paper stencils on which the patterns were pricked.   Those needles which entered the punched holes passed through freely and did not raise their connected warp threads, while the others were active.   Thus the pattern could be picked out, weft thread by weft thread, one to each line of holes. Only three years later another Lyons inventor, one M. Falcon, extended the scheme to allow the use of several long rows of needles, one row above the other.   Obviously this could be done more neatly by using a rectangular card instead of a paper roll.   Using a perforated platen held in the assistant's hand, a card was pressed against the needles.   This is certainly the full punched-card principle; the obvious extension to a long series of cards, all strung together, was already Falcon's.

It was not until 1801 that J. M. Jacquard of Lyons made a fully successful automatic drawloom.   He took over the cards from Falcon, but very much elaborated the mechanics of the apparatus, so that the loom became a one-man operation.   Jacquard looms were made by the tens of thousands in the early decades of the nineteenth century; indeed, they are even more common today.   We know directly from Babbage that it was the Jacquard string of punched cards actuating the lifting cords of the silk loom which inspired his use of the same principle for the Analytical Engine.   He was very proud of a remarkable woven silk portrait which he owned, showing the inventor Jacquard surrounded by the machines of his trade. This work was woven with about 1,000 threads to the inch and resembled a line engraving in fineness of detail. A total of 24,000 cards, each one capable of receiving 1,050 punch-holes, was used to weave its five square feet. This truly extraordinary display fascinated Babbage and surely did much to convince him of the practicality of complex information storage through the mechanical perusal of punched cards.

Babbage never built a punched-card machine, but the inventor, patentee, and co-founder of the earliest practical punched-card tabulating machine may have been influenced by Babbage's work, perhaps through the summary report of the Committee of the British Association, published in 1878; more probably he took his ideas directly from the Jacquard loom. Herman Hollerith went to the U. S. Patent Office to begin serious work on punch-card tabulation in the early 1880's. By 1890 crude Hollerith machines (looking not unlike the Falcon drawloom mechanism) were in practical use at the Census Bureau, and Hollerith and other punch-card machines were extensively employed thereafter both in the United States and abroad. Contemporary punch-card machines of all manufacturers stem from one or another of these early devices.

P.M.

E.M.

BIBLIOGRAPHY

I Standard biographical works:

Dictionary of National Biography, Vol. II. London: Smith, Elder & Co., 1885.

Eminent Persons, Vol. I, 1871–75. London: Macmillan and Co. (Biographies reprinted from *Times* of London.)

Encyclopedia Britannica, 11th ed.; Cambridge, 1910. (Vols. III, XVI, XXVIII, etc. In addition to biographies, articles on "Logarithms," "Weaving," etc.)

II Nineteenth-century works on Babbage:

Maj.-Gen. H. P. Babbage (ed.), *Babbage's Calculating Engines, Being a Collection of Papers Relating to Them; Their History and Construction*, London: E. and F. N. Spon, 1889. Usually referred to in the Introduction and notes as *Engines*.

W. W. Rouse Ball, *A History of the Study of Mathematics at Cambridge*, Cambridge, 1889.

Mathematical and Scientific Library of Late Charles Babbage, London, 1872. A bookseller's catalogue, privately printed.

U.S. Lighthouse Board, *Experiments on Mr. Babbage's Method of Distinguishing Lighthouses*, J. H. Alexander, 1861.

III Twentieth-century works on Babbage and Machines:

Books:

B. V. Bowden, *Faster than Thought*, London: Sir Isaac Pitman & Son, 1953.

R. T. Gunther, *Early Science in Cambridge*, Cambridge, 1937.

A. Schuster and A. E. Shipley, *Britain's Heritage of Science*, London: Constable and Co., 1917.

Articles:

Maj.-Gen. H. P. Babbage, "Babbage's Analytical Engine," *Monthly Notices* of Royal Astronomical Society, Vol. LXX, No. 6, April, 1910. This gives the history of the construction of the "mill" of the frontispiece of this volume and includes a specimen page of multiples to 29 places. We find as

Errata in the same volume (page 645) that π was given to the machine with an error in the fourteenth decimal so that all the multiples are wrong, and in 8π, 9π, and 11π there were errors "caused by weakness in certain springs." The printing mechanism did not correctly print what was calculated by the machine!

L. H. D. Buxton, "Charles Babbage and His Difference Engines," in *The Newcomen Society Transactions*, Vol. XIV, London: Courier Press, 1935.

L. J. Comrie, "Babbage's Dream Comes True," *Nature*, October 26, 1946.

P. and E. Morrison, "Charles Babbage," in *Scientific American*: April, 1952. (Reprinted in *Lives in Science*, New York: Simon and Schuster, 1958.)

Charles F. Mullett, "Charles Babbage, a Scientific Gadfly," in *Scientific Monthly*, November, 1948.

Works on Scheutz Machine:

Annals of the Dudley Observatory, Vol. 1, Albany, 1866.

Anecdotes:

Nora Barlow (ed.), *The Autobiography of Charles Darwin*, London: Collins, 1958.

J. Churton Collins (ed.), *Early Poems of A. Tennyson*, London: Methuen, 1901.

L. A. and B. L. Tollemache, *Safe Studies*, London, 1884. Privately printed.

Works on Punch Cards:

Dictionary of American Biography, Supplement to Vol. XXI, New York, 1944.

Singer *et al.*, *History of Technology*, Vols. 2–5, Oxford, 1955–58.

W. Willets, *Chinese Art*, Vol. 1, Middlesex: Penguin, 1958.

PART I

Chapters from

PASSAGES FROM THE LIFE
OF A PHILOSOPHER

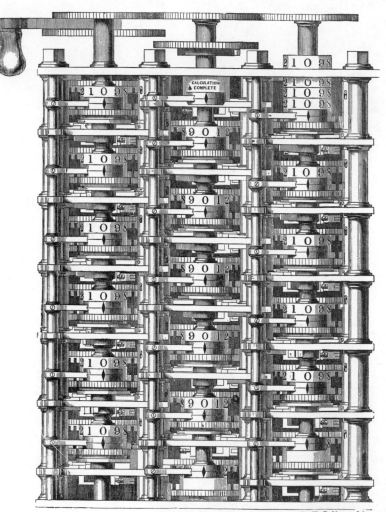

Impression from a woodcut ot a small portion of Mr. Babbage's Difference
Engine No. 1, the property of Government, at present deposited in the Museum
at South Kensington.

It was commenced 1823.
This portion put together 1833.
The construction abandoned 1842.
This plate was printed June, 1853.
This portion was in the Exhibition 1862.

Facsimile of frontispiece from *Passages from the Life of a Philosopher* published in 1864.

PASSAGES

FROM

THE LIFE OF A PHILOSOPHER.

BY

CHARLES BABBAGE, ESQ., M.A.,

F.R.S., F.R.S.E., F.R.A.S., F. STAT. S., HON. M.R.I.A., M.C.P.S.,

COMMANDER OF THE ITALIAN ORDER OF ST. MAURICE AND ST. LAZARUS,

INST. IMP. (ACAD. MORAL.) PARIS CORR., ACAD. AMER. ART. ET SC. BOSTON, REG. ŒCON. BORUSS.,
PHYS. HIST. NAT. GENEV., ACAD. REG. MONAC., HAFN., MASSIL., ET DIVION., SOCIUS.
ACAD. IMP. ET REG. PETROP., NEAP., BRUX., PATAV., GEORG. FLOREN, LYNCEI ROM., MUT., PHILOMATH.
PARIS, SOC. CORR., ETC.

"I'm a philosopher. Confound them all—
Birds, beasts, and men; but no, not womankind."—*Don Juan.*

"I now gave my mind to philosophy: the great object of my ambition was to make out a complete system of the universe, including and comprehending the origin, causes, consequences, and termination of all things. Instead of countenance, encouragement, and applause, which I should have received from every one who has the true dignity of an oyster at heart, I was exposed to calumny and misrepresentation. While engaged in my great work on the universe, some even went so far as to accuse me of infidelity;—such is the malignity of oysters."—"*Autobiography of an Oyster*" *deciphered by the aid of photography in the shell of a philosopher of that race,—recently scolloped.*

LONDON:

LONGMAN, GREEN, LONGMAN, ROBERTS, & GREEN.

1864.

[The right of Translation is reserved.]

Title page from *Passages from the Life of a Philosopher* published in 1864.

DEDICATION

TO VICTOR EMMANUEL II, KING OF ITALY

Sire,

In dedicating this volume to your Majesty, I am also doing an act of justice to the memory of your illustrious father.

In 1840, the King, Charles Albert, invited the learned of Italy to assemble in his capital. At the request of her most gifted Analyst, I brought with me the drawings and explanations of the Analytical Engine. These were thoroughly examined and their truth acknowledged by Italy's choicest sons.

To the King, your father, I am indebted for the first public and official acknowledgment of this invention.

I am happy in thus expressing my deep sense of that obligation to his son, the Sovereign of united Italy, the country of Archimedes and of Galileo.

I am, Sire,
With the highest respect,
Your Majesty's faithful Servant,
Charles Babbage

PREFACE

SOME men write their lives to save themselves from *ennui*, careless of the amount they inflict on their readers.

Others write their personal history, lest some kind friend should survive them, and, in showing off his own talent, unwittingly show them up.

Others, again, write their own life from a different motive—from fear that the vampires of literature might make it their prey.

I have frequently had applications to write my life, both from my countrymen and from foreigners. Some caterers for the public offered to pay me for it. Others required that I should pay them for its insertion; others offered to insert it without charge. One proposed to give me a quarter of a column gratis, and as many additional lines of eloge as I chose to write and pay for at ten-pence per line. To many of these I sent a list of my works, with the remark that they formed the best life of an author; but nobody cared to insert them.

I have no desire to write my own biography, as long as I have strength and means to do better work.

The remarkable circumstances attending those Calculating Machines on which I have spent so large a portion of my life, make me wish to place on record some account of their past history. As, however, such a work would be utterly uninteresting to the greater part of my countrymen, I thought it might be rendered less unpalatable by relating some of my experience amongst various classes of society, widely differing from each other, in which I have occasionally mixed.

This volume does not aspire to the name of an autobiography. It relates a variety of isolated circumstances in which I have taken part—some of them arranged in the order of time, and others grouped together in separate chapters, from similarity of subject.

The selection has been made in some cases from the importance of the matter. In others, from the celebrity of the persons concerned; whilst several of them furnish interesting illustrations of human character.

CONTENTS*

* [A complete Table of Contents, including subheads, from *Passages from the Life of a Philosopher* has been reproduced as Appendix V, p. 385.]

CHAPTER II

CHILDHOOD

The Prince of Darkness is a gentleman.—*Hamlet*

Early Passion for inquiry and inquisition into Toys—Lost on London Bridge—
Supposed value of the young Philosopher—Found again—Strange Coincidence
in after-years—Poisoned—Frightened a Schoolfellow by a Ghost—Frightened
himself by trying to raise the Devil—Effect of Want of Occupation for the Mind—
Treasure-trove—Death and Non-appearance of a Schoolfellow.

FROM MY earliest years I had a great desire to inquire into the causes
of all those little things and events which astonish the childish mind.
At a later period I commenced the still more important inquiry into
those laws of thought and those aids which assist the human mind in
passing from received knowledge to that other knowledge then unknown to our race.   I now think it fit to record some of those views to
which, at various periods of my life, my reasoning has led me.   Truth
only has been the object of my search, and I am not conscious of ever
having turned aside in my inquiries from any fear of the conclusions to
which they might lead.

As it may be interesting to some of those who will hereafter read
these lines, I shall briefly mention a few events of my earliest, and even
of my childish years.   My parents being born at a certain period of
history, and in a certain latitude and longitude, of course followed the
religion of their country.   They brought me up in the Protestant form of
the Christian faith.   My excellent mother taught me the usual forms of
my daily and nightly prayer; and neither in my father nor my mother
was there any mixture of bigotry and intolerance on the one hand, nor on
the other of that unbecoming and familiar mode of addressing the
Almighty which afterwards so much disgusted me in my youthful years.

My invariable question on receiving any new toy, was "Mamma,
what is inside of it?"   Until this information was obtained those
around me had no repose, and the toy itself, I have been told, was
generally broken open if the answer did not satisfy my own little ideas
of the "fitness of things."

Earliest Recollections

Two events which impressed themselves forcibly on my memory
happened, I think, previously to my eighth year.

When about five years old, I was walking with my nurse, who had in her arms an infant brother of mine, across London Bridge, holding, as I thought, by her apron. I was looking at the ships in the river. On turning round to speak to her, I found that my nurse was not there, and that I was alone upon London Bridge. My mother had always impressed upon me the necessity of great caution in passing any street-crossing: I went on, therefore, quietly until I reached Tooley Street, where I remained watching the passing vehicles, in order to find a safe opportunity of crossing that very busy street.

In the mean time the nurse, having lost one of her charges, had gone to the crier, who proceeded immediately to call, by the ringing of his bell, the attention of the public to the fact that a young philosopher was lost, and to the still more important fact that five shillings would be the reward of his fortunate discoverer. I well remember sitting on the steps of the door of the linendraper's shop on the opposite corner of Tooley Street, when the gold-laced crier was making proclamation of my loss; but I was too much occupied with eating some pears to attend to what he was saying.

The fact was, that one of the men in the linendraper's shop, observing a little child by itself, went over to it, and asked what it wanted. Finding that it had lost its nurse, he brought it across the street, gave it some pears, and placed it on the steps at the door: having asked my name, the shopkeeper found it to be that of one of his own customers. He accordingly sent off a messenger, who announced to my mother the finding of young Pickle before she was aware of his loss.

Those who delight in observing coincidences may perhaps account for the following singular one. Several years ago when the houses in Tooley Street were being pulled down, I believe to make room for the new railway terminus, I happened to pass along the very spot on which I had been lost in my infancy. A slate of the largest size, called a Duchess,* was thrown from the roof of one of the houses, and penetrated into the earth close to my feet.

The other event, which I believe happened some time after the one just related, is as follows. I give it from memory, as I have always repeated it.

I was walking with my nurse and my brother in a public garden, called Montpelier Gardens, in Walworth. On returning through the private road leading to the gardens, I gathered and swallowed some dark berries very like black currants:—these were poisonous.

* There exists an aristocracy even amongst slates, perhaps from their occupying the most *elevated* position in every house. Small ones are called Ladies, a larger size Countesses, and the biggest of all are Duchesses.

On my return home, I recollect being placed between my father's knees, and his giving me a glass of castor oil, which I took from his hand.

My father at that time possessed a collection of pictures. He sat on a chair on the right hand side of the chimney-piece in the breakfast room, under a fine picture of our Saviour taken down from the cross. On the opposite wall was a still-celebrated "Interior of Antwerp Cathedral."

In after-life I several times mentioned the subject both to my father and to my mother; but neither of them had the slightest recollection of the matter.

Having suffered in health at the age of five years, and again at that of ten by violent fevers, from which I was with difficulty saved, I was sent into Devonshire and placed under the care of a clergyman (who kept a school at Alphington, near Exeter), with instructions to attend to my health; but, not to press too much knowledge upon me: a mission which he faithfully accomplished. Perhaps great idleness may have led to some of my childish reasonings.

Relations of ghost stories often circulate amongst children, and also of visitations from the devil in a *personal* form. Of course I shared the belief of my comrades, but still had some doubts of the existence of these personages, although I greatly feared their appearance. Once, in conjunction with a companion, I frightened another boy, bigger than myself, with some pretended ghost; how prepared or how represented by natural objects I do not now remember: I believe it was by the accidental passing shadows of some external objects upon the walls of our common bedroom.

The effect of this on my playfellow was painful; he was much frightened for several days; and it naturally occurred to me, after some time, that as I had deluded him with ghosts, I might myself have been deluded by older persons, and that, after all, it might be a doubtful point whether ghost or devil ever really existed. I gathered all the information I could on the subject from the other boys, and was soon informed that there was a peculiar process by which the devil might be raised and become personally visible. I carefully collected from the traditions of different boys the visible forms in which the Prince of Darkness had been recorded to have appeared. Amongst them were—

A rabbit,
An owl,
A black cat, very frequently,
A raven,
A man with a cloven foot, also frequent.

After long thinking over the subject, although checked by a belief that the inquiry was wicked, my curiosity at length over-balanced my fears, and I resolved to attempt to raise the devil. Naughty people, I was told, had made written compacts with the devil, and had signed them with their names written in their own blood. These had become very rich and great men during their life, a fact which might be well known. But, after death, they were described as having suffered and continuing to suffer physical torments throughout eternity, another fact which, to my uninstructed mind, it seemed difficult to prove.

As I only desired an interview with the gentleman in black simply to convince my senses of his existence, I declined adopting the legal forms of a bond, and preferred one more resembling that of leaving a visiting card, when, if not at home, I might expect the satisfaction of a return of the visit by the devil in person.

Accordingly, having selected a promising locality, I went one evening towards dusk up into a deserted garret. Having closed the door, and I believe opened the window, I proceeded to cut my finger and draw a circle on the floor with the blood which flowed from the incision.

I then placed myself in the centre of the circle, and either said or read the Lord's Prayer backwards. This I accomplished at first with some trepidation and in great fear towards the close of the scene. I then stood still in the centre of that magic and superstitious circle, looking with intense anxiety in all directions, especially at the window and at the chimney. Fortunately for myself, and for the reader also, if he is interested in this narrative, no owl or black cat or unlucky raven came into the room.

In either case my then weakened frame might have expiated this foolish experiment by its own extinction, or by the alienation of that too curious spirit which controlled its feeble powers.

After waiting some time for my expected but dreaded visitor, I, in some degree, recovered my self-possession, and leaving the circle of my incantation, I gradually opened the door and gently closing it, descended the stairs, at first slowly, and by degrees much more quickly. I then rejoined my companions, but said nothing whatever of my recent attempt. After supper the boys retired to bed. When we were in bed and the candle removed, I proceeded as usual to repeat my prayers silently to myself. After the few first sentences of the Lord's Prayer, I found that I had forgotten a sentence, and could not go on to the conclusion. This alarmed me very much, and having repeated another prayer or hymn, I remained long awake, and very unhappy. I thought that this forgetfulness was a punishment inflicted upon me by the Almighty, and that I was a wicked little boy for having attempted to satisfy myself about the existence of a devil. The next night my memory was more faithful, and my prayers went on as usual. Still, however, I was unhappy, and continued to brood over the inquiry. My uninstructed faculties led me from doubts of the existence of a devil to doubts of the book and the religion which asserted him to be a living being. My sense of justice (whether it be innate or acquired) led me to believe that it was impossible that an almighty and all-merciful God could punish me, a poor little boy, with eternal torments because I had anxiously taken the only means I knew of to verify the truth or falsehood of the religion I had been taught. I thought over these things for a long time, and, in my own childish mind, wished and prayed that God would tell me what was true. After long meditation, I resolved to make an experiment to settle the question. I thought, if it was really of such immense importance to me here and hereafter to believe rightly, that the Almighty would not consign me to eternal misery because, after trying all means that I could devise, I was unable to know the truth. I took an odd mode of making the experiment; I resolved that at a certain hour of a certain day I would go to a certain room in the house, and that if I found the door open, I would believe the Bible; but that if it were closed, I should conclude that it was not true. I remember well that the observation was made, but I have no recollection as to the state of the door. I presume it was found open from the circumstance that, for many years after, I was no longer troubled by doubts, and indeed went through the usual religious forms with very little thought about their origin.

At length, as time went on, my bodily health was restored by my native air: my mind, however, receiving but little instruction, began, I imagine, to prey upon itself—such at least I infer to have been the case from the following circumstance. One day, when uninterested in the sports of my little companions, I had retired into the shrubbery and was leaning my head, supported by my left arm, upon the lower branch of a thorn-tree. Listless and unoccupied, I *imagined* I had a head-ache. After a time I perceived, lying on the ground just under me, a small bright bit of metal. I instantly seized the precious discovery, and turning it over, examined both sides. I immediately concluded that I had discovered some valuable treasure, and running away to my deserted companions, showed them my golden coin. The little company became greatly excited, and declared that it must be gold, and that it was a piece of money of great value. We ran off to get the opinion of the usher; but whether he partook of the delusion, or we acquired our knowledge from the higher authority of the master, I know not. I only recollect the entire dissipation of my head-ache, and then my ultimate great disappointment when it was pronounced, upon the undoubted authority of the village doctor, that the square piece of brass I had found was a half-dram weight which had escaped from the box of a pair of medical scales. This little incident had an important effect upon my after-life. I reflected upon the extraordinary fact, that my head-ache had been entirely cured by the discovery of the piece of brass. Although I may not have put into words the principle, *that occupation of the mind is such a source of pleasure that it can relieve even the pain of a head-ache*; yet I am sure it practically gave an additional stimulus to me in many a difficult inquiry. Some few years after, when suffering under a form of tooth-ache, not acute though tediously wearing, I often had recourse to a volume of Don Quixote, and still more frequently to one of Robinson Crusoe. Although at first it required a painful effort of attention, yet it almost always happened, after a time, that I had forgotten the moderate pain in the overpowering interest of the novel.

My most intimate companion and friend was a boy named Dacres, the son of Admiral Richard Dacres. We had often talked over such questions as those I have mentioned in this chapter, and we had made an agreement that whichever died first should, if possible, appear to the other after death, in order to satisfy the survivor about their solution.

After a year or two my young friend entered the navy, but we kept up our friendship, and when he was ashore I saw him frequently. He was in a ship of eighty guns at the passage of the Dardanelles, under the command of Sir Thomas Duckworth. Ultimately he was sent home in charge of a prize-ship, in which he suffered the severest hardships during a long and tempestuous voyage, and then died of consumption.

I saw him a few days before his death, at the age of about eighteen. We talked of former times, but neither of us mentioned the compact. I believed it occurred to his mind: it was certainly strongly present to my own.

He died a few days after. On the evening of that day I retired to my own room, which was partially detached from the house by an intervening conservatory. I sat up until after midnight, endeavouring to read, but found it impossible to fix my attention on any subject, except the overpowering feeling of curiosity, which absorbed my mind. I then undressed and went into bed; but sleep was entirely banished. I had previously carefully examined whether any cat, bird, or living animal might be accidentally concealed in my room, and I had studied the forms of the furniture lest they should in the darkness mislead me.

I passed a night of perfect sleeplessness. The distant clock and a faithful dog, just outside my own door, produced the only sounds which disturbed the intense silence of that anxious night.

CHAPTER III

BOYHOOD

DURING MY boyhood my mother took me to several exhibitions of machinery. I well remember one of them in Hanover Square, by a man who called himself Merlin. I was so greatly interested in it, that the Exhibitor remarked the circumstance, and after explaining some of the objects to which the public had access, proposed to my mother to take me up to his workshop, where I should see still more wonderful automata. We accordingly ascended to the attic. There were two uncovered female figures of silver, about twelve inches high.

One of these walked or rather glided along a space of about four feet, when she turned round and went back to her original place. She used an eye-glass occasionally, and bowed frequently, as if recognizing her acquaintances. The motions of her limbs were singularly graceful.

The other silver figure was an admirable *danseuse*, with a bird on the fore finger of her right hand, which wagged its tail, flapped its wings, and opened its beak. This lady attitudinized in a most fascinating manner. Her eyes were full of imagination, and irresistible.

These silver figures were the chef-d'œuvres of the artist: they had cost him years of unwearied labour, and were not even then finished.

After I left Devonshire I was placed at a school in the neighbourhood of London, in which there were about thirty boys.

My first experience was unfortunate, and probably gave an unfavourable turn to my whole career during my residence of three years.

After I had been at school a few weeks, I went with one of my companions into the play-ground in the dusk of the evening. We heard a noise, as of people talking in an orchard at some distance, which belonged to our master. As the orchard had recently been robbed, we thought that thieves were again at work. We accordingly climbed over the boundary wall, ran across the field, and saw in the orchard beyond a couple of fellows evidently running away. We pursued as fast as our legs could carry us, and just got up to the supposed thieves at the ditch on the opposite side of the orchard.

A roar of laughter then greeted us from two of our own companions, who had entered the orchard for the purpose of getting some manure for their flowers out of a rotten mulberry-tree. These boys were aware of our mistake, and had humoured it.

We now returned all together towards the play-ground, when we met our master, who immediately pronounced that we were each fined one shilling for being out of bounds. We two boys who had gone out of bounds to protect our master's property, and who if thieves had really been there would probably have been half-killed by them, attempted to remonstrate and explain the case; but all remonstrance was vain, and we were accordingly fined. I never forgot that injustice.

The school-room adjoined the house, but was not directly connected with it. It contained a library of about three hundred volumes on various subjects, generally very well selected; it also contained one or two works on subjects which do not usually attract at that period of life. I derived much advantage from this library; and I now mention it because I think it of great importance that a library should exist in every school-room.

Amongst the books was a treatise on Algebra, called "Ward's Young Mathematician's Guide." I was always partial to my arithmetical lessons, but this book attracted my particular attention. After I had been at this school for about a twelvemonth, I proposed to one of my school-fellows, who was of a studious habit, that we should get up every morning at three o'clock, light a fire in the school-room, and work until five or half-past five. We accomplished this pretty regularly for several months. Our plan had, however, become partially known to a few of our companions. One of these, a tall boy, bigger than ourselves, having heard of it, asked me to allow him to get up with us, urging that his sole object was to study, and that it would be of great importance to him in after-life. I had the cruelty to refuse this very reasonable request. The subject has often recurred to my memory, but never without regret.

Another of my young companions, Frederick Marryat,* made the same request, but not with the same motive. I told him we got up in order to work; that he would only play, and that we should then be

* Afterwards Captain Marryat.

found out. After some time, having exhausted all his arguments, Marryat told me he was determined to get up, and would do it whether I liked it or not.

Marryat slept in the same room as myself: it contained five beds. Our room opened upon a landing, and its door was exactly opposite that of the master. A flight of stairs led up to a passage just over the room in which the master and mistress slept. Passing along this passage, another flight of stairs led down, on the other side of the master's bed-room, to another landing, from which another flight of stairs led down to the external door of the house, leading by a long passage to the school-room.

Through this devious course I had cautiously threaded my way, calling up my companion in his room at the top of the last flight of stairs, almost every night for several months.

One night on trying to open the door of my own bed-room, I found Marryat's bed projecting a little before the door, so that I could not open it. I perceived that this was done purposely, in order that I might awaken him. I therefore cautiously, and by degrees, pushed his bed back without awaking him, and went as usual to my work. This occurred two or three nights successively.

One night, however, I found a piece of pack-thread tied to the door lock, which I traced to Marryat's bed, and concluded it was tied to his arm or hand. I merely untied the cord from the lock, and passed on.

A few nights after I found it impossible to untie the cord, so I cut it with my pocket-knife. The cord then became thicker and thicker for several nights, but still my pen-knife did its work.

One night I found a small chain fixed to the lock, and passing thence into Marryat's bed. This defeated my efforts for that night, and I retired to my own bed. The next night I was provided with a pair of plyers, and unbent one of the links, leaving the two portions attached to Marryat's arm and to the lock of the door. This occurred several times, varying by stouter chains, and by having a padlock which I could not pick in the dark.

At last one morning I found a chain too strong for the tools I possessed; so I retired to my own bed, defeated. The next night, however, I provided myself with a ball of packthread. As soon as I heard by his breathing that Marryat was asleep, I crept over to the door, drew one end of my ball of packthread through a link of the too-powerful chain, and bringing it back with me to bed, gave it a sudden jerk by pulling both ends of the packthread passing through the link of the chain.

Marryat jumped up, put out his hand to the door, found his chain all right, and then lay down. As soon as he was asleep again, I repeated the operation. Having awakened him for the third time, I let go one end of the string, and drew it back by the other, so that he was unable at daylight to detect the cause.

At last, however, I found it expedient to enter into a treaty of peace, the basis of which was that I should allow Marryat to join the night party; but that nobody else should be admitted. This continued for a short time; but, one by one, three or four other boys, friends of Marryat, joined our party, and, as I had anticipated, no work was done. We all got to play; we let off fire-works in the play-ground, and were of course discovered.

Our master read us a very grave lecture at breakfast upon the impropriety of this irregular system of turning night into day, and pointed out its injurious effects upon the health. This, he said, was so remarkable that he could distinguish by their pallid counten-ances those who had taken part in it. Now he certainly did point out every boy who had been up on the night we were detected. But it appeared to me very odd that the same means of judging had not enabled him long before to discover the two boys who had for several months habitually practised this system of turning night into day.

Another of our pranks never received its solution in our master's mind; indeed I myself scarcely knew its early history. Somehow or other, a Russian young gentleman, who was a parlour-boarder, had I believe, expatiated to Marryat on the virtues of Cognac.

One evening my friend came to me with a quart bottle of what he called excellent stuff. A council was held amongst a few of us boys to decide how we should dispose of this treasure. I did not myself much admire the liquid, but suggested that it might be very good when mixed up with a lot of treacle. This thought was unanimously adopted, and a subscription made to purchase the treacle. Having no vessel sufficiently large to hold the intended mixture, I proposed to take one of our garden-pots, stopping up the hole in its bottom with a cork.

A good big earthen vessel, thus extemporised, was then filled with this wonderful mixture. A spoon or two, an oystershell, and various other contrivances delivered it to its numerous consumers, and all the boys got a greater or less share, according to their taste for this extra-ordinary liqueur.

The feast was over, the garden-pot was restored to its owner, and the treacled lips of the boys had been wiped with their handkerchiefs or on their coat-sleeves, when the bell announced that it was prayer-time. We all knelt in silence at our respective desks. As soon as the prayers were over, one of the oddest scenes occurred.

Many boys rose up from their knees—but some fell down again. Some turned round several times, and then fell. Some turned round so often that they resembled spinning dervishes. Others were only more stupid than usual; some complained of being sick; many were very sleepy; others were sound asleep, and had to be carried to bed; some talked fast and heroically, two attempted psalmody, but none listened.

All investigation at the time was useless: we were sent off to bed as quickly as possible. It was only known that Count Cognac had married the sweet Miss Treacle, whom all the boys knew and loved, and who lodged at the grocer's, in the neighbouring village. But I believe neither the pedigree of the bridegroom nor his domicile were ever discovered. It is probable that he was of French origin, and dwelt in a cellar.

After I left this school I was for a few years under the care of an excellent clergyman in the neighbourhood of Cambridge. There were only six boys; but I fear I did not derive from it all the advantage that I might have done. I came into frequent contact with the Rev. Charles Simeon, and with many of his enthusiastic disciples. Every Sunday I had to write from memory an abstract of the sermon he preached in our village. Even at that period of my life I had a taste for generalization. Accordingly, having generalized some of Mr. Simeon's sermons up to a kind of skeleton form, I tried, by way of experiment, to fill up such a form in a sermon of my own composing from the text of "Alexander the coppersmith hath done us much harm." As well as I remember, there were in my sermon some queer deductions from this text; but then they fulfilled all the usual conditions of our sermons: so thought also two of my companions to whom I communicated *in confidence* this new manufacture.

By some unexplained circumstance my sermon relating to copper being isomorphous with Simeon's own productions, got by substitution into the hands of our master as the recollections of one of the other boys. Thereupon arose an awful explosion which I decline to paint.

I did, however, learn something at this school, for I observed a striking illustration of the Economy of Manufactures. Mr. Simeon had the cure of a very wicked parish in Cambridge, whilst my instructor held that of a tolerably decent country village. If each minister had stuck to the instruction of his own parish, it would have necessitated the manufacture of four sermons per week, whilst, by this beneficial interchange of duties, only two were required.

Each congregation enjoyed also another advantage from this arrangement—the advantage of variety, which, when moderately indulged in, excites the appetite.

CHAPTER IV

CAMBRIDGE

Universal Language—Purchase Lacroix's Quarto Work on the Integral Calculus—
Disappointment on getting no explanation of my Mathematical Difficulties—Origin
of the Analytical Society—The Ghost Club—Chess—Sixpenny Whist and Guinea
Whist—Boating—Chemistry—Elected Lucasian Professor of Mathematics in 1828.

MY FATHER, with a view of acquiring some information which might
be of use to me at Cambridge, had consulted a tutor of one of the
colleges, who was passing his long vacation at the neighbouring
watering-place, Teignmouth. He dined with us frequently. The
advice of the Rev. Doctor was quite sound, but very limited. It
might be summed up in one short sentence: "Advise your son not to
purchase his wine in Cambridge."

Previously to my entrance at Trinity College, Cambridge, I resided
for a time at Totnes, under the guidance of an Oxford tutor, who
undertook to superintend my classical studies only.

During my residence at this place I accidentally heard, for the
first time, of an idea of forming a universal language. I was much
fascinated by it, and, soon after, proceeded to write a kind of grammar,
and then to devise a dictionary. Some trace of the former, I think, I
still possess: but I was stopped in my idea of making a universal
dictionary by the apparent impossibility of arranging signs in any
consecutive order, so as to find, as in a dictionary, the meaning of each
when wanted. It was only after I had been some time at Cambridge
that I became acquainted with the work of "Bishop Wilkins on
Universal Language."

Being passionately fond of algebra, I had instructed myself by means
of Ward's "Young Mathematician's Guide," which had casually
fallen into my hands at school. I now employed all my leisure in
studying such mathematical works as accident brought to my knowledge. Amongst these were Humphrey Ditton's "Fluxions," of which
I could make nothing; Madame Agnesi's "Analytical Institutions,"
from which I acquired some knowledge; Woodhouse's "Principles of
Analytical Calculation," from which I learned the notation of Leibnitz;
and Lagrange's "Théorie des Fonctions." I possessed also the
Fluxions of Maclaurin and of Simpson.

22

Thus it happened that when I went to Cambridge I could work out such questions as the very moderate amount of mathematics which I then possessed admitted, with equal facility, in the dots of Newton, the d's of Leibnitz, or the dashes of Lagrange. I had, however, met with many difficulties, and looked forward with intense delight to the certainty of having them all removed on my arrival at Cambridge. I had in my imagination formed a plan for the institution amongst my future friends of a chess club, and also of another club for the discussion of mathematical subjects.

In 1811, during the war, it was very difficult to procure foreign books. I had heard of the great work of Lacroix, on the "Differential and Integral Calculus," which I longed to possess, and being mis-informed that its price was two guineas, I resolved to purchase it in London on my passage to Cambridge. As soon as I arrived I went to the French bookseller, Dulau, and to my great surprise found that the price of the book was seven guineas. After much thought I made the costly purchase, went on immediately to Cambridge, saw my tutor Hudson, got lodgings, and then spent the greater part of the night in turning over the pages of my newly-acquired purchase. After a few days, I went to my public tutor Hudson, to ask the explanation of one of my mathematical difficulties. He listened to my question, said it would not be asked in the Senate House, and was of no sort of conse-quence, and advised me to get up the earlier subjects of the university studies.

After some little while I went to ask the explanation of another difficulty from one of the lecturers. He treated the question just in the same way. I made a third effort to be enlightened about what was really a doubtful question, and felt satisfied that the person I addressed knew nothing of the matter, although he took some pains to disguise his ignorance.

I thus acquired a distaste for the routine of the studies of the place, and devoured the papers of Euler and other mathematicians, scattered through innumerable volumes of the academies of Petersburgh, Berlin, and Paris, which the libraries I had recourse to contained.

Under these circumstances it was not surprising that I should perceive and be penetrated with the superior power of the notation of Leibnitz.

At an early period, probably at the commencement of the second year of my residence at Cambridge, a friend of mine, Michael Slegg, of Trinity, was taking wine with me, discussing mathematical subjects, to which he also was enthusiastically attached. Hearing the chapel bell ring, he took leave of me, promising to return for a cup of coffee.

At this period Cambridge was agitated by a fierce controversy. Societies had been formed for printing and circulating the Bible. One party proposed to circulate it with notes, in order to make it intelligible; whilst the other scornfully rejected all explanations of the word of God as profane attempts to mend that which was perfect.

The walls of the town were placarded with broadsides, and posters were sent from house to house. One of the latter form of advertisement was lying upon my table when Slegg left me. Taking up the paper, and looking through it, I thought it, from its exaggerated tone, a good subject for a parody.

I then drew up the sketch of a society to be instituted for translating the small work of Lacroix on the Differential and Integral Calculus. It proposed that we should have periodical meetings for the propagation of d's; and consigned to perdition all who supported the heresy of dots. It maintained that the work of Lacroix was so perfect that any comment was unnecessary.

On Slegg's return from chapel I put the parody into his hands. My friend enjoyed the joke heartily, and at parting asked my permission to show the parody to a mathematical friend of his, Mr. Bromhead.*

The next day Slegg called on me, and said that he had put the joke into the hand of his friend, who, after laughing heartily, remarked that it was too good a joke to be lost, and proposed seriously that we should form a society for the cultivation of mathematics.

The next day Bromhead called on me. We talked the subject over, and agreed to hold a meeting at his lodgings for the purpose of forming a society for the promotion of analysis.

At that meeting, besides the projectors, there were present Herschel, Peacock, D'Arblay,† Ryan,‡ Robinson,§ Frederick Maule,‖ and several others. We constituted ourselves "The Analytical Society"; hired a meeting-room, open daily; held meetings, read papers, and discussed them. Of course we were much ridiculed by the Dons; and, not being put down, it was darkly hinted that we were young infidels, and that no good would come of us.

In the meantime we quietly pursued our course, and at last resolved to publish a volume of our Transactions. Owing to the illness of one of the number, and to various other circumstances, the volume which was published was entirely contributed by Herschel and myself.

---

* Afterwards Sir Edward French Bromhead, Bart., the author of an interesting paper in the Transactions of the Royal Society.
† The only son of Madame D'Arblay.
‡ Now the Right Honourable Sir Edward Ryan.
§ The Rev. Dr. Robinson, Master of the Temple.
‖ A younger brother of the late Mr. Justice Maule.

At last our work was printed, and it became necessary to decide upon a title. Recalling the slight imputation which had been made upon our faith, I suggested that the most appropriate title would be—

The Principles of pure D-ism in opposition to the Dot-age of the University.*

In thus reviving this wicked pun, I ought at the same time to record an instance of forgiveness unparalleled in history. Fourteen years after, being then at Rome, I accidentally read in Galignani's newspaper the following paragraph, dated Cambridge:—"Yesterday the bells of St. Mary rang on the election of Mr. Babbage as Lucasian Professor of Mathematics."

If this event had happened during the lifetime of my father, it would have been most gratifying to myself, because, whilst it would have given him much pleasure, it would then also have afforded intense delight to my mother.

I concluded that the next post would bring me the official confirmation of this report, and after some consideration I sketched the draft of a letter, in which I proposed to thank the University sincerely for the honour they had done me, but to decline it.

This sketch of a letter was hardly dry when two of my intimate friends, the Rev. Mr. Lunn and Mr. Beilby Thompson,† who resided close to me in the Piazza del Populo, came over to congratulate me on the appointment. I showed them my proposed reply, against which they earnestly protested. Their first, and as they believed their strongest, reason was that it would give so much pleasure to my mother. To this I answered that my mother's opinion of her son had been confirmed by the reception he had met with in every foreign country he had visited, and that this, in her estimation, would add but little to it. To their next argument I had no satisfactory answer. It was that this election could not have occurred unless some friends of mine in England had taken active measures to promote it; that some of these might have been personal friends, but that many others might have exerted themselves entirely upon principle, and that it would be harsh to disappoint such friends, and reject such a compliment.

My own feelings were of a mixed nature. I saw the vast field that the Difference Engine had opened out; for, before I left England in the previous year, I had extended its mechanism to the tabulation of functions having no constant difference, and more particularly I had arrived at the knowledge of the entire command it would have over

* Leibnitz indicated fluxions by a *d*, Newton by a dot.
† Afterwards Lord Wenlock.

the computation of the most important classes of tables, those of astronomy and of navigation. I was also most anxious to give my whole time to the completion of the mechanism of the Difference Engine No. 1 which I had then in hand. Small as the admitted duties of the Lucasian Chair were, I felt that they would absorb time which I thought better devoted to the completion of the Difference Engine. If I had then been aware that the lapse of a few years would have thrown upon me the enormous labour which the Analytical Engine absorbed, no motive short of absolute necessity would have induced me to accept any office which might, in the slightest degree, withdraw my attention from its contrivance.

The result of this consultation with my two friends was that I determined to accept the Chair of Newton, and to hold it for a few years. In 1839 the demands of the Analytical Engine upon my attention had become so incessant and so exhausting, that even the few duties of the Lucasian Chair had a sensible effect in impairing my bodily strength. I therefore sent in my resignation.

In January, 1829, I visited Cambridge, to fulfil one of the first duties of my new office, the examination for Dr. Smith's prizes.

These two prizes, of twenty-five pounds each, exercise a very curious and important influence. Usually three or four hundred young men are examined previously to taking their degree. The University officers examine and place them in the order of their mathematical merit. The class called Wranglers is the highest; of these the first is called the senior wrangler, the others the second and third, &c., wranglers.

All the young men who have just taken their degree, whether with or without honours, are qualified to compete for the Smith's prizes by sending in notice to the electors, who consist of the three Professors of Geometry, Astronomy, and Physics, assisted occasionally by two official electors, the Vice-Chancellor and the Master of Trinity College. However, in point of fact, generally three, and rarely above six young men compete.

It is manifest that the University officers, who examine several hundred young men, cannot bestow the same minute attention upon each as those who, at the utmost, only examine six. Nor is this of any importance, except to the few first wranglers, who usually are candidates for these prizes. The consequence is that the examiners of the Smith's prizes constitute, as it were, a court of appeal from the decision of the University officers. The decision of the latter is thus therefore, necessarily appealed against upon every occasion. Perhaps in one out of five or six cases the second or third wrangler obtains the first Smith's prize. I may add that in the few cases known to me previously to my becoming an examiner, the public opinion of the University always approved those decisions, without implying any censure on the officers of the University.

In forming my set of questions, I consulted the late Dean of Ely and another friend, in order that I might not suddenly deviate too much from the usual style of examinations.

After having examined the young men, I sat up the whole night, carefully weighing the relative merits of their answers. I found, with some mortification, that, according to my marks, the second wrangler ought to have the first prize. I therefore put aside the papers until the day before the decision. I then took an unmarked copy of my questions, and put new numbers for their respective values. After very carefully going over the whole of the examination-papers again, I arrived almost exactly at my former conclusion.

On our meeting at the Vice-Chancellor's, that functionary asked me, as the senior professor, what was my decision as to the two prizes. I stated that the result of my examination obliged me to award the first prize to the second wrangler. Professor Airy was then asked the same question. He made the same reply. Professor Lax being then asked, said he had arrived at the same conclusion as his two colleagues.

The Vice-Chancellor remarked that when we altered the arrangement of the University Examiners, it was very satisfactory that we should be unanimous. Professor Airy observed that this satisfaction was enhanced by the fact of the remarkable difference in the tastes of the three examiners.

The Vice-Chancellor, turning to me, asked whether it might be permitted to inquire the numbers we had respectively assigned to each candidate.

I and my colleagues immediately mentioned our numbers, which Professor Airy at once reduced to a common scale. On this it appeared that the number of marks assigned to each by Professor Airy and myself very nearly agreed, whilst that of Professor Lax differed but little.

On this occasion the first Smith's prize was assigned to the second wrangler, Mr. Cavendish, now Duke of Devonshire, the present Chancellor of the University.

The result of the whole of my after-experience showed that amongst the highest men the peculiar tastes of the examiners had no effect in disturbing the proper decision.

I held the Chair of Newton for some few years, and still feel deeply grateful for the honour the University conferred upon me—the only honour I ever received in my own country.*

I must now return to my pursuits during my residence at Cambridge, the account of which has been partially interrupted by the history of my appointment to the Chair of Newton.

Whilst I was an undergraduate, I lived probably in a greater variety of sets than any of my young companions. But my chief and choicest consisted of some ten or a dozen friends who usually breakfasted with me every Sunday after chapel; arriving at about nine, and remaining to between twelve and one o'clock. We discussed all knowable and many unknowable things.

At one time we resolved ourselves into a Ghost Club, and proceeded to collect evidence, and entered into a considerable correspondence upon the subject. Some of this was both interesting and instructive.

At another time we resolved ourselves into a Club which we called The Extractors. Its rules were as follows,—

1st. Every member shall communicate his address to the Secretary once in six months.

2nd. If this communication is delayed beyond twelve months, it shall be taken for granted that his relatives had shut him up as insane.

3rd. Every effort legal and illegal shall be made to get him out of the madhouse. Hence the name of the club—The Extractors.

4th. Every candidate for admission as a member shall produce six certificates. Three that he is sane and three others that he is insane.

It has often occurred to me to inquire of my legal friends whether, if the sanity of any member of the club had been questioned in after-life, he would have adduced the fact of membership of the Club of Extractors as an indication of sanity or of insanity.

During the first part of my residence at Cambridge, I played at chess very frequently, often with D'Arblay and with several other good players. There was at that period a fellow-commoner at Trinity named Brande, who devoted almost his whole time to the study of chess. I was invited to meet him one evening at the rooms of a common friend for the purpose of trying our strength.

On arriving at my friend's rooms, I found a note informing me that he had gone to Newmarket, and had left coffee and the chessmen for us. I was myself tormented by great shyness, and my yet unseen

---

* This professorship is not in the gift of the Government. The electors are the masters of the various colleges. It was founded in 1663 by Henry Lucas, M.P. for the University, and was endowed by him with a small estate in Bedfordshire. During my tenure of that office my net receipts were between 80*l.* and 90*l.* a year. I am glad to find that the estate is now improved, and that the University have added an annual salary to the Chair of Newton.

adversary was, I understood, equally diffident.   I was sitting before the chess-board when Brande entered.   I rose, he advanced, sat down, and took a white and a black pawn from the board, which he held, one in either hand.   I pointed with my finger to the left hand and won the move.

The game then commenced; it was rather a long one, and I won it: but not a word was exchanged until the end: when Brande uttered the first word.   "Another?"   To this I nodded assent.

How that game was decided I do not now remember; but the first sentence pronounced by either of us, was a remark by Brande, that he had lost the first game by a certain move of his white bishop.   To this I replied, that I thought he was mistaken, and that the real cause of his losing the game arose from the use I had made of my knight two moves previously to his white bishop's move.

We then immediately began to replace the men on the board in the positions they occupied at that particular point of the game when the white bishop's move was made.   Each took up any piece indiscriminately, and placed it without hesitation on the exact square on which it had stood.   It then became apparent that the effective move to which I had referred was that of my knight.

Brande, during his residence at Cambridge, studied chess regularly several hours each day, and read almost every treatise on the subject. After he left college he travelled abroad, took lessons from every celebrated teacher, and played with all the most eminent players on the Continent.

At intervals of three or four years I occasionally met him in London. After the usual greeting he always proposed that we should play a game of chess.

I found on these occasions, that if I played any of the ordinary openings, such as are found in the books, I was sure to be beaten. The only way in which I had a chance of winning, was by making early in the game a move so bad that it had not been mentioned in any treatise.   Brande possessed, and had read, almost every book upon the subject.

Another set which I frequently joined were addicted to sixpenny whist.   It consisted of Higman, afterwards Tutor of Trinity; Follet, afterwards Attorney-General; of a learned and accomplished Dean still living, and I have no doubt still playing an excellent rubber, and myself.   We not unfrequently sat from chapel-time in the evening until the sound of the morning chapel bell again called us to our religious duties.

I mixed occasionally with a different set of whist players at Jesus

College.   They played high: guinea points, and five guineas on the
rubber.   I was always a most welcome visitor, not from my skill at the
game; but because I never played more than shilling points and five
shillings on the rubber.   Consequently my partner had what they
considered an advantage: namely, that of playing guinea points with
one of our adversaries and pound points with the other.

Totally different in character was another set in which I mixed.   I
was very fond of boating, not of the manual labour of rowing, but the
more intellectual art of sailing.   I kept a beautiful light, London-
built boat, and occasionally took long voyages down the river, beyond
Ely into the fens.   To accomplish these trips, it was necessary to
have two or three strong fellows to row when the wind failed or was
contrary.   These were useful friends upon my aquatic expeditions, but
not being of exactly the same calibre as my friends of the Ghost Club,
were very cruelly and disrespectfully called by them "my Tom fools."

The plan of our voyage was thus:—I sent my servant to the apothe-
cary for a thing called an ægrotat, which I understood, for I never saw
one, meant a certificate that I was indisposed, and that it would be
injurious to my health to attend chapel, or hall, or lectures.   This was
forwarded to the college authorities.

I also directed my servant to order the cook to send me a large
well-seasoned meat pie, a couple of fowls, &c.   These were packed
in a hamper with three or four bottles of wine and one of noyeau.   We
sailed when the wind was fair, and rowed when there was none.
Whittlesea Mere was a very favourite resort for sailing, fishing, and
shooting.   Sometimes we reached Lynn.   After various adventures
and five or six days of hard exercise in the open air, we returned with
our health more renovated than if the best physician had prescribed
for us.

During my residence at Cambridge, Smithson Tennant was the
Professor of Chemistry, and I attended his lectures.   Having a spare
room, I turned it into a kind of laboratory, in which Herschel worked
with me, until he set up a rival one of his own.   We both occasionally
assisted the Professor in preparing his experiments.   The science of
chemistry had not then assumed the vast development it has now
attained.   I gave up its practical pursuit soon after I resided in
London, but I have never regretted the time I bestowed upon it at the
commencement of my career.   I had hoped to have long continued
to enjoy the friendship of my entertaining and valued instructor, and
to have profited by his introducing me to the science of the metropolis,
but his tragical fate deprived me of that advantage.   Whilst riding

with General Bulow across a drawbridge at Boulogne, the bolt having been displaced, Smithson Tennant was precipitated to the bottom, and killed on the spot. The General, having an earlier warning, set spurs to his horse, and just escaped a similar fate.

My views respecting the notation of Leibnitz now (1812) received confirmation from an extensive course of reading. I became convinced that the notation of fluxions must ultimately prove a strong impediment to the progress of English science. But I knew, also, that it was hopeless for any young and unknown author to attempt to introduce the notation of Leibnitz into an elementary work. This opinion naturally suggested to me the idea of translating the smaller work of Lacroix. It is possible, although I have no recollection of it, that the same idea may have occurred to several of my colleagues of the Analytical Society, but most of them were so occupied, first with their degree, and then with their examination for fellowships, that no steps were at that time taken by any of them on that subject.

Unencumbered by these distractions, I commenced the task, but at what period of time I do not exactly recollect. I had finished a portion of the translation, and laid it aside, when, some years afterwards, Peacock called on me in Devonshire Street, and stated that both Herschel and himself were convinced that the change from the dots to the d's would not be accomplished until some foreign work of eminence should be translated into English. Peacock then proposed that I should either finish the translation which I had commenced, or that Herschel and himself should complete the remainder of my translation. I suggested that we should toss up which alternative to take. It was determined by lot that we should make a joint translation. Some months after, the translation of the small work of Lacroix was published.

For several years after, the progress of the notation of Leibnitz at Cambridge was slow. It is true that the tutors of the two largest colleges had adopted it, but it was taught at none of the other colleges.

It is always difficult to think and reason in a new language, and this difficulty discouraged all but men of energetic minds. I saw, however, that, by making it their interest to do so, the change might be accomplished. I therefore proposed to make a large collection of examples of the differential and integral calculus, consisting merely of the statement of each problem and its final solution. I foresaw that if such a publication existed, all those tutors who did not approve of the change of the Newtonian notation would yet, in order to save their own time and trouble, go to this collection of examples to find problems to set to their pupils. After a short time the use of the new signs would become familiar, and I anticipated their general adoption at Cambridge as a matter of course.

I commenced by copying out a large portion of the work of Hirsch. I then communicated to Peacock and Herschel my view, and proposed that they should each contribute a portion.

Peacock considerably modified my plan by giving the process of solution to a large number of the questions. Herschel prepared the questions in finite differences, and I supplied the examples to the calculus of functions. In a very few years the change was completely established; and thus at last the English cultivators of mathematical science, untrammelled by a limited and imperfect system of signs, entered on equal terms into competition with their continental rivals.

CHAPTER V

DIFFERENCE ENGINE No. 1

"Oh no! we never mention it,
Its name is never heard."

Difference Engine No. 1—First Idea at Cambridge, 1812—Plan for Dividing Astronomical Instruments—Idea of a Machine to calculate Tables by Differences—Illustrations by Piles of Cannon-balls.

CALCULATING MACHINES comprise various pieces of mechanism for assisting the human mind in executing the operations of arithmetic. Some few of these perform the whole operation without any mental attention when once the given numbers have been put into the machine.

Others require a moderate portion of mental attention: these latter are generally of much simpler construction than the former, and it may also be added, are less useful.

The simplest way of deciding to which of these two classes any calculating machine belongs is to ask its maker—Whether, when the numbers on which it is to operate are placed in the instrument, it is capable of arriving at its result by the mere motion of a spring, a descending weight, or any other constant force? If the answer be in the affirmative, the machine is really automatic; if otherwise, it is not self-acting.

Of the various machines I have had occasion to examine, many of those for Addition and Subtraction have been found to be automatic. Of machines for Multiplication and Division, which have fully come under my examination, I cannot at present recall one to my memory as absolutely fulfilling this condition.

The earliest idea that I can trace in my own mind of calculating arithmetical Tables by machinery arose in this manner:—

One evening I was sitting in the rooms of the Analytical Society, at Cambridge, my head leaning forward on the Table in a kind of dreamy mood, with a Table of logarithms lying open before me. Another member, coming into the room, and seeing me half asleep, called out, "Well, Babbage, what are you dreaming about?" to which I replied, "I am thinking that all these Tables (pointing to the logarithms) might be calculated by machinery."

I am indebted to my friend, the Rev. Dr. Robinson, the Master of the Temple, for this anecdote.   The event must have happened either in 1812 or 1813.

About 1819 I was occupied with devising means for accurately dividing astronomical instruments, and had arrived at a plan which I thought was likely to succeed perfectly.   I had also at that time been speculating about making machinery to compute arithmetical Tables.

One morning I called upon the late Dr. Wollaston, to consult him about my plan for dividing instruments.   On talking over the matter, it turned out that my system was exactly that which had been described by the Duke de Chaulnes, in the Memoirs of the French Academy of Sciences, about fifty or sixty years before.   I then mentioned my other idea of computing Tables by machinery, which Dr. Wollaston thought a more promising subject.

I considered that a machine to execute the mere isolated operations of arithmetic, would be comparatively of little value, unless it were very easily set to do its work, and unless it executed not only accurately, but with great rapidity, whatever it was required to do.

On the other hand, the method of differences supplied a general principle by which *all* Tables might be computed through limited intervals, by one uniform process.   Again, the method of differences required the use of mechanism for Addition only.   In order, however, to insure accuracy in the printed Tables, it was necessary that the machine which computed Tables should also set them up in type, or else supply a mould in which stereotype plates of those Tables could be cast.

I now began to sketch out arrangements for accomplishing the several partial processes which were required.   The arithmetical part must consist of two distinct processes—the power of adding one digit to another, and also of carrying the tens to the next digit, if it should be necessary.

The first idea was, naturally, to add each digit successively.   This, however, would occupy much time if the numbers added together consisted of many places of figures.

The next step was to add all the digits of the two numbers each to each at the same instant, but reserving a certain mechanical memorandum, wherever a carriage became due.   These carriages were then to be executed successively.

Having made various drawings, I now began to make models of some portions of the machine, to see how they would act.   Each number was to be expressed upon wheels placed upon an axis; there being one wheel for each figure in the number operated upon.

Having arrived at a certain point in my progress, it became necessary to have teeth of a peculiar form cut upon these wheels. As my own lathe was not fit for this job, I took the wheels to a wheel-cutter at Lambeth, to whom I carefully conveyed my instructions, leaving with him a drawing as his guide.

These wheels arrived late one night, and the next morning I began putting them in action with my other mechanism, when, to my utter astonishment, I found they were quite unfit for their task. I examined the shape of their teeth, compared them with those in the drawings, and found they agreed perfectly; yet they could not perform their intended work. I had been so certain of the truth of my previous reasoning, that I now began to be somewhat uneasy. I reflected that, if the reasoning about which I had been so certain should prove to have been really fallacious, I could then no longer trust the power of my own reason. I therefore went over with my wheels to the artist who had formed the teeth, in order that I might arrive at some explanation of this extraordinary contradiction.

On conferring with him, it turned out that, when he had understood fully the peculiar form of the teeth of wheels, he discovered that his wheel-cutting engine had not got amongst its divisions that precise number which I had required. He therefore had asked me whether another number, which his machine possessed, would not equally answer my object. I had inadvertently replied in the affirmative. He then made arrangements for the precise number of teeth I required; and the new wheels performed their expected duty perfectly.

The next step was to devise means for printing the tables to be computed by this machine. My first plan was to make it put together moveable type. I proposed to make metal boxes, each containing 3,000 types of one of the ten digits. These types were to be made to pass out one by one from the bottom of their boxes, when required by the computing part of the machine.

But here a new difficulty arose. The attendant who put the types into the boxes might, by mistake, put a wrong type in one or more of them. This cause of error I removed in the following manner:— There are usually certain notches in the side of the type. I caused these notches to be so placed that all the types of any given digit possessed the same characteristic notches, which no other type had. Thus, when the boxes were filled, by passing a small wire down these peculiar notches, it would be impeded in its passage, if there were included in the row a single wrong figure. Also, if any digit were accidentally turned upside down, it would be indicated by the stoppage of the testing wire.

One notch was reserved as common to every species of type. The object of this was that, before the types which the Difference Engine had used for its computation were removed from the iron platform on which they were placed, a steel wire should be passed through this common notch, and remain there. The tables, composed of moveable types, thus interlocked, could never have any of their figures drawn out by adhesion to the inking-roller, and then by possibility be restored in an inverted order. A small block of such figures tied together by a bit of string, remained unbroken for several years, although it was rather roughly used as a plaything by my children. One such box was finished, and delivered its type satisfactorily.

Another plan for printing the tables, was to place the ordinary printing type round the edges of wheels. Then, as each successive number was produced by the arithmetical part, the type-wheels would move down upon a plate of soft composition, upon which the tabular number would be impressed. This mould was formed of a mixture of plaster-of-Paris with other materials, so as to become hard in the course of a few hours.

The first difficulty arose from the impression of one tabular number on the mould being distorted by the succeeding one.

I was not then aware that a very slight depth of impression from the type would be quite sufficient. I surmounted the difficulty by previously passing a roller, having longitudinal wedge-shaped projections, over the plastic material. This formed a series of small depressions in the matrix between each line. Thus the expansion arising from the impression of one line partially filled up the small depression or ditch which occurred between each successive line.

The various minute difficulties of this kind were successively overcome; but subsequent experience had proved that the depth necessary for stereotype moulds is very small, and that even thick paper, prepared in a peculiar manner, is quite sufficient for the purpose.

Another series of experiments were, however, made for the purpose of punching the computed numbers upon copper plate. A special machine was contrived and constructed, which might be called a co-ordinate machine, because it moved the copper plate and steel punches in the direction of three rectangular co-ordinates. This machine was afterwards found very useful for many other purposes. It was, in fact, a general shaping machine, upon which many parts of the Difference Engine were formed.

Several specimens of surface and copper-plate printing, as well as of the copper plates, produced by these means, were exhibited at the Exhibition of 1862.

I have proposed and drawn various machines for the purpose of calculating a series of numbers forming Tables by means of a certain system called "The Method of Differences," which it is the object of this sketch to explain.

The first Difference Engine with which I am acquainted comprised a few figures, and was made by myself, between 1820 and June 1822. It consisted of from six to eight figures. A much larger and more perfect engine was subsequently commenced in 1823 for the Government.

It was proposed that this latter Difference Engine should have six orders of differences, each consisting of about twenty places of figures, and also that it should print the Tables it computed.

The small portion of it which was placed in the International Exhibition of 1862 was put together nearly thirty years ago. It was accompanied by various parts intended to enable it to print the results it calculated, either as a single copy on paper—or by putting together moveable types—or by stereotype plates taken from moulds punched by the machine—or from copper plates impressed by it. The parts necessary for the execution of each of these processes were made, but these were not at that time attached to the calculating part of the machine.

A considerable number of the parts by which the printing was to be accomplished, as also several specimens of portions of tables punched on copper, and of stereotype moulds, were exhibited in a glass case adjacent to the Engine.

In 1834 Dr. Lardner published, in the 'Edinburgh Review,'* a very elaborate description of this portion of the machine, in which he explained clearly the method of Differences.

It is very singular that two persons, one resident in London, the other in Sweden, should both have been struck, on reading this review, with the simplicity of the mathematical principle of differences as applied to the calculation of Tables, and should have been so fascinated with it as to have undertaken to construct a machine of the kind.

Mr. Deacon, of Beaufort House, Strand, whose mechanical skill is well known, made, for his own satisfaction, a small model of the calculating part of such a machine, which was shown only to a few friends, and of the existence of which I was not aware until after the Swedish machine was brought to London.

Mr. Scheutz, an eminent printer at Stockholm, had far greater difficulties to encounter. The construction of mechanism, as well as

* 'Edinburgh Review,' No. CXX, July, 1834. [*See* p. 163.]

the mathematical part of the question, was entirely new to him. He, however, undertook to make a machine having four differences, and fourteen places of figures, and capable of printing its own Tables.

After many years' indefatigable labour, and an almost ruinous expense, aided by grants from his Government, by the constant assistance of his son, and by the support of many enlightened members of the Swedish Academy, he completed his Difference Engine. It was brought to London, and some time afterwards exhibited at the great Exhibition at Paris. It was then purchased for the Dudley Observatory at Albany by an enlightened and public-spirited merchant of that city, John F. Rathbone, Esq.

An exact copy of this machine was made by Messrs. Donkin and Co., for the English Government, and is now in use in the Registrar-General's Department at Somerset House. It is very much to be regretted that this specimen of English workmanship was not exhibited in the International Exhibition.

Explanation of the Difference Engine

Those who are only familiar with ordinary arithmetic may, by following out with the pen some of the examples which will be given, easily make themselves acquainted with the simple principles on which the Difference Engine acts.

It is necessary to state distinctly at the outset, that the Difference Engine is not intended to answer special questions. Its object is to calculate and print a *series* of results formed according to given laws. These are called Tables—many such are in use in various trades. For example—there are collections of Tables of the amount of any number of pounds from 1 to 100 lbs. of butchers' meat at various prices per lb. Let us examine one of these Tables: viz.—the price of meat 5*d.* per lb., we find

Number Lbs.	Table Price
	s. *d.*
1	0 5
2	0 10
3	1 3
4	1 8
5	2 1

There are two ways of computing this Table:—

1st. We might have multiplied the number of lbs. in each line by 5,

the price per lb., and have put down the result in *l. s. d.*, as in the
2nd column: or,

2nd. We might have put down the price of 1 lb., which is 5*d.*, and
have added five pence for each succeeding lb.

Let us now examine the relative advantages of each plan.   We shall
find that if we had multiplied each number of lbs. in the Table by 5,
and put down the resulting amount, then every number in the Table
would have been computed independently.   If, therefore, an error
had been committed, it would not have affected any but the single
tabular number at which it had been made.   On the other hand, if a
single error had occurred in the system of computing by adding five
at each step, any such error would have rendered the whole of the rest
of the Table untrue.

Thus the system of calculating by differences, which is the easiest,
is much more liable to error.   It has, on the other hand, this great
advantage: viz., that when the Table has been so computed, if we
calculate its last term directly, and if it agree with the last term found
by the continual addition of 5, we shall then be quite certain that every
term throughout is correct.   In the system of computing each term
directly, we possess no such check upon our accuracy.

Now the Table we have been considering is, in fact, merely a Table
whose first difference is constant and equal to five.   If we express it in
pence it becomes—

	Table	1st Difference
1	5	5
2	10	5
3	15	5
4	20	5
5	25	

Any machine, therefore, which could add one number to another,
and at the same time retain the original number called the first
difference for the next operation, would be able to compute all such
Tables.

Let us now consider another form of Table which might readily
occur to a boy playing with his marbles, or to a young lady with the
balls of her solitaire board.

The boy may place a row of his marbles on the sand, at equal
distances from each other, thus—

●   ●   ●   ●   ●

He might then, beginning with the second, place two other marbles

under each, thus—

He might then, beginning with the third, place three other marbles under each group, and so on; commencing always one group later, and making the addition one marble more each time.   The several groups would stand thus arranged—

   He will not fail to observe that he has thus formed a series of triangular groups, every group having an equal number of marbles in each of its three sides.   Also that the side of each successive group contains one more marble than that of its preceding group.
   Now an inquisitive boy would naturally count the numbers in each group and he would find them thus—

        1   3   6   10   15   21

   He might also want to know how many marbles the thirtieth or any other distant group might contain.   Perhaps he might go to papa to obtain this information; but I much fear papa would snub him, and would tell him that it was nonsense—that it was useless—that nobody knew the number, and so forth.   If the boy is told by papa, that he is not able to answer the question, then I recommend him to pay careful attention to whatever that father may at any time say, for he has overcome two of the greatest obstacles to the acquisition of knowledge—inasmuch as he possesses the consciousness that he does not know—and he has the moral courage to avow it.*
   If papa fail to inform him, let him go to mamma, who will not fail to find means to satisfy her darling's curiosity.   In the meantime the author of this sketch will endeavour to lead his young friend to make use of his own common sense for the purpose of becoming better acquainted with the triangular figures he has formed with his marbles.
   In the case of the Table of the price of butchers' meat, it was obvious that it could be formed by adding the same *constant* difference continually to the first term.   Now suppose we place the numbers of our groups of marbles in a column, as we did our prices of various weights

   * The most remarkable instance I ever met with of the distinctness with which any individual perceived the exact boundary of his own knowledge, was that of the late Dr. Wollaston.

of meat. Instead of adding a certain difference, as we did in the former case, let us subtract the figures representing each group of marbles from the figures of the succeeding group in the Table. The process will stand thus:—

TABLE

Number of the Group	Number of Marbles in each Group	1st Difference Difference between the number of Marbles in each Group and that in the next	2nd Difference
1	1	1	1
2	3	2	1
3	6	3	1
4	10	4	1
5	15	5	1
6	21	6	
7	28	7	

It is usual to call the third column thus formed *the column of first differences*. It is evident in the present instance that that column represents the natural numbers. But we already know that the first difference of the natural numbers is constant and equal to unity. It appears, therefore, that a Table of these numbers, representing the group of marbles, might be constructed to any extent by mere addition —using the number 1 as the first number of the Table, the number 1 as the first Difference, and also the number 1 as the second Difference, which last always remains constant.

Now as we could find the value of any given number of pounds of meat directly, without going through all the previous part of the Table, so by a somewhat different rule we can find at once the value of any group whose number is given.

Thus, if we require the number of marbles in the fifth group, proceed thus:—

Take the number of the group 5
Add 1 to this number, it becomes 6

Multiply these numbers together . . . . . . 2)30

Divide the product by 2 15
This gives 15, the number of marbles in the 5th group.

If the reader will take the trouble to calculate with his pencil the five groups given above, he will soon perceive the general truth of this rule.

We have now arrived at the fact that this Table—like that of the price of butchers' meat—can be calculated by two different methods. By the first, each number of the Table is calculated independently: by the second, the truth of each number depends upon the truth of all the previous numbers.

Perhaps my young friend may now ask me, What is the use of such Tables? Until he has advanced further in his arithmetical studies, he must take for granted that they are of some use. The very Table about which he has been reasoning possesses a special name—it is called a Table of Triangular Numbers. Almost every general collection of Tables hitherto published contains portions of it of more or less extent.

Above a century ago, a volume in small quarto, containing the first 20,000 triangular numbers, was published at the Hague by E. De Joncourt, A.M., and Professor of Philosophy.* I cannot resist quoting the author's enthusiastic expression of the happiness he enjoyed in composing his celebrated work:

"The Trigonals here to be found, and nowhere else, are exactly elaborate. Let the candid reader make the best of these numbers, and feel (if possible) in perusing my work the pleasure I had in composing it.

"That sweet joy may arise from such contemplations cannot be denied. Numbers and lines have many charms, unseen by vulgar eyes, and only discovered to the unwearied and respectful sons of Art. In features the serpentine line (who starts not at the name) produces beauty and love; and in numbers, high powers, and humble roots, give soft delight.

"Lo! the raptured arithmetician! Easily satisfied, he asks no Brussels lace, nor a coach and six. To calculate, contents his liveliest desires, and obedient numbers are within his reach."

I hope my young friend is acquainted with the fact—that the product of any number multiplied by itself is called the square of that number. Thus 36 is the product of 6 multiplied by 6, and 36 is called the square of 6. I would now recommend him to examine the series of square numbers

$$1, \quad 4, \quad 9, \quad 16, \quad 25, \quad 36, \quad 49, \quad 64, \quad \&c.,$$

and to make, for his own instruction, the series of their first and

* 'On the Nature and Notable Use of the most Simple Trigonal Numbers.' By E. De Joncourt, at the Hague. 1762.

second differences, and then to apply to it the same reasoning which has been already applied to the Table of Triangular Numbers.

When he feels that he has mastered that Table, I shall be happy to accompany mamma's darling to Woolwich or to Portsmouth, where he will find some practical illustrations of the use of his newly-acquired numbers. He will find scattered about in the Arsenal various heaps of cannon balls, some of them triangular, others square or oblong pyramids.

Looking on the simplest form—the triangular pyramid—he will observe that it exactly represents his own heaps of marbles placed each successively above one another until the top of the pyramid contains only a single ball.

The new series thus formed by the addition of his own triangular numbers is—

Number	Table	1st Difference	2nd Difference	3rd Difference
1	1	3	3	1
2	4	6	4	1
3	10	10	5	1
4	20	15	6	
5	35	21		
6	56			

He will at once perceive that this Table of the number of cannon balls contained in a triangular pyramid can be carried to any extent by simply adding successive differences, the third of which is constant.

The next step will naturally be to inquire how any number in this Table can be calculated by itself. A little consideration will lead him to a fair guess; a little industry will enable him to confirm his conjecture.

It will be observed at p. 38 that in order to find independently any number of the Table of the price of butchers' meat, the following rule was observed:—

Take the number whose tabular number is required.

Multiply it by the first difference.

This product is equal to the required tabular number.

Again, at p. 41, the rule for finding any triangular number was:—

Take the number of the group   5

Add 1 to this number, it becomes   6

Multiply these numbers together .   .   .   . 2)30

Divide the product by 2   15

This is the number of marbles in the 5th group.

Now let us make a bold conjecture respecting the Table of cannon balls, and try this rule:—

Take the number whose tabular number is required, say . 5
Add 1 to that number 6
Add 1 more to that number 7

Multiply all three numbers together . . . . . 2)210

Divide by 2 105

The real number in the 5th pyramid is 35. But the number 105 at which we have arrived is exactly three times as great. If, therefore, instead of dividing by 2 we had divided by 2 and also by 3, we should have arrived at a true result in this instance.

The amended rule is therefore—

Take the number whose tabular number is required, say . n
Add 1 to it . . . . $n+1$
Add 1 to this . . . . $n+2$
Multiply these three numbers together . . $n \times (n+1) \times (n+2)$
Divide by $1 \times 2 \times 3$.

The result is . . . . $\dfrac{n(n+1)\ (n+2)}{6}$

This rule will, upon trial, be found to give correctly every tabular number.

By similar reasoning we might arrive at the knowledge of the number of cannon balls in square and rectangular pyramids. But it is presumed that enough has been stated to enable the reader to form some general notion of the method of calculating arithmetical Tables by differences which are constant.

It may now be stated that mathematicians have discovered that all the Tables most important for practical purposes, such as those relating to Astronomy and Navigation, can, although they may not possess any constant differences, still be calculated in detached portions by that method.

Hence the importance of having machinery to calculate by differences, which, if well made, cannot err; and which, if carelessly set, presents in the last term it calculates the power of verification of every antecedent term.

Of the Mechanical Arrangements necessary for computing
Tables by the Method of Differences

From the preceding explanation it appears that all Tables may be calculated, to a greater or less extent, by the method of Differences.

That method requires, for its successful execution, little beyond mechanical means of performing the arithmetical operation of Addition. Subtraction can, by the aid of a well-known artifice, be converted into Addition.

The process of Addition includes two distinct parts—1st. The first consists of the addition of any one digit to another digit; 2nd. The second consists in carrying the tens to the next digit above.

Let us take the case of the addition of the two following numbers, in which no carriages occur:—

$$6023$$
$$1970$$
$$\overline{}$$
$$7993$$

It will be observed that, in making this addition, the mind acts by successive steps.   The person adding says to himself—

0 and 3 make three,
7 and 2 make nine,
9 and 0 make nine,
1 and 6 make seven.

In the following addition there are several carriages:—

$$2648$$
$$4564$$
$$\overline{}$$
$$7212$$

The person adding says to himself—

4 and 8 make 12: put down 2 and carry one.
1 and 6 are 7 and 4 make 11: put down 1 and carry one.
1 and 5 are 6 and 6 make 12: put down 2 and carry one.
1 and 4 are 5 and 2 make   7: put down 7.

Now, the length of time required for adding one number to another is mainly dependent upon the number of figures to be added.   If we could tell the average time required by the mind to add two figures together, the time required for adding any given number of figures to another equal number would be found by multiplying that average time by the number of digits in either number.

When we attempt to perform such additions by machinery we might follow exactly the usual process of the human mind.   In that case we might take a series of wheels, each having marked on its edges the digits 0, 1, 2, 3, 4, 5, 6, 7, 8, 9.   These wheels might be placed above each other upon an axis.   The lowest would indicate the units' figure, the next above the tens, and so on, as in the Difference Engine at the Exhibition, a woodcut of which faces the title-page.*

Several such axes, with their figure wheels, might be placed around a system of central wheels, with which the wheels of any one or more axes might at times be made to gear. Thus the figures on any one axis might, by means of those central wheels, be added to the figure wheels of any other axis.

But it may fairly be expected, and it is indeed of great importance that calculations made by machinery should not merely be exact, but that they should be done in a much shorter time than those performed by the human mind. Suppose there were no tens to carry, as in the first of the two cases; then, if we possessed mechanism capable of adding any one digit to any other in the units' place of figures, a similar mechanism might be placed above it to add the tens' figures, and so on for as many figures as might be required.

But in this case, since there are no carriages, each digit might be added to its corresponding digit at the same time. Thus, the time of adding by means of mechanism, any two numbers, however many figures they might consist of, would not exceed that of adding a single digit to another digit. If this could be accomplished it would render additions and subtractions with numbers having ten, twenty, fifty, or any number of figures, as rapid as those operations are with single figures.

Let us now examine the case in which there were several carriages. Its successive stages may be better explained, thus—

	2648
Stages.	4584
1 Add units' figure = 4	2642
2 Carry	1
	2652
3 Add tens' figure = 8	8
	2632
4 Carry	1
	2732
5 Add hundreds' figure = 5	5
	2232
6 Carry	1
	3232
7 Add thousands' figure = 4	4
	7232

8 Carry 0. There is no carr.

* [See page 2 of this edition.]

Now if, as in this case, all the carriages were known, it would then be possible to make all the additions of digits at the same time, provided we could also record each carriage as it became due. We might then complete the addition by adding, at the same instant, each carriage in its proper place. The process would then stand thus:—

$$
\begin{array}{l}
2648 \\
4564 \\
\hline
\end{array}
$$

Stages 1 $\left\{\begin{array}{l} 6102 \quad \text{Add each digit to the digit above.} \\ 111 \quad \text{Record the carriages.} \end{array}\right.$

2    $\{$ 7212   Add the above carriages.

Now, whatever mechanism is contrived for adding any one digit to any other must, of course, be able to add the largest digit, nine, to that other digit. Supposing, therefore, one unit of number to be passed over in one second of time, it is evident that any number of pairs of digits may be added together in nine seconds, and that, when all the consequent carriages are known, as in the above case, it will cost one second more to make those carriages. Thus, addition and carriage would be completed in ten seconds, even though the numbers consisted each of a hundred figures.

But, unfortunately, there are multitudes of cases in which the carriages that become due are only known in successive periods of time. As an example, add together the two following numbers:—

8473
1528

Stages
1 Add all the digits   9991
2 Carry on tens and warn next car.   1

9901
3 Carry on hundreds, and ditto   1

9001
4 Carry on thousands, and ditto   1

00001
5 Carry on ten thousands   1

10001

In this case the carriages only become known successively, and they amount to the number of figures to be added; consequently, the mere addition of two numbers, each of fifty places of figures, would require only nine seconds of time, whilst the possible carriages would consume fifty seconds.

The mechanical means I employed to make these carriages bears some slight analogy to the operation of the faculty of memory. A toothed wheel had the ten digits marked upon its edge; between the nine and the zero a projecting tooth was placed. Whenever any wheel, in receiving addition, passed from nine to zero, the projecting tooth pushed over a certain lever. Thus, as soon as the nine seconds of time required for addition were ended, every carriage which had *become due* was indicated by the altered position of its lever. An arm now went round, which was so contrived that the act of replacing that lever caused the carriage which its position indicated to be made to the next figure above. But this figure might be a nine, in which case, in passing to zero, it would put over its lever, and so on. By placing the arms spirally round an axis, these successive carriages were accomplished.

Multitudes of contrivances were designed, and almost endless drawings made, for the purpose of economizing the time and simplifying the mechanism of carriage. In that portion of the Difference Engine in the Exhibition of 1862 the time of carriage has been reduced to about one-fourth part of what was at first required.

At last having exhausted, during years of labour, the principle of successive carriages, it occurred to me that it might be possible to teach mechanism to accomplish another mental process, namely—to foresee. This idea occurred to me in October, 1834. It cost me much thought, but the principle was arrived at in a short time. As soon as that was attained, the next step was to teach the mechanism which could foresee to act upon that foresight. This was not so difficult: certain mechanical means were soon devised which, although very far from simple, were yet sufficient to demonstrate the possibility of constructing such machinery.

The process of simplifying this form of carriage occupied me, at intervals, during a long series of years. The demands of the Analytical Engine, for the mechanical execution of arithmetical operations, were of the most extensive kind. The multitude of similar parts required by the Analytical Engine, amounting in some instances to upwards of fifty thousand, rendered any, even the simplest, improvement of each part a matter of the highest importance, more especially as regarded the diminished amount of expenditure for its construction.

Description of the existing portion of Difference Engine No. 1

That portion of Difference Engine, No. 1, which during the last twenty years has been in the museum of King's College, at Somerset House, is represented in the woodcut opposite the title page.*

It consists of three columns; each column contains six cages; each cage contains one figure-wheel.

The column on the right hand has its lowest figure-wheel covered by a shade which is never removed, and to which the reader's attention need not be directed.

The figure-wheel next above may be placed by hand at any one of the ten digits.   In the woodcut it stands at zero.

The third, fourth, and fifth cages are exactly the same as the second.

The sixth cage contains exactly the same as the four just described. It also contains two other figure-wheels, which with a similar one above the frame, may also be dismissed from the reader's attention.   Those wheels are entirely unconnected with the moving part of the engine, and are only used for memoranda.

It appears, therefore, that there are in the first column on the right hand five figure-wheels, each of which may be set by hand to any of the figures 0, 1, 2, 3, 4, 5, 6, 7, 8, 9.

The lowest of these figure-wheels represents the unit's figure of any number; the next above the ten's figure, and so on.   The highest figure-wheel will therefore represent tens of thousands.

Now, as each of these figure-wheels may be set by hand to any digit, it is possible to place on the first column any number up to 99999.   It is on these wheels that the Table to be calculated by the engine is expressed.   This column is called the Table column, and the axis of the wheels the Table axis.

The second or middle column has also six cages, in each of which a figure-wheel is placed.   It will be observed that in the lowest cage, the figure on the wheel is concealed by a shade.   It may therefore be dismissed from the attention.   The five other figure-wheels are exactly like the figure-wheels on the Table axis, and can also represent any number up to 99999.

This column is called the First Difference column, and the axis is called the First Difference axis.

The third column, which is that on the left hand, has also six cages, in each of which is a figure-wheel capable of being set by hand to any digit.

The mechanism is so contrived that whatever may be the numbers

* [See page 2 of this edition.]

placed respectively on the figure-wheels of each of the three columns, the following succession of operations will take place as long as the handle is moved:—

1st. Whatever number is found upon the column of first differences will be added to the number found upon the Table column.

2nd. The same first difference remaining upon its own column, the number found upon the column of second differences will be added to that first difference.

It appears, therefore, that with this small portion of the Engine any Table may be computed by the method of differences, provided neither the Table itself, nor its first and second differences, exceed five places of figures.

If the whole Engine had been completed it would have had six orders of differences, each of twenty places of figures, whilst the three first columns would each have had half a dozen additional figures.

This is the simplest explanation of that portion of the Difference Engine No. 1, at the Exhibition of 1862. There are, however, certain modifications in this fragment which render its exhibition more instructive, and which even give a mechanical insight into those higher powers with which I had endowed it in its complete state.

As a matter of convenience in exhibiting it, there is an arrangement by which the *three* upper figures of the second difference are transformed into a small engine which counts the natural numbers.

By this means it can be set to compute any Table whose second difference is constant and less than 1000, whilst at the same time it thus shows the position in the Table of each tabular number.

In the existing portion there are three bells; they can be respectively ordered to ring when the Table, its first difference and its second difference, pass from positive to negative. Several weeks after the machine had been placed in my drawing-room, a friend came by appointment to test its power of calculating Tables. After the Engine had computed several Tables, I remarked that it was evidently finding the root of a quadratic equation; I therefore set the bells to watch it. After some time the proper bell sounded twice, indicating, and giving the two positive roots to be 28 and 30. The Table thus calculated related to the barometer and really involved a quadratic equation, although its maker had not previously observed it. I afterwards set the Engine to tabulate a formula containing impossible roots, and of course the other bell warned me when it had attained those roots. I had never before used these bells, simply because I did not think the power it thus possessed to be of any practical utility.

Again, the lowest cages of the Table, and of the first difference, have been made use of for the purpose of illustrating three important faculties of the finished engine.

1st. The portion exhibited can calculate any Table whose third difference is constant and less than 10.

2nd. It can be used to show how much more rapidly astronomical Tables can be calculated in an engine in which there is no constant difference.

3rd. It can be employed to illustrate those singular laws which might continue to be produced through ages, and yet after an enormous interval of time change into other different laws; each again to exist for ages, and then to be superseded by new laws. These views were first proposed in the "Ninth Bridgewater Treatise."

Amongst the various questions which have been asked respecting the Difference Engine, I will mention a few of the most remarkable:— One gentleman addressed me thus: "Pray, Mr. Babbage, can you explain to me in two words what is the principle of this machine?" Had the querist possessed a moderate acquaintance with mathematics I might in four words have conveyed to him the required information by answering, "The method of differences." The question might indeed have been answered with six characters thus—

$$\Delta^7 u_x = 0$$

but such information would have been unintelligible to such inquirers.

On two occasions I have been asked,—"Pray, Mr. Babbage, if you put into the machine wrong figures, will the right answers come out?" In one case a member of the Upper, and in the other a member of the Lower, House put this question. I am not able rightly to apprehend the kind of confusion of ideas that could provoke such a question. I did, however, explain the following property, which might in some measure approach towards an answer to it.

It is possible to construct the Analytical Engine in such a manner that after the question is once communicated to the engine, it may be stopped at any turn of the handle and set on again as often as may be desired. At each stoppage every figure-wheel throughout the Engine, which is capable of being moved without breaking, may be moved on to any other digit. Yet after each of these apparent falsifications the engine will be found to make the next calculation with perfect truth.

The explanation is very simple, and the property itself useless. The whole of the mechanism ought of course to be enclosed in glass, and kept under lock and key, in which case the mechanism necessary to give it the property alluded to would be useless.

CHAPTER VIII

OF THE ANALYTICAL ENGINE

Man wrongs, and Time avenges.

BYRON—*The Prophecy of Dante*

Built Workshops for constructing the Analytical Engine—Difficulties about carrying the Tens—Unexpectedly solved—Application of the Jacquard Principle—Treatment of Tables—Probable Time required for Arithmetical Operations—Conditions it must fulfil—Unlimited in Number of Figures, or in extent of Analytical Operations—The Author invited to Turin in 1840—Meetings for Discussion—Plana, Menabrea, MacCullagh, Mossotti—Difficulty proposed by the latter—Observations on the Errata of Astronomical Tables—Suggestions for a Reform of Analytical Signs.

THE circular arrangement of the axes of the Difference Engine round large central wheels led to the most extended prospects. The whole of arithmetic now appeared within the grasp of mechanism. A vague glimpse even of an Analytical Engine at length opened out, and I pursued with enthusiasm the shadowy vision. The drawings and the experiments were of the most costly kind. Draftsmen of the highest order were necessary to economize the labour of my own head; whilst skilled workmen were required to execute the experimental machinery to which I was obliged constantly to have recourse.

In order to carry out my pursuits successfully, I had purchased a house with above a quarter of an acre of ground in a very quiet locality. My coach-house was now converted into a forge and a foundry, whilst my stables were transformed into a workshop. I built other extensive workshops myself, and had a fire-proof building for my drawings and draftsmen. Having myself worked with a variety of tools, and having studied the art of constructing each of them, I at length laid it down as a principle—that, except in rare cases, I would never do anything myself if I could afford to hire another person who could do it for me.

The complicated relations which then arose amongst the various parts of the machinery would have baffled the most tenacious memory. I overcame that difficulty by improving and extending a language of signs, the Mechanical Notation, which in 1826 I had explained in a paper printed in the "Phil. Trans." By such means I succeeded in mastering trains of investigation so vast in extent that no length of years ever allotted to one individual could otherwise have enabled me to control. By the aid of the Mechanical Notation, the Analytical Engine became a reality: for it became susceptible of demonstration.

Such works could not be carried on without great expenditure. The fluctuations in the demand and supply of skilled labour were considerable. The railroad mania withdrew from other pursuits the most intellectual and skilful draftsmen. One who had for some years been my chief assistant was tempted by an offer so advantageous that in justice to his own family he could scarcely have declined it. Under these circumstances I took into consideration the plan of advancing his salary to one guinea per day. Whilst this was in abeyance, I consulted my venerable surviving parent. When I had fully explained the circumstances, my excellent mother replied: "My dear son, you have advanced far in the accomplishment of a great object, which is worthy of your ambition. You are capable of completing it. My advice is—pursue it, even if it should oblige you to live on bread and cheese."

This advice entirely accorded with my own feelings. I therefore retained my chief assistant at his advanced salary.

The most important part of the Analytical Engine was undoubtedly the mechanical method of carrying the tens. On this I laboured incessantly, each succeeding improvement advancing me a step or two. The difficulty did not consist so much in the more or less complexity of the contrivance as in the reduction of the *time* required to effect the carriage. Twenty or thirty different plans and modifications had been drawn. At last I came to the conclusion that I had exhausted the principle of successive carriage. I concluded also that nothing but teaching the Engine to foresee and then to act upon that foresight could ever lead me to the object I desired, namely, to make the whole of any unlimited number of carriages in one unit of time. One morning, after I had spent many hours in the drawing-office in endeavouring to improve the system of successive carriages, I mentioned these views to my chief assistant, and added that I should retire to my library, and endeavour to work out the new principle. He gently expressed a doubt whether the plan was *possible*, to which I replied that, not being able to prove its impossibility, I should follow out a slight glimmering of light which I thought I perceived.

After about three hours' examination, I returned to the drawing-office with much more definite ideas upon the subject. I had discovered a principle that proved the possibility, and I had contrived mechanism which, I thought, would accomplish my object.

I now commenced the explanation of my views, which I soon found were but little understood by my assistant; nor was this surprising, since in the course of my own attempt at explanation, I found several defects in my plan, and was also led by his questions to perceive others. All these I removed one after another, and ultimately terminated at a late hour my morning's work with the conviction that *anticipating* carriage was not only within my power, but that I had devised one mechanism at least by which it might be accomplished.

Many years after, my assistant, on his return from a long residence abroad, called upon me, and we talked over the progress of the Analytical Engine. I referred back to the day on which I had made that most important step, and asked him if he recollected it. His reply was that he perfectly remembered the circumstance; for that on retiring to my library, he seriously thought that my intellect was beginning to become deranged. The reader may perhaps be curious to know how I spent the rest of that remarkable day.

After working, as I constantly did, for ten or eleven hours a day, I had arrived at this satisfactory conclusion, and was revising the rough sketches of the new contrivance, when my servant entered the drawing-office, and announced that it was seven o'clock—that I dined in Park Lane—and that it was time to dress. I usually arrived at the house of my friend about a quarter of an hour before the appointed time, in order that we might have a short conversation on subjects on which we were both much interested. Having mentioned my recent success, in which my host thoroughly sympathized, I remarked that it had produced an exhilaration of the spirits which not even his excellent champagne could rival. Having enjoyed the society of Hallam, of Rogers, and of some few others of that delightful circle, I retired, and joined one or perhaps two much more extensive reunions. Having thus forgotten science, and enjoyed society for four or five hours, I returned home. About one o'clock I was asleep in my bed, and thus continued for the next five hours.

This new and rapid system of carrying the tens when two numbers are added together, reduced the actual time of the addition of any number of digits, however large, to nine units of time for the addition, and one unit for the carriage. Thus in ten's units of time, any two numbers, however large, might be added together. A few more units of time, perhaps five or six, were required for making the requisite previous arrangements.

Having thus advanced as nearly as seemed possible to the minimum of time requisite for arithmetical operations, I felt renewed power and increased energy to pursue the far higher object I had in view.

To describe the successive improvements of the Analytical Engine would require many volumes. I only propose here to indicate a few of its more important functions, and to give to those whose minds are duly prepared for it some information which will remove those vague notions of wonder, and even of its impossibility, with which it is surrounded in the minds of some of the most enlightened.

To those who are acquainted with the principles of the Jacquard loom, and who are also familiar with analytical formulæ, a general idea of the means by which the Engine executes its operations may be obtained without much difficulty. In the Exhibition of 1862 there were many splendid examples of such looms.

It is known as a fact that the Jacquard loom is capable of weaving any design which the imagination of man may conceive. It is also the constant practice for skilled artists to be employed by manufacturers in designing patterns. These patterns are then sent to a peculiar artist, who, by means of a certain machine, punches holes in a set of pasteboard cards in such a manner that when those cards are placed in a Jacquard loom, it will then weave upon its produce the exact pattern designed by the artist.

Now the manufacturer may use, for the warp and weft of his work, threads which are all of the same colour; let us suppose them to be unbleached or white threads. In this case the cloth will be woven all of one colour; but there will be a damask pattern upon it such as the artist designed.

But the manufacturer might use the same cards, and put into the warp threads of any other colour. Every thread might even be of a different colour, or of a different shade of colour; but in all these cases the *form* of the pattern will be precisely the same—the colours only will differ.

The analogy of the Analytical Engine with this well-known process is nearly perfect.

The Analytical Engine consists of two parts:—

1st. The store in which all the variables to be operated upon, as well as all those quantities which have arisen from the result of other operations, are placed.

2nd. The mill into which the quantities about to be operated upon are always brought.

Every formula which the Analytical Engine can be required to compute consists of certain algebraical operations to be performed upon given letters, and of certain other modifications depending on the numerical value assigned to those letters.

There are therefore two sets of cards, the first to direct the nature of the operations to be performed—these are called operation cards: the other to direct the particular variables on which those cards are required to operate—these latter are called variable cards. Now the symbol of each variable or constant, is placed at the top of a column capable of containing any required number of digits.

Under this arrangement, when any formula is required to be computed, a set of operation cards must be strung together, which contain the series of operations in the order in which they occur. Another set of cards must then be strung together, to call in the variables into the mill, the order in which they are required to be acted upon. Each operation card will require three other cards, two to represent the variables and constants and their numerical values upon which the previous operation card is to act, and one to indicate the variable on which the arithmetical result of this operation is to be placed.

But each variable has below it, on the same axis, a certain number of figure-wheels marked on their edges with the ten digits: upon these any number the machine is capable of holding can be placed. Whenever variables are ordered into the mill, these figures will be brought in, and the operation indicated by the preceding card will be performed upon them. The result of this operation will then be replaced in the store.

The Analytical Engine is therefore a machine of the most general nature. Whatever formula it is required to develop, the law of its development must be communicated to it by two sets of cards. When these have been placed, the engine is special for that particular formula. The numerical value of its constants must then be put on the columns of wheels below them, and on setting the Engine in motion it will calculate and print the numerical results of that formula.

Every set of cards made for any formula will at any future time recalculate that formula with whatever constants may be required.

Thus the Analytical Engine will possess a library of its own. Every set of cards once made will at any future time reproduce the calculations for which it was first arranged. The numerical value of its constants may then be inserted.

It is perhaps difficult to apprehend these descriptions without a familiarity both with analytical forms and mechanical structures. I will now, therefore, confine myself to the mathematical view of the Analytical Engine, and illustrate by example some of its supposed difficulties.

An excellent friend of mine, the late Professor MacCullagh, of Dublin, was discussing with me, at breakfast, the various powers of the Analytical Engine. After a long conversation on the subject, he inquired what the machine could do if, in the midst of algebraic operations, it was required to perform logarithmic or trigonometric operations.

My answer was, that whenever the Analytical Engine should exist, all the developments of formula would be directed by this condition— that the machine should be able to compute their numerical value in the shortest possible time.   I then added that if this answer were not satisfactory, I had provided means by which, with equal accuracy, it might compute by logarithmic or other Tables.

I explained that the Tables to be used must, of course, be computed and punched on cards by the machine, in which case they would undoubtedly be correct.   I then added that when the machine wanted a tabular number, say the logarithm of a given number, that it would ring a bell and then stop itself.   On this, the attendant would look at a certain part of the machine, and find that it wanted the logarithm of a given number, say of 2303.   The attendant would then go to the drawer containing the pasteboard cards representing its table of logarithms.   From amongst these he would take the required logarithmic card, and place it in the machine.   Upon this the engine would first ascertain whether the assistant had or had not given him the correct logarithm of the number; if so, it would use it and continue its work.   But if the engine found the attendant had given him a wrong logarithm, it would then ring a louder bell, and stop itself.   On the attendant again examining the engine, he would observe the words, "Wrong tabular number," and then discover that he really had given the wrong logarithm, and of course he would have to replace it by the right one.

Upon this, Professor MacCullagh naturally asked why, if the machine could tell whether the logarithm was the right one, it should have asked the attendant at all?   I told him that the means employed were so ridiculously simple that I would not at that moment explain them; but that if he would come again in the course of a few days, I should be ready to explain it.   Three or four days after, Bessel and Jacobi, who had just arrived in England, were sitting with me, in-quiring about the Analytical Engine, when fortunately my friend MacCullagh was announced.   The meeting was equally agreeable to us all, and we continued our conversation.   After some time Bessel put to me the very same question which MacCullagh had previously asked.   On this Jacobi remarked that he, too, was about to make the same inquiry when Bessel had asked the question.   I then explained to them the following very simple means by which that verification was accomplished.

Besides the sets of cards which direct the nature of the operations to be performed, and the variables or constants which are to be operated upon, there is another class of cards called number cards. These are much less general in their uses than the others, although they are necessarily of much larger size.

Any number which the Analytical Engine is capable of using or of producing can, if required, be expressed by a card with certain holes in it; thus—

Number				Table						
2	3	0	3	3	6	2	2	9	3	9
●	●	○	●	●	●	●	●	●	●	●
●	●	○	●	●	●	●	●	●	●	●
○	●	○	●	●	●	○	○	●	●	●
○	○	○	○	○	●	○	○	●	○	●
○	○	○	○	○	●	○	○	●	○	●
○	○	○	○	○	●	○	○	●	○	●
○	○	○	○	○	○	○	○	●	○	●
○	○	○	○	○	○	○	○	●	○	●
○	○	○	○	○	○	○	○	●	○	●

The above card contains eleven vertical rows for holes, each row having nine or any less number of holes.   In this example the tabular number is 3 6 2 2 9 3 9, whilst its number in the order of the table is 2 3 0 3.   In fact, the former number is the logarithm of the latter.

The Analytical Engine will contain,

1st. Apparatus for printing on paper, one, or, if required, two copies of its results.

2nd. Means for producing a stereotype mould of the tables or results it computes.

3rd. Mechanism for punching on blank pasteboard cards or metal plates the numerical results of any of its computations.

Of course the Engine will compute all the Tables which it may itself be required to use.   These cards will therefore be entirely free from error.   Now when the Engine requires a tabular number, it will stop, ring a bell, and ask for such number.   In the case we have assumed, it asks for the logarithm of 2 3 0 3.

When the attendant has placed a tabular card in the Engine, the first step taken by it will be to verify the *number* of the card given it by subtracting its number from 2 3 0 3, the number whose logarithm it asked for.   If the remainder is zero, then the engine is certain that the logarithm must be the right one, since it was computed and punched by itself.

Thus the Analytical Engine first computes and punches on cards its own tabular numbers. These are brought to it by its attendant when demanded. But the Engine itself takes care that the *right* card is brought to it by verifying the *number* of that card by the number of the card which it demanded. The Engine will always reject a wrong card by continually ringing a loud bell and stopping itself until supplied with the precise intellectual food it demands.

It will be an interesting question, which time only can solve, to know whether such tables of cards will ever be required for the Engine. Tables are used for saving the time of continually computing individual numbers. But the computations to be made by the Engine are so rapid that it seems most probable that it will make shorter work by computing directly from proper formulæ than by having recourse even to its own Tables.

The Analytical Engine I propose will have the power of expressing every number it uses to fifty places of figures. It will multiply any two such numbers together, and then, if required, will divide the product of one hundred figures by number of fifty places of figures.

Supposing the velocity of the moving parts of the Engine to be not greater than forty feet per minute, I have no doubt that

> Sixty additions or subtractions may be completed and printed in one minute.
>
> One multiplication of two numbers, each of fifty figures, in one minute.
>
> One division of a number having 100 places of figures by another of 50 in one minute.

In the various sets of drawings of the modifications of the mechanical structure of the Analytical Engines, already numbering upwards of thirty, two great principles were embodied to an unlimited extent.

1st. The entire control over *arithmetical* operations, however large, and whatever might be the number of their digits.

2nd. The entire control over the *combinations* of algebraic symbols, however lengthened those processes may be required. The possibility of fulfilling these two conditions might reasonably be doubted by the most accomplished mathematician as well as by the most ingenious mechanician.

The difficulties which naturally occur to those capable of examining the question, as far as they relate to arithmetic, are these,—

> (*a*). The number of digits in *each constant* inserted in the Engine must be without limit.

(b). The number of constants to be inserted in the Engine must also be without limit.

(c). The number of operations necessary for arithmetic is only four, but these four may be repeated an *unlimited* number of times.

(d). These operations may occur in any order, or follow an *unlimited* number of laws.

The following conditions relate to the algebraic portion of the Analytical Engine:—

(e). The number of *literal* constants must be *unlimited*.

(f). The number of *variables* must be *without limit*.

(g). The combinations of the algebraic signs must *be unlimited*.

(h). The number of *functions* to be employed must be *without limit*.

This enumeration includes eight conditions, each of which is absolutely *unlimited* as to the number of its combinations.

Now it is obvious that no *finite* machine can include infinity. It is also certain that no question *necessarily* involving infinity can ever be converted into any other in which the idea of infinity under some shape or other does not enter.

It is impossible to construct machinery occupying unlimited space; but it is possible to construct finite machinery, and to use it through unlimited time. It is this substitution of the *infinity of time* for the *infinity of space* which I have made use of, to limit the size of the engine and yet to retain its unlimited power.

(a). I shall now proceed briefly to point out the means by which I have effected this change.

Since every calculating machine must be constructed for the calculation of a definite number of figures, the first datum must be to fix upon that number. In order to be somewhat in advance of the greatest number that may ever be required, I chose fifty places of figures as the standard for the Analytical Engine. The intention being that in such a machine two numbers, each of fifty places of figures, might be multiplied together and the resultant product of one hundred places might then be divided by another number of fifty places. It seems to me probable that a long period must elapse before the demands of science will exceed this limit. To this it may be added that the addition and subtraction of numbers in an engine constructed for n places of figures would be equally rapid whether n were equal to five or five thousand digits. With respect to multiplication and division, the time required is greater:—

Thus if $a \cdot 10^{50} + b$ and $a' \cdot 10^{50} + b'$ are two numbers each of less than a hundred places of figures, then each can be expressed upon two columns of fifty figures, and a, b, a', b' are each less than fifty places of figures: they can therefore be added and subtracted upon any column holding fifty places of figures.

The product of two such numbers is—

$$aa'\, 10^{100} + (ab' + a'b)10^{50} + bb'.$$

This expression contains four pairs of factors, aa', ab', $a'b$, bb', each factor of which has less than fifty places of figures. Each multiplication can therefore be executed in the Engine. The time, however, of multiplying two numbers, each consisting of any number of digits between fifty and one hundred, will be nearly four times as long as that of two such numbers of less than fifty places of figures.

The same reasoning will show that if the numbers of digits of each factor are between one hundred and one hundred and fifty, then the time required for the operation will be nearly nine times that of a pair of factors having only fifty digits.

Thus it appears that whatever may be the number of digits the Analytical Engine is capable of holding, if it is required to make all the computations with k times that number of digits, then it can be executed by the same Engine, but in an amount of time equal to k^2 times the former. Hence the condition (a), or the unlimited number of digits contained in each constant employed, is fulfilled.

It must, however, be admitted that this advantage is gained at the expense of diminishing the number of the constants the Engine can hold. An engine of fifty digits, when used as one of a hundred digits, can only contain half the number of variables. An engine containing m columns, each holding n digits, if used for computations requiring kn digits, can only hold $\dfrac{m}{k}$ constants or variables.

(b). The next step is therefore to prove (b), viz.: to show that a finite engine can be used as if it contained an unlimited number of constants. The method of punching cards for tabular numbers has already been alluded to. Each Analytical Engine will contain one or more apparatus for printing any numbers put into it, and also an apparatus for punching on pasteboard cards the holes corresponding to those numbers. At another part of the machine a series of number cards, resembling those of Jacquard, but delivered to and computed by the machine itself, can be placed. These can be called for by the Engine itself in any order in which they may be placed, or according to *any law* the Engine may be directed to use. Hence the condition (b) is fulfilled, namely: an *unlimited number of constants* can be inserted in the machine in an *unlimited* time.

I propose in the Engine I am constructing to have places for only a thousand constants, because I think it will be more than sufficient.   But if it were required to have ten, or even a hundred times that number, it would be quite possible to make it, such is the simplicity of its structure of that portion of the Engine.

(c). The next stage in the arithmetic is the number of times the four processes of addition, subtraction, multiplication, and division can be repeated.   It is obvious that four different cards thus punched

would give the orders for the four rules of arithmetic.

Now there is no limit to the number of such cards which may be strung together according to the nature of the operations required. Consequently the condition (c) is fulfilled.

(d). The fourth arithmetical condition (d), that the order of succession in which these operations can be varied, is itself *unlimited*, follows as a matter of course.

The four remaining conditions which must be fulfilled, in order to render the Analytical Engine as general as the science of which it is the powerful executive, relate to algebraic quantities with which it operates.

The thousand columns, each capable of holding any number of less than fifty-one places of figures, may each represent a constant or a variable quantity.   These quantities I have called by the comprehensive title of variables, and have denoted them by V_n, with an index below.   In the machine I have designed, n may vary from 0 to 999. But after any one or more columns have been used for variables, if those variables are not required afterwards, they may be printed upon paper, and the columns themselves again used for other variables.   In such cases the variables must have a new index; thus, $^mV^n$. I propose to make n vary from 0 to 99.   If more variables are required, these may be supplied by Variable Cards, which may follow each other in unlimited succession.   Each card will cause its symbol to be printed with its proper indices.

For the sake of uniformity, I have used V with as many indices as may be required throughout the Engine. This, however, does not prevent the printed result of a development from being represented by any letters which may be thought to be more convenient.   In that part in which the results are printed, type of any form may be used, according to the taste of the proposer of the question.

It thus appears that the two conditions, (e) and (f), which require that the number of constants and of variables should be unlimited, are both fulfilled.

The condition (g) requiring that the number of combinations of the four algebraic signs shall be unlimited, is easily fulfilled by placing them on cards in any order of succession the problem may require.

The last condition (h), namely, that the number of functions to be employed must be without limit, might seem at first sight to be difficult to fulfil. But when it is considered that any function of any number of operations performed upon any variables is but a combination of the four simple signs of operation with various quantities, it becomes apparent that any function whatever may be represented by two groups of cards, the first being signs of operation, placed in the order in which they succeed each other, and the second group of cards representing the variables and constants placed in the order of succession in which they are acted upon by the former.

Thus it appears that the whole of the conditions which enable a *finite* machine to make calculations of *unlimited* extent are fulfilled in the Analytical Engine. The means I have adopted are uniform. I have converted the infinity of space, which was required by the conditions of the problem, into the infinity of time. The means I have employed are in daily use in the art of weaving patterns. It is accomplished by systems of cards punched with various holes strung together to any extent which may be demanded. Two large boxes, the one empty and the other filled with perforated cards, are placed before and behind a polygonal prism, which revolves at intervals upon its axis, and advances through a short space, after which it immediately returns.

A card passes over the prism just before each stroke of the shuttle; the cards that have passed hang down until they reach the empty box placed to receive them, into which they arrange themselves one over the other. When the box is full, another empty box is placed to receive the coming cards, and a new full box on the opposite side replaces the one just emptied. As the suspended cards on the entering side are exactly equal to those on the side at which the others are delivered, they are perfectly balanced, so that whether the formulæ to be computed be excessively complicated or very simple, the force to be exerted always remains nearly the same.

In 1840 I received from my friend M. Plana a letter pressing me strongly to visit Turin at the then approaching meeting of Italian philosophers. In that letter M. Plana stated that he had inquired anxiously of many of my countrymen about the power and mechanism of the Analytical Engine. He remarked that from all the information he could collect the case seemed to stand thus:—

"Hitherto the *legislative* department of our analysis has been all powerful—the *executive* all feeble.

"Your engine seems to give us the same control over the executive which we have hitherto only possessed over the legislative department."

Considering the exceedingly limited information which could have reached my friend respecting the Analytical Engine, I was equally surprised and delighted at his exact prevision of its powers. Even at the present moment I could not express more clearly, and in fewer terms, its real object. I collected together such of my models, drawings, and notations as I conceived to be best adapted to give an insight into the principles and mode of operating of the Analytical Engine. On mentioning my intention to my excellent friend the late Professor MacCullagh, he resolved to give up a trip to the Tyrol, and join me at Turin.

We met at Turin at the appointed time, and as soon as the first bustle of the meeting had a little abated, I had the great pleasure of receiving at my own apartments, for several mornings, Messrs. Plana, Menabrea, Mossotti, MacCullagh, Plantamour, and others of the most eminent geometers and engineers of Italy.

Around the room were hung the formula, the drawings, notations, and other illustrations which I had brought with me. I began on the first day to give a short outline of the idea. My friends asked from time to time further explanations of parts I had not made sufficiently clear. M. Plana had at first proposed to make notes, in order to write an outline of the principles of the engine. But his own laborious pursuits induced him to give up this plan, and to transfer the task to a younger friend of his, M. Menabrea, who had already established his reputation as a profound analyst.

These discussions were of great value to me in several ways. I was thus obliged to put into language the various views I had taken, and I observed the effect of my explanations on different minds. My own ideas became clearer, and I profited by many of the remarks made by my highly-gifted friends.

One day Mossotti, who had been unavoidably absent from the previous meeting, when a question of great importance had been discussed, again joined the party. Well aware of the acuteness and rapidity of my friend's intellect, I asked my other friends to allow me five minutes to convey to Professor Mossotti the substance of the preceding sitting. After putting a few questions to Mossotti himself, he placed before me distinctly his greatest difficulty.

He remarked that he was now quite ready to admit the power of mechanism over numerical, and even over algebraical relations, to any extent. But he added that he had no conception how the machine could perform the act of judgment sometimes required during an analytical inquiry, when two or more different courses presented themselves, especially as the proper course to be adopted could not be known in many cases until all the previous portion had been gone through.

I then inquired whether the solution of a numerical equation of any degree by the usual, but very tedious proceeding of approximation would be a type of the difficulty to be explained. He at once admitted that it would be a very eminent one.

For the sake of perspicuity and brevity I shall confine my present explanation to possible roots.

I then mentioned the successive stages:—

Number of Operation
    Cards used

1   *a.* Ascertain the number of possible roots by applying Sturm's theorem to the coefficients.

2   *b.* Find a number greater than the greatest root.

3   *c.* Substitute the powers of ten (commencing with that next greater than the greatest root, and diminishing the powers by unity at each step) for the value of x in the given equation.

      Continue this until the sign of the resulting number changes from positive to negative.

      The index of the last power of ten (call it n), which is positive, expresses the number of digits in that part of the root which consists of whole numbers. Call this index $n+1$.

4   *d.* Substitute successively for x in the original equation 0×10^n, 1×10^n, 2×10^n, 3×10^n, ... 9×10^n, until a change of sign occurs in the result. The digit previously substituted will be the first figure of the root sought.

5   *e.* Transform the original equation into another whose roots are less by the number thus found.

      The transformed equation will have a real root, the digit, less than 10^n.

6   *f.* Substitute $1 \times 10^{n-1}$, $2 \times 10^{n-1}$, $3 \times 10^{n-1}$, &c., successively for the root of this equation, until a change of sign occurs in the result, as in process 4.

      This will give the second figure of the root.

      This process of alternately finding a new figure in the

Number of Operation
    Cards used

                  root, and then transforming the equation into
                  another (as in process 4 and 5), must be carried on
                  until as many figures as are required, whether
                  whole numbers or decimals, are arrived at.

7   *g*. The root thus found must now be used to reduce the original
        equation to one dimension lower.

8   *h*. This new equation of one dimension lower must now be
        treated by sections 3, 4, 5, 6, and 7, until the new root is
        found.

9   *i*. The repetition of sections 7 and 8 must go on until all the
        roots have been found.

Now it will be observed that Professor Mossotti was quite ready to admit at once that each of these different processes could be performed by the Analytical Machine through the medium of properly-arranged sets of Jacquard cards.

His real difficulty consisted in teaching the engine to know when to change from one set of cards to another, and back again repeatedly, at intervals not known to the person who gave the orders.

The dimensions of the algebraic equation being known, the number of arithmetical processes necessary for Sturm's theorem is consequently known. A set of operation cards can therefore be prepared. These must be accompanied by a corresponding set of variable cards, which will represent the columns in the store, on which the several coefficients of the given equation, and the various combinations required amongst them, are to be placed.

The next stage is to find a number greater than the greatest root of the given equation. There are various courses for arriving at such a number. Any one of these being selected, another set of operation and variable cards can be prepared to execute this operation.

Now, as this second process invariably follows the first, the second set of cards may be attached to the first set, and the engine will pass on from the first to the second process, and again from the second to the third process.

But here a difficulty arises: successive powers of ten are to be substituted for x in the equation, until a certain event happens. A set of cards may be provided to make the substitution of the highest power of ten, and similarly for the others; but on the occurrence of a certain event, namely, the change of a sign from $+$ to $-$, this stage of the calculation is to terminate.

Now at a very early period of the inquiry I had found it necessary to teach the engine to know when any numbers it might be computing passed through zero or infinity.

The passage through zero can be easily ascertained, thus: Let the continually-decreasing number which is being computed be placed upon a column of wheels in connection with a carrying apparatus. After each process this number will be diminished, until at last a number is subtracted from it which is greater than the number expressed on those wheels.

Thus let it be   .   00000,00000,00000,00423
Subtract .   .   00000,00000,00000,00511
                    99999,99999,99999,99912

Now in every case of a carriage becoming due, a certain lever is transferred from one position to another in the cage next above it.

Consequently in the highest cage of all (say the fiftieth in the Analytical Engine), an arm will be moved or not moved accordingly as the carriages do or do not run up beyond the highest wheel.

This arm can, of course, make any change which has previously been decided upon. In the instance we have been considering it would order the cards to be turned on to the next set.

If we wish to find when any number, which is increasing, exceeds in the number of its digits the number of wheels on the columns of the machine, the same carrying arm can be employed. Hence any directions may be given which the circumstances require.

It will be remarked that this does not actually prove, even in the Analytical Engine of fifty figures, that the number computed has passed through infinity; but only that it has become greater than any number of fifty places of figures.

There are, however, methods by which any machine made for a given number of figures may be made to compute the same formulæ with double or any multiple of its original number. But the nature of this work prevents me from explaining that method.

It may here be remarked that in the process, the cards employed to make the substitutions of the powers of ten are *operation* cards. They are, therefore, quite independent of the numerical values substituted. Hence the same set of operation cards which order the substitutions 1×10^n will, if backed, order the substitution of 2×10^n, &c. We may, therefore, avail ourselves of mechanism for backing these cards, and call it into action whenever the circumstances themselves require it.

The explanation of M. Mossotti's difficulty is this:—Mechanical means have been provided for backing or advancing the operation cards to any extent.   There exist means of expressing the conditions under which these various processes are required to be called into play. It is not even necessary that two courses only should be possible.   Any number of courses may be possible at the same time; and the choice of each may depend upon any number of conditions.

It was during these meetings that my highly valued friend, M. Menabrea, collected the materials for that lucid and admirable description which he subsequently published in the Bibli. Univ. de Genève, t. xli. Oct. 1842.

The elementary principles on which the Analytical Engine rests were thus in the first instance brought before the public by General Menabrea.*

Some time after the appearance of his memoir on the subject in the "Bibliothèque Universelle de Genève," the late Countess of Lovelace† informed me that she had translated the memoir of Menabrea.   I asked why she had not herself written an original paper on a subject with which she was so intimately acquainted?   To this Lady Lovelace replied that the thought had not occurred to her.   I then suggested that she should add some notes to Menabrea's memoir; an idea which was immediately adopted.

We discussed together the various illustrations that might be introduced: I suggested several, but the selection was entirely her own.   So also was the algebraic working out of the different problems, except, indeed, that relating to the numbers of Bernoulli, which I had offered to do to save Lady Lovelace the trouble.   This she sent back to me for an amendment, having detected a grave mistake which I had made in the process.

The notes of the Countess of Lovelace extend to about three times the length of the original memoir.   Their author has entered fully into almost all the very difficult and abstract questions connected with the subject.

These two memoirs taken together furnish, to those who are capable of understanding the reasoning, a complete demonstration—*That the whole of the developments and operations of analysis are now capable of being executed by machinery.*

There are various methods by which these developments are arrived at:—1. By the aid of the Differential and Integral Calculus. 2. By the Combinatorial Analysis of Hindenburg.   3. By the Calculus of Derivations of Arbogast.

Each of these systems professes to expand any function according

* [See p. 225.]
† Ada Augusta, Countess of Lovelace, only child of the Poet Byron.

to any laws. Theoretically each method may be admitted to be perfect; but practically the time and attention required are, in the greater number of cases, more than the human mind is able to bestow. Consequently, upon several highly interesting questions relative to the Lunar theory, some of the ablest and most indefatigable of existing analysts are at variance.

The Analytical Engine is capable of executing the laws prescribed by each of these methods. At one period I examined the Combinatorial Analysis, and also took some pains to ascertain from several of my German friends, who had had far more experience of it than myself, whether it could be used with greater facility than the Differential system. They seemed to think that it was more readily applicable to all the usual wants of analysis.

I have myself worked with the system of Arbogast, and if I were to decide from my own limited use of the three methods, I should, for the purposes of the Analytical Engine, prefer the Calcul des Derivations.

As soon as an Analytical Engine exists, it will necessarily guide the future course of the science. Whenever any result is sought by its aid, the question will then arise—By what course of calculation can these results be arrived at by the machine in the *shortest time*?

In the drawings I have prepared I proposed to have a thousand variables, upon each of which any number not having more than fifty figures can be placed. This machine would multiply 50 figures by other 50, and print the product of 100 figures. Or it would divide any number having 100 figures by any other of 50 figures, and print the quotient of 50 figures. Allowing but a moderate velocity for the machine, the time occupied by either of these operations would be about one minute.

The whole of the *numerical* constants throughout the works of Laplace, Plana, Le Verrier, Hansen, and other eminent men whose indefatigable labours have brought astronomy to its present advanced state, might easily be recomputed. They are but the numerical coefficients of the various terms of functions developed according to certain series. In all cases in which these numerical constants can be calculated by more than one method, it might be desirable to compute them by several processes until frequent practice shall have confirmed our belief in the infallibility of mechanism.

The great importance of having accurate Tables is admitted by all who understand their uses; but the multitude of errors really occurring is comparatively little known. Dr. Lardner, in the "Edinburgh Review," has made some very instructive remarks on this subject.

I shall mention two within my own experience: these are selected because they occurred in works where neither care nor expense were spared on the part of the Government to insure perfect accuracy. It is, however, but just to the eminent men who presided over the preparation of these works for the press to observe, that the real fault lay not in them but in *the nature of things*.

In 1828 I lent the Government an original MS. of the table of Logarithmic Sines, Cosines, &c., computed to every second of the quadrant, in order that they might have it compared with Taylor's Logarithms, 4to., 1792, of which they possessed a considerable number of copies. Nineteen errors were thus detected, and a list of these errata was published in the Nautical Almanac for 1832: these may be called

Nineteen errata of the first order 1832

An error being detected in one of these errata, in the following Nautical Almanac we find an

Erratum of the errata in N. Alm. 1832 1833

But in this very erratum of the second order a new mistake was introduced larger than any of the original mistakes. In the year next following there ought to have been found

Erratum in the erratum of the errata in N. Alm. 1832 . 1834

In the "Tables de la Lune," by M. P. A. Hansen, 4to, 1857, published at the expense of the English Government, under the direction of the Astronomer Royal, is to be found a list of errata amounting to 155. In the 21st of these original errata there have been found *three* mistakes. These are duly noted in a newly-printed list of errata discovered during computations made with them in the "Nautical Almanac"; so that we now have the errata of an erratum of the original work.

This list of errata from the office of the "Nautical Almanac" is larger than the original list. The total number of errors at present (1862) discovered in Hansen's "Tables of the Moon" amounts to above three hundred and fifty. In making these remarks I have no intention of imputing the slightest blame to the Astronomer Royal, who, like other men, cannot avoid submitting to inevitable fate. The only circumstance which is really extraordinary is that, when it was demonstrated that all tables are capable of being computed by machinery, and even when a machine existed which computed certain tables, that the Astronomer Royal did not become the most enthusiastic supporter of an instrument which could render such invaluable service to his own science.

In the Supplementary Notices of the Astronomical Society, No. 9, vol. xxiii., p. 259, 1863, there occurs a Paper by M. G. de Ponteculant, in which forty-nine numerical coefficients relative to the Longitude, Latitude, and Radius vector of the Moon are given as computed by Plana, Delaunay, and Ponteculant.   The computations of Plana and Ponteculant agree in thirteen cases; those of Delaunay and Ponteculant in two; and in the remaining thirty-four cases they all three differ.

I am unwilling to terminate this chapter without reference to another difficulty now arising, which is calculated to impede the progress of Analytical Science.   The extension of analysis is so rapid, its domain so unlimited, and so many inquirers are entering into its fields, that a variety of new symbols have been introduced, formed on no common principles.   Many of these are merely new ways.of expressing well-known functions.   Unless some philosophical principles are generally admitted as the basis of all notation, there appears a great probability of introducing the confusion of Babel into the most accurate of all languages.

A few months ago I turned back to a paper in the Philosophical Transactions, 1844, to examine some analytical investigations of great interest by an author who has thought deeply on the subject.   It related to the separation of symbols of operation from those of quantity, a question peculiarly interesting to me, since the Analytical Engine contains the embodiment of that method.   There was no ready, sufficient and simple mode of distinguishing letters which represented quantity from those which indicated operation.   To understand the results the author had arrived at, it became necessary to read the whole Memoir.

Although deeply interested in the subject, I was obliged, with great regret, to give up the attempt; for it not only occupied much time, but placed too great a strain on the memory.

Whenever I am thus perplexed it has often occurred to me that the very simple plan I have adopted in my *Mechanical Notation* for lettering drawings might be adopted in analysis.

On the geometrical drawings of machinery every piece of matter which represents framework is invariably denoted by an *upright* letter; whilst all letters indicating moveable parts are marked by *inclined* letters.

The analogous rule would be—

Let all letters indicating operations or modifications be expressed by *upright* letters;

Whilst all letters representing quantity should be represented by *inclined* letters.

The subject of the principles and laws of notation is so important that it is desireable, before it is too late, that the scientific academies of the world should each contribute the results of their own examination and conclusions, and that some congress should assemble to discuss them.   Perhaps it might be still better if each academy would draw up its own views, illustrated by examples, and have a sufficient number printed to send to all other academies.

CHAPTER XIII

RECOLLECTIONS OF WOLLASTON, DAVY, AND ROGERS

In 1826, one of the secretaryships of the Royal Society became vacant. Dr. Wollaston and several others of the leading members of the Society and of the Council wished that I should be appointed. This would have been the more agreeable to me, because my early friend Herschel was at that time the senior Secretary.

This arrangement was agreed to by Sir H. Davy, and I left town with the full assurance that I was to have the appointment. In the mean time Sir H. Davy summoned a Council at an unusual hour—eight o'clock in the evening—for a special purpose, namely, some arrangement about the Treasurer's accounts.

After the business relating to the Treasurer was got through, Sir H. Davy observed that there was a secretaryship vacant, and he proposed to fill it up.

Dr. Wollaston then asked Sir Humphry Davy if he claimed the nomination as a right of the President, to which Sir H. Davy replied that he did, and then nominated Mr. Children. The President, as president, has no such right; and even if he had possessed it, he had promised Mr. Herschel that I should be his colleague. There were upright and eminent men on that council; yet no one of them had the moral courage to oppose the President's dictation, or afterwards to set it aside on the ground of its irregularity.

A few years after, whilst I was on a visit at Wimbledon Park, Dr. and Mrs. Somerville came down to spend the day. Dr. Somerville mentioned a very pleasant dinner he had had with the late Mr. John Murray of Albemarle Street, and also a conversation relating to my book "On the Decline of Science in England." Mr. Murray felt hurt

at a remark I had made on himself whilst criticizing a then unexplained
job of Sir Humphry Davy's. Dr. Somerville assured Mr. Murray
that he knew me intimately, and that if I were convinced that I had
done him an injustice, nobody would be more ready to repair it. A
few days after, Mr. Murray put into Dr. Somerville's hands papers
explaining the whole of the transaction. These papers were now
transferred to me. On examining them I found ample proof of what
I had always suspected. The observation I had made which pained
Mr. Murray fell to the ground as soon as the real facts were known, and
I offered to retract it in any suitable manner. One plan I proposed
was to print a supplemental page, and have it bound up with all the
remaining copies of the "Decline of Science."

Mr. Murray was satisfied with my explanation, but did not wish me
to take the course I proposed, at least, not at that time. Various
objections may have presented themselves to his mind, but the affair
was adjourned with the understanding that at some future time I
should explain the real state of the facts which had led to this mis-
interpretation of Mr. Murray's conduct.

The true history of the affair was this: Being on the Council of the
Royal Society in 1827, I observed in our accounts a charge of 381*l*. 5*s*.
as paid to Mr. Murray for 500 copies of Sir Humphry Davy's Discourses.

I asked publicly at the Council for an explanation of this item. The
answer given by Dr. Young and others was—

"That the Council had agreed to purchase these volumes at that
price, in order to *induce* Mr. Murray to print the President's speeches."

To this I replied that such an explanation was entirely inadmissible.
I then showed that even allowing a very high price for composing,
printing, and paper, if the Council had wished to print 500 copies
of those Discourses they could have done it themselves for 150*l*. at the
outside. I could not extract a single word to elucidate this mystery,
about which, however, I had my own ideas.

It appeared by the papers put into my hands that Sir Humphry
Davy had applied to Mr. Murray, and had sold him the copyright of
the Discourses for 500 guineas, one of the conditions being that the
Royal Society should purchase of him 500 copies at the trade price.

Mr. Murray paid Sir H. Davy the 500 guineas in three bills at six,
twelve, and eighteen months. These bills passed through Drummond's (Sir H. Davy's banker), and I have had them in my own
hands for examination.

Thus it appears that Mr. Murray treated the whole affair as a
matter of business, and acted in this purchase in his usual liberal
manner. I have had in my hand a statement of the winding-up of

that account copied from Mr. Murray's books, and I find that he was
a considerable loser by his purchase.   Sir H. Davy, on the other hand,
contrived to transfer between three and four hundred pounds from
the funds of the Royal Society into his own pocket.*

It was my determination to have called for an explanation of this
affair at the election of our President and officers at our anniversary
on the 30th November if Sir H. Davy had been again proposed as
President in 1827.

The Thaumatrope

One day Herschel, sitting with me after dinner, amusing himself by
spinning a pear upon the table, suddenly asked whether I could show
him the two sides of a shilling at the same moment.

I took out of my pocket a shilling, and holding it up before the
looking-glass, pointed out *my* method.   "No," said my friend, "that
won't do"; then spinning my shilling upon the table, he pointed out
his method of seeing both sides at once.   The next day I mentioned the
anecdote to the late Dr. Fitton, who a few days after brought me a
beautiful illustration of the principle.   It consisted of a round disc
of card suspended between the two pieces of sewing-silk.   These
threads being held between the finger and thumb of each hand, were
then made to turn quickly, when the disc of card, of course, revolved
also.

Upon one side of this disc of card was painted a bird; upon the other
side, an empty bird-cage.   On turning the thread rapidly, the bird
appeared to have got inside the cage.   We soon made numerous
applications, as a rat on one side and a trap upon the other, &c.   It
was shown to Captain Kater, Dr. Wollaston, and many of our friends,
and was, after the lapse of a short time, forgotten.

Some months after, during dinner at the Royal Society Club, Sir
Joseph Banks being in the chair, I heard Mr. Barrow, then Secretary
to the Admiralty, talking very loudly about a wonderful invention of
Dr. Paris, the object of which I could not quite understand.   It was
called the thaumatrope, and was said to be sold at the Royal Institution,
in Albermarle-street.   Suspecting that it had some connection with
our unnamed toy, I went the next morning and purchased, for seven
shillings and sixpence, a thaumatrope, which I afterwards sent down
to Slough to the late Lady Herschel.   It was precisely the thing which
her son and Dr. Fitton had contributed to invent, which amused all
their friends for a time and had then been forgotten.   There was
however *one* additional thaumatrope made afterwards.   It consisted

* See "Decline of Science in England," p. 105.   8vo.   1830.

of the usual disc of paper. On one side was represented a thaumatrope (the design upon it being a penny-piece) with the motto, "How
to turn a penny."

On the other side was a gentleman in black, with his hands held
out in the act of spinning a thaumatrope, the motto being, "A new
trick from Paris."

After my contest for Finsbury was decided, Mr. Rogers the banker,
and the brother of the poet, who had been one of my warmest sup-
porters, proposed accompanying me to the hustings at the declaration
of the poll. He had also invited a party of some of the most influential
electors of his district to dine with him in the course of the week, in
order that they might meet me, and consider about measures for
supporting me at the next opportunity.

On a cold drizzling rainy day in November the final state of the poll
was declared. Mr. Rogers took me in his carriage to the hustings, and
caught a cold, which seemed at first unimportant. On the day of the
dinner, when we met at Mr. Rogers's, who resided at Islington, he
was unable to leave his bed. Miss Rogers, his sister, who lived with
him, and his brother the poet, received us, quite unconscious of the
dangerous condition of their relative, who died the next day.

Thus commenced a friendship with both of my much-valued friends
which remained unruffled by the slightest wave until their lamented
loss. Miss Rogers removed to a house in the Regent's Park, in which
the paintings by modern artists collected by her elder brother, and
increased by her own judicious taste, were arranged. The society at
that house comprised all that was most eminent in literature and in art.
The adjournment after her breakfasts to the delightful verandah
overlooking the Park still clings to my fading memory, and the voices
of her poet brother, of Jeffrey, and of Sidney Smith still survive in the
vivid impressions of their wisdom and their wit.

I do not think the genuine kindness of the poet's character was
sufficiently appreciated. I occasionally walked home with him from
parties during the first years of our acquaintance. In later years, when
his bodily strength began to fail, I always accompanied him, though
sometimes not without a little contest.

I have frequently walked with him from his sister's house, in the
Regent's Park, to his own in St. James's Place, and he has sometimes
insisted upon returning part of the way home with me.

On one of those occasions we were crossing a street near Cavendish
Square: a cart coming rapidly round the corner, I almost dragged
him over. As soon as we were safe, the poet said, very much as a
child would, "There, now, that was all your fault; you would come

with me, and so I was nearly run over." However, I found less and less resistance to my accompanying him, and only regretted that I could not be constantly at his side on those occasions.

Soon after the publication of the "Economy of Manufactures," Mr. Rogers told me that he had met one evening, at a very fashionable party, a young dandy, with whom he had had some conversation. The poet had asked him whether he had read that work. To this his reply was, "Yes: it is a very nice book—just the kind of book that anybody could have written."

One day, when I was in great favour with the poet, we were talking about the preservation of health. He told me he would teach me how to live for ever; for which I thanked him in a compliment after his own style, rather than in mine. I answered, "Only embalm me in your poetry, and it is done." Mr. Rogers invited me to breakfast with him the next morning, when he would communicate the receipt. We were alone, and I enjoyed a very entertaining breakfast. The receipt consisted mainly of cold ablutions and the frequent use of the flesh brush. Mr. Rogers himself used the latter to a moderate extent regularly, three times every day—before he dressed himself, when he dressed for dinner, and before he got into bed. About six or eight strokes of the flesh-brush completed each operation. We then adjourned to a shop, where I purchased a couple of the proper brushes, which I used for several years, and still use occasionally, with, I believe, considerable advantage.

Once, at Mr. Rogers's table, I was talking with one of his guests about the speed with which some authors composed, and the slowness of others. I then turned to our host, and, much to his surprise, inquired how many lines a-day on the average a poet usually wrote. My friend, when his astonishment had a little subsided, very good-naturedly gave us the result of his own experience. He said that he had never written more than four* lines of verse in any one day of his life. This I can easily understand; for Mr. Rogers' taste was the most fastidious, as well as the most just, I ever met with. Another circumstance also, I think, contributed to this slowness of composition.

An author may adopt either of two modes of composing. He may write off the whole of his work roughly, so as to get upon paper the plan and general outline, without attending at all to the language. He may afterwards study minutely every clause of each sentence, and then every word of each clause.

* I am not quite certain that the number was four; but I am absolutely certain that it was either four or six.

Or the author may finish and polish each sentence as soon as it is written.

This latter process was, I think, employed by Mr. Rogers, at least in his poetry.

He then told us that Southey composed with much greater rapidity than himself, as well in poetry as in prose. Of the latter Southey frequently wrote a great many pages before breakfast.

Once, at a large dinner party, Mr. Rogers was speaking of an inconvenience arising from the custom, then commencing, of having windows formed of one large sheet of plate-glass. He said that a short time ago he sat at dinner with his back to one of these single panes of plate-glass: it appeared to him that the window was wide open, and such was the force of imagination, that he actually caught cold.

It so happened that I was sitting just opposite to the poet. Hearing this remark, I immediately said, "Dear me, how odd it is, Mr. Rogers, that you and I should make such a very different use of the faculty of imagination. When I go to the house of a friend in the country, and unexpectedly remain for the night, having no night-cap, I should naturally catch cold. But by tying a bit of pack-thread tightly round my head, I go to sleep imagining that I have a night-cap on; consequently I catch no cold at all." This sally produced much amusement in all around, who supposed I had improvised it; but, odd as it may appear, it is a practice I have often resorted to. Mr. Rogers, who knew full well the respect and regard I had for him, saw at once that I was relating a simple fact, and joined cordially in the merriment it excited.

In the latter part of Mr. Rogers's life, when, being unable to walk, he was driven in his carriage round the Regent's Park, he frequently called at my door, and, when I was able, I often accompanied him in his drive. On some one of these occasions, when I was unable to accompany him, I put into his hands a parcel of proof-sheets of a work I was then writing, thinking they might amuse him during his drive, and that I might profit by his criticism. Some years before, I had consulted him about a novel I had proposed to write solely for the purpose of making money to assist me in completing the Analytical Engine. I breakfasted alone with the poet, who entered fully into the subject. I proposed to give up a twelvemonth to writing the novel, but I determined not to commence it unless I saw pretty clearly that I could make about 5,000l. by the sacrifice of my time. The novel was to have been in three volumes, and there would probably have been reprints of another work in two volumes. Both of these works

would have had graphic illustrations. The poet gave me much
information on all the subjects connected with the plan, and amongst
other things, observed that when he published his beautifully illustrated
work on Italy, that he had paid 9,000*l.* out of his own pocket before he
received any return for that work.

CHAPTER XIV

RECOLLECTIONS OF LAPLACE, BIOT, AND HUMBOLDT

My first visit to Paris was made in company with my friend John Herschel. On reaching Abbeville, we wanted breakfast, and I undertook to order it. Each of us usually required a couple of eggs. I preferred having mine moderately boiled, but my friend required his to be boiled quite hard. Having explained this matter to the waiter, I concluded by instructing him that each of us required two eggs thus cooked, concluding my order with the words, "pour chacun deux."

The garçon ran along the passage half way towards the kitchen, and then called out in his loudest tone—

"Il faut faire bouillir cinquante-deux œufs pour Messieurs les Anglais." I burst into such a fit of uncontrollable laughter at this absurd misunderstanding of *chacun deux*, for *cinquante-deux*, that it was some time before I could explain it to Herschel, and but for his running into the kitchen to countermand it, the half hundred of eggs would have assuredly been simmering over the fire.

A few days after our arrival in Paris, we dined with Laplace, where we met a large party, most of whom were members of the Institut. The story had already arrived at Paris, having rapidly passed through several editions.

To my great amusement, one of the party told the company that, a few days before, two young Englishmen being at Abbeville, had ordered fifty-two eggs to be boiled for their breakfast, and that they ate up every one of them, as well as a large pie which was put before them.

My next neighbour at dinner asked me if I thought it probable. I replied, that there was no absurdity a young Englishman would not occasionally commit.

One morning Herschel and I called on Laplace, who spoke to us of various English works on mathematical subjects. Amongst others, he mentioned with approbation, "Un ouvrage de vous deux." We were both quite at a loss to know to what work he referred. Herschel and I had not written any joint work, although we had together translated the work of Lacroix. The volume of the "Memoirs of the Analytical Society," though really our joint production, was not known to be such, and it was also clear that Laplace did not refer to that work. Perceiving that we did not recognise the name of the author to whom he referred, Laplace varied the pronunciation by calling him *vous deux*; the first word being pronounced as the French word "vous," and the second as the English word "deuce."

Upon further explanation, it turned out that Laplace meant to speak of a work published by Woodhouse, whose name is in the pronunciation of the French so very like *vous deux*.

Poisson, Fourier, and Biot were amongst my earliest friends in Paris. Fourier, then Secretary of the Institute, had accompanied the first Napoleon in his expedition to Egypt. His profound acquaintance with analysis remains recorded in his works. His unaffected and genial manner, the vast extent of his acquirements, and his admirable taste conspicuous even in the apartments he inhabited, were most felt by those who were honoured by his friendship.

With M. Biot I became acquainted in early life; he was then surrounded by a happy family. In my occasional visits to Paris I never omitted an opportunity of paying my respects to him: when deprived of those supports and advanced in life, he still earnestly occupied himself in carrying out the investigations of his earlier years.

His son, M. Biot, a profound oriental scholar, who did me the honour of translating the 'Economy of Manufactures,' died many years before his father.

In one of my visits to Paris, at a period when beards had become fashionable amongst a certain class of my countrymen, I met Biot. After our first greeting, looking me full in the face, he said, "My dear friend, you are the best shaved man in Europe."

At a later period I took with me to Paris the complete drawings of Difference Engine No. 2. As soon as I had hung them up round my own apartments to explain them to my friends I went to the Collège de France, where M. Biot resided. I mentioned to him the fact, and said that if it was a subject in which he was interested, and had leisure to look at these drawings, I should have great pleasure in bringing them to him, and giving him any explanation that he might desire. I told him, however, that I was fully aware how much the time of every man who really adds to science must be occupied, and that I made this proposal rather to satisfy my own mind that I had not neglected one of my oldest friends than in the expectation that he had time for the examination of this new subject.

The answer of my friend was remarkable. After thanking me in the warmest terms for this mark of friendship, he explained to me that the effect of age upon his own mind was to render the pursuit of any new inquiry a matter of slow and painful effort; but that in following out the studies of his youth he was not so much impeded. He added that in those subjects he could still study with satisfaction, and even make advances in them, assisted in the working out of his views experimentally by the aid of his younger friends.

I was much gratified by this unreserved expression of the state of the case, and I am sure those younger men who so kindly assisted the aged philosopher will be glad to know that their assistance was duly appreciated.

The last time during M. Biot's life that I visited Paris I went, as usual, to the Collège de France. I inquired of the servant who opened the door after the state of M. Biot's health, which was admitted to be feeble. I then asked whether he was well enough to see an old friend. Biot himself had heard the latter part of this conversation. Coming into the passage he seized my hand and said "My dear friend, I would see you even if I were dying."

Alexander Humboldt

One of the most remarkable characteristics of Humboldt's mind was, that he not merely loved and pursued science for its own sake, but that he derived pleasure from assisting with his information and advice any other inquirer, however humble, who might need it.

In one of my visits to Paris, Humboldt was sitting with me when a friend of mine, an English clergyman, who had just arrived in Paris, and had only two days to spare for it, called upon me to ask my assistance about getting access to certain MSS. Putting into Humboldt's hand a tract lying on my table, I asked him to excuse me for a few minutes whilst I gave what advice I could to my countryman.

My friend told me that he wanted to examine a MS., which he was informed was in a certain library in a certain street in Paris: that he knew nobody in the city to help him in his mission.

Humboldt having heard this statement, came over to us and said, "If you will introduce me to your friend, I can put him in the way of seeing the MSS. he is in search of." He then explained that the MSS. had been removed to another library in Paris, and proposed to give my friend a note of introduction to the librarian, and mentioned other MSS. and other libraries in which he would find information upon the same subject.

Many years after, being at Vienna, I heard that Humboldt was at Töplitz, a circumstance which induced me to visit that town. On my arrival I found he had left it a few days before on his return to Berlin. In the course of a few days, I followed him to that city, and having arrived in the middle of the day, I took apartments in the Linden Walk, and got all my travelling apparatus in order; I then went out to call on Humboldt. Finding that he had gone to dine with his brother William, who resided at a short distance from Berlin, I therefore merely left my card.

The next morning at seven o'clock, before I was out of bed, I received a very kind note from Humboldt, to ask me to breakfast with him at nine. In a postscript he added, "What are the moving molecules of Robert Brown?" These atoms of dead matter in rapid motion, when examined under the microscope, were then exciting great attention amongst philosophers.

I met at breakfast several of Humboldt's friends, with whose names and reputation I was well acquainted.

Humboldt himself expressed great pleasure that I should have visited Berlin to attend the great meeting of German philosophers, who in a few weeks were going to assemble in that capital. I assured him that I was quite unaware of the intended meeting, and had directed my steps to Berlin merely to enjoy the pleasure of his society. I soon perceived that this meeting of philosophers on a very large scale, supported by the King and by all the science of Germany, might itself have a powerful influence upon the future progress of human knowledge. Amongst my companions at the breakfast-table were Derichlet and Magnus. In the course of the morning Humboldt mentioned to me that his own duties required his attendance on the King every day at three o'clock, and having also in his hands the organization of the great meeting of philosophers, it would not be in his power to accompany me as much as he wished in seeing the various institutions in Berlin. He said that, under these circumstances, he had asked his two young friends, Derichlet and Magnus, to supply his place. During many weeks of my residence in Berlin, I felt the daily advantage of this thoughtful kindness of Humboldt. Accompanied by one or other, and frequently by both, of my young friends, I saw everything to the best advantage, and derived an amount of information and instruction which under less favourable circumstances it would have been impossible to have obtained.

The next morning, I again breakfasted with Humboldt. On the previous day I had mentioned that I was making a collection of the signs employed in map-making. I now met Von Buch and General Ruhl, both of whom were profoundly acquainted with that subject. I had searched in vain for any specimen of a map shaded upon the principle of lines of equal elevation. Von Buch the next morning gave me an engraving of a small map upon that principle, which was, I believe, at that time the only one existing.

After breakfast we went into Humboldt's study to look at something he wished to show us. In turning over his papers, which, like my own, were lying apparently in great disorder upon the table, he picked up the cover of a letter on which was written a number of names in different parallel columns. "That," he observed incidentally, "is for you." After he had shown us the object of our visit to his sanctum, he reverted to the envelop which he put into my hands, explaining that he had grouped roughly together for my use all the remarkable men then in Berlin, and several of those who were expected.

These he had arranged in classes:—Men of science, men of letters, sculptors, painters, and artists generally, instrument-makers, &c. This list I found very convenient for reference.

When the time of the great meeting approached, it became necessary to prepare the arrangements for the convenience of the assembled science of Europe. One of the first things, of course, was the important question, how they were to dine? A committee was therefore appointed to make experiment by dining successively at each of the three or four hotels competing for the honour of providing a table d'hôte for the savans.

Humboldt put me on that committee, remarking, that an Englishman always appreciates a good dinner. The committee performed their agreeable duty in a manner quite satisfactory to themselves, and I hope, also, to the digestions of the Naturforschers.

During the meeting much gaiety was going on at Berlin. One evening previous to our parties, I was walking in the Linden Walk with Humboldt, discussing the singularities of several of our learned acquaintance. My companion made many acute and very amusing remarks; some of these were a little caustic, but not one was ill-natured. I had contributed a very small and much less brilliant share to this conversation, when the clock striking, warned us that the hour for our visits had arrived. I never shall forget the expression of archness which lightened up Humboldt's countenance when shaking my hand he said, in English, "My dear friend, I think it may be as well that we should not speak of each other until we meet again." We then each kept our respective engagements, and met again at the most recherché of all, a concert at Mendelssohn's.

Of the Buonaparte Family

From my father's house on the coast, near Teignmouth, we could, with a telescope, see every ship which entered Torbay. When the "Bellerophon" anchored, the news was rapidly spread that Napoleon was on board. On hearing the rumour, I put a small telescope into my pocket, and, mounting my horse, rode over to Torbay. A crowd of boats surrounded the ship, then six miles distant; but, by the aid of my glass, I saw upon the quarter-deck that extraordinary man, with many members of whose family I subsequently became acquainted. Of those who are no more I may without impropriety say a few words.

My first acquaintance with several branches of the family of Napoleon Buonaparte arose under the following circumstances:—

When his elder brother Lucien, to avoid the necessity of accepting a kingdom, fled from his imperial brother, and took refuge in England, his position was either not well understood, or, perhaps, was entirely mistaken. Lucien seems to have been looked upon with suspicion by our Government, and was placed in the middle of England under a species of espionage.

Political parties then ran high, and he did not meet with those attentions which his varied and highly-cultivated tastes, especially in the fine arts, entitled him to receive, as a stranger in a foreign land.

A family connection of mine, residing in Worcestershire, was in the habit of visiting Lucien Buonaparte. Thus, in my occasional visits to my brother-in-law's place, I became acquainted with the Prince of Canino. In after-years, when he occasionally visited London, I had generally the pleasure of seeing him.

In 1828 I met at Rome the eldest son of Lucien, who introduced me to his sisters, Lady Dudley Stuart and the Princess Gabrielli.

In the same year I became acquainted, at Bologna, with the Princess d'Ercolano, another daughter of Lucien, whom I afterwards met at Florence, at the palace of her uncle Louis, the former king of Holland. During a residence of several months in that city I was a frequent guest at the family table of the Comte St. Leu. One of his sons had married the Princess Charlotte, the second daughter of the King of Spain, a most accomplished, excellent, and charming person. They reminded me much of a sensible English couple, in the best class of English society. Both had great taste in the fine arts. The prince had a workshop at the top of the palace, in which he had a variety of tools and a lithographic printing press. Occasionally, in the course of their morning drives, some picturesque scene, in that beautiful country, would arrest their attention. Stopping the carriage, they would select a favourable spot, and the princess would then make a sketch of it.

At other times they would spend the evening, the prince in extemporizing an imaginary scene, which he described to his wife, who, with admirable skill, embodied upon paper the tasteful conceptions of her husband. These sketches then passed up to the workshop of the Prince, were transferred to stone, and in a few days lithographic impressions descended to the drawing-room. I fortunately possess some of these impressions, which I value highly, not only as the productions of an amiable and most accomplished lady, but of one who did not shrink from the severer duties of life, and died in fulfilling them.

After the melancholy loss of her husband, the Princess Charlotte remained with her father, who resided at one period in the Regent's Park, where I from time to time paid my respects to them. Occasionally I received them at my own house. One summer letters from Florence reached them, announcing the dangerous illness of the Comte de St. Leu. The daughter of Joseph immediately set out alone for Florence to minister to the comfort of her uncle and father-in-law. On her return from Italy she was attacked by cholera and died in the south of France.

CHAPTER XV

EXPERIENCE BY WATER

Shooting Sea-birds—Walking on the Water—A Screw being loose—The Author nearly drowned—Adventure in the Thames Tunnel—Descent in a Diving-bell— Plan for Submarine Navigation.

THE grounds surrounding my father's house, near Teignmouth, extended to the sea. The cliffs, though lofty, admitted at one point of a descent to the beach, of which I very frequently availed myself for the purpose of bathing. One Christmas when I was about sixteen I determined to see if I could manage a gun. I accordingly took my father's fowling-piece, and climbing with it down to the beach, I began to look about for the large sea-birds which I thought I might have a chance of hitting.

I fired several charges in vain. At last, however, I was fortunate enough to hit a sea-bird called a diver; but it fell at some distance into the sea: I had no dog to get it out for me; the sea was rough, and no boat was within reach; also it was snowing.

So I took advantage of a slight recess in the rock to protect my clothes from the snow, undressed, and swam out after my game, which I succeeded in capturing. The next day, having got the cook to roast it, I tried to eat it; but this was by no means an agreeable task, so for the future I left the sea-birds to the quiet possession of their own dominion.

Shortly after this, whilst residing on the beautiful banks of the Dart, I constantly indulged in swimming in its waters. One day an idea struck me, that it was possible, by the aid of some simple mechanism, to walk upon the water, or at least to keep in a vertical position, and have head, shoulders, and arms above water.

My plan was to attach to each foot two boards closely connected together by hinges themselves fixed to the sole of the shoe. My theory was, that in lifting up my leg, as in the act of walking, the two boards would close up towards each other; whilst on pushing down my foot, the water would rush between the boards, cause them to open out into a flat surface, and thus offer greater resistance to my sinking in the water.

I took a pair of boots for my experiment, and cutting up a couple of old useless volumes with very thick binding, I fixed the boards by hinges in the way I proposed. I placed some obstacle between the two flaps of each book to prevent them from approaching too nearly to each other so as to impede their opening by the pressure of the water.

I now went down to the river, and thus prepared, walked into the water. I then struck out to swim as usual, and found little difficulty. Only it seemed necessary to keep the feet farther apart. I now tried the grand experiment. For a time, by active exertion of my legs, I kept my head and shoulders above water and sometimes also my arms. I was now floating down the river with the receding tide, sustained in a vertical position with a very slight exertion of force.

But unfortunately one pair of my hinges got out of order, and refused to perform its share of the propulsion. The result was that I became lop-sided. I was therefore obliged to swim, which I now did with considerable exertion; but another difficulty soon occurred,—the instrument on the disabled side refused to do its share in propelling me. The tide was rapidly carrying me down the river; my own exertions alone would have made me revolve in a small circle, consequently I was obliged to swim in a spiral. It was very difficult to calculate the curve I was describing upon the surface of the water, and still more so to know at what point, if at any, I might hope to reach its banks again. I became very much fatigued by my efforts, and endeavoured to relieve myself for a time by resuming the vertical position.

After floating, or rather struggling for some time, my feet at last touched the bottom. With some difficulty and much exertion I now gained the bank, on which I lay down in a state of great exhaustion.

This experiment satisfied me of the danger as well as of the practicability of my plan, and ever after, when in the water, I preferred trusting to my own unassisted powers.

At the close of the year 1827, as I anticipated a long absence from England, I paid a visit to the Thames Tunnel, in the construction of which I took a great interest. My eldest son, then about twelve years of age, accompanied me in this visit. I fortunately found the younger Brunel at the works, who kindly took us with him into the workings.

We stood upon a timber platform, distant about fifty feet from the shield, which was full of busy workmen, each actively employed in his own cell. As we were conversing together, I observed some commotion in the upper cell on the right hand side. From its higher corner there entered a considerable stream of liquid mud. Brunel ran directly to the shield, a line of workmen was instantly formed, and whatever tools or timber was required was immediately conveyed to the spot.

I observed the progress with some anxiety, since but a short time before a similar occurrence had been the prelude to the inundation of the whole tunnel. I remained watching the fit time, if necessary, to run away; but also noticing what effect the apparent danger had on my son. After a short time it was clear that the ingress of liquid mud had been checked, and in a few minutes more Brunel returned to me, having this time succeeded in stopping up the breach. I then inquired what was really the nature of the danger we had escaped. Brunel told me that unless himself or Gravatt had been present, the whole tunnel would in less than ten minutes have been full of water. The next day I embarked for Holland, and in about a week after I read in Galignani's newspaper, that the Thames had again broken into the tunnel; that five or six of the workmen had been drowned, and that Brunel himself had escaped with great difficulty by swimming.

In 1818, during a visit to Plymouth, I had an opportunity of going down in a diving-bell: I was accompanied by two friends and the usual director of that machine.

The diving-bell in which I descended was a cast-iron vessel about six feet long by four feet and a half wide, and five feet eight inches high. In the top of the bell there were twelve circular apertures, each about six inches in diameter, filled by thick plate-glass fixed by water-tight cement. Exactly in the centre there were a number of small holes through which the air was continually pumped in from above.

At the ends of the bell are two seats, placed at such a height, that the top of the head is but a few inches below the top of the bell; these will conveniently hold two persons each. Exactly in the middle of the bell, and about six inches above its lower edge, is placed a narrow board, on which the feet of the divers rest. On one side, nearly on a level with the shoulders, is a small shelf, with a ledge to contain a few tools, chalk for writing messages, and a ring to which a small rope is tied. A board is connected with this rope; and after writing any orders on the board with a piece of chalk, on giving it a pull, the superintendent above, round whose arm the other end of the rope is fastened, will draw it up to the surface, and, if necessary, return an answer by the same conveyance.

In order to enter the bell, it is raised about three or four feet above the surface of the water; and the boat, in which the persons who propose descending are seated, is brought immediately under it; the bell is then lowered, so as to enable them to step upon the foot-board within it; and having taken their seats, the boat is removed, and the bell gradually descends to the water.

On touching the surface, and thus cutting off the communication with the external air, a peculiar sensation is perceived in the ears; it is not, however, painful. The attention is soon directed to another object. The air rushing in through the valve at the top of the bell overflows, and escapes with a considerable bubbling noise under the sides. The motion of the bell proceeds slowly, and almost imperceptibly; and, on looking at the glass lenses close to the head, when the top of the machine just reaches the surface of the water, it may be perceived, by means of the little impurities which float about in it, flowing into the recesses containing the glasses. A pain now begins to be felt in the ears, arising from the increased external pressure; this may sometimes be removed by the act of yawning, or by closing the nostrils and mouth, and attempting to force air through the ears. As soon as the equilibrium is established the pain ceases, but recommences almost immediately by the continuance of the descent. On returning, the same sensation of pain is felt in the ears; but it now arises from the dense air which had filled them endeavouring, as the pressure is removed, to force its way out.

If the water is clear, and not much disturbed, the light in the bell is very considerable; and, even at the depth of twenty feet, was more than is usual in many sitting-rooms. Within the distance of eight or ten feet, the stones at the bottom began to be visible. The pain in the ears still continues to occur at intervals, until the descent of the bell terminates by its resting on the ground. The light is sufficient, after passing through twenty feet of sea water, even for delicate experiments; and a far less quantity is enough for the work which is usually performed in those situations.

The temperatures of the hand and of the mouth, under the tongue, were measured by a thermometer, but they did not seem to differ from those which had been determined by the same instrument previous to the descent; at least, the difference did not amount to one-sixth of a degree of Fahrenheit's scale. The pulse was more frequent.

A small magnetic needle did not appear to have entirely lost its directive power, when placed on the footboard in the middle of the bell; but its direction was not the same as that which it indicated on shore. This was determined by directing, by means of signals, the workmen above to move the bell in the direction of one of the co-ordinates; a stick then being pressed against the bottom drew a line parallel to that co-ordinate, its direction by compass was ascertained in the bell, and the direction of the co-ordinate was determined on returning to the surface after leaving the bell.

Signals are communicated by the workmen in the bell to those above, by striking against the side of the bell with a hammer. Those most frequently wanted are indicated by the fewest number of blows; thus a single stroke is to require more air. The sound is heard very distinctly by those above; but, it must be confessed, that to persons unaccustomed to it, the force with which a weighty hammer is driven against so brittle a material as cast iron is a little alarming.

After ascending a few inches from the bottom, the air in the bell became slightly obscured. At the distance of a few feet this appearance increased. Before it had half reached the surface, it was evident that the whole atmosphere it contained was filled with a mist or cloud, which at last began to condense in large drops on the whole of the internal surface.

The explanation of this phenomenon seems to be, that on the rising of the bell the pressure on the air within being diminished by a weight equal to several feet of water, it began to expand; and some portion of it escaping under the edges of the bell, reduced the temperature of that which remained so much, that it was unable to retain, in the state of invisible vapour, the water which it had previously held in solution. Thus the same principle which constantly produces clouds in the atmosphere filled the diving-bell with mist.

This first led me to consider the much more extensive question of submarine navigation. I was aware that Fulton had already descended in a diving-vessel, and remained under water during several hours. He also carried down a copper sphere containing one cubic foot of space into which he had forced two hundred atmospheres. With these means he remained under water and moved about at pleasure during four hours.

But a closed vessel is obviously of little use for the most important purposes to which submarine navigation would be applied in case of war. In the article Diving Bell, published in 1826, in the 'Encyclopedia Metropolitana,' I gave a description and drawings of an *open* submarine vessel which would contain sufficient air for the consumption of four persons during more than two days. A few years ago, I understand, experiments were made in the Seine at Paris, on a similar kind of open diving-vessel. Such a vessel could be propelled by a screw, and might enter, without being suspected, any harbour, and place any amount of explosive matter under the bottoms of ships at anchor.

Such means of attack would render even iron and iron-clad ships unsafe when blockading a port. For though chains were kept constantly passing under their keels, it would yet be possible to moor explosive magazines at some distance below, which would effectually destroy them.

CHAPTER XVI

EXPERIENCE BY FIRE

Baked in an Oven—A Living Volcano—Vesuvius in action—Carried up the Cone of Ashes in a Chair—View of the Crater in a Dark Night—Sunrise—Descent by Ropes and Rolling into the great Crater—Watched the small Crater in active eruption at intervals—Measured a Base of 330 feet—Depth of great Crater 570 feet—Descent into small Crater—A Lake of red-hot Boiling Lava—Regained the great Crater with the sacrifice of my Boots—Lunched on Biscuits and Irish Whisky—Visit to the Hot Springs of Ischia—Towns destroyed by Earthquake—Coronets of Smoke projected by Vesuvius—Artificial Mode of producing them—Fire-damp visited in Welsh Coal-mine in company with Professor Moll.

Baked in an Oven

CALLING one morning upon Chantrey, I met Captain Kater and the late Sir Thomas Lawrence, the President of the Royal Academy. Chantrey was engaged at that period in casting a large bronze statue. An oven of considerable size had been built for the purpose of drying the moulds. I made several inquiries about it, and Chantrey kindly offered to let me pay it a visit, and thus ascertain by my own feelings the effects of high temperature on the human body.

I willingly accepted the proposal, and Captain Kater offered to accompany me. Sir Thomas Lawrence, who was suffering from indisposition, did not think it prudent to join our party. In fact, he died on the second or third day after our experiment.

The iron folding-doors of the small room or oven were opened. Captain Kater and myself entered, and they were then closed upon us. The further *corner* of the room, which was paved with squared stones, was visibly of a dull-red heat. The thermometer marked, if I recollect rightly, 265°. The pulse was quickened, and I ought to have counted but did not count the number of inspirations per minute. Perspiration commenced immediately and was very copious. We remained, I believe, about five or six minutes without very great discomfort, and I experienced no subsequent inconvenience from the result of the experiment.

A Living Volcano

I have never been so fortunate as to be *conscious* of having experienced the least shock of an earthquake, although, when a town had been destroyed in Ischia I hastened on from Rome in the hope of getting a slight shake. My passion was disappointed, so I consoled myself by a flirtation with a volcano.

The situation of my apartments during my residence at Naples enabled me constantly to see the cone of Vesuvius, and the continual projections of matter from its crater. Amongst these were occasionally certain globes of air, or of some gas, which, being shot upwards to a great height above the cone, spread out into huge coronets of smoke, having a singular motion amongst their particles.

A similar phenomenon sometimes occurs on a small scale during the firing of heavy ordnance. I have frequently seen such at Plymouth and elsewhere; but I was not satisfied about the cause of this phenomenon. I was told that it occurred more frequently if the muzzle of the gun were rubbed with grease; but this did not always succeed.

Soon after my return to London I made a kind of drum, by stretching wet parchment over a large tin funnel. On directing the point of the funnel at a candle placed a few feet distant, and giving a smart blow upon the parchment, it is observed that the candle is immediately extinguished.

This arises from what is called an air shot. In fact, the air in the tubular part is projected bodily forward, and so blows out the candle. The statements about persons being killed by cannon balls passing close to but not touching them, if true, are probably the results of air shots.

Wishing to trace the motions of such air shots, I added two small tubes towards the large end of the tin funnel, in order that I might fill it with smoke, and thus trace more distinctly the progress of the ball of air.

To my great delight the first blow produced a beautiful coronet of smoke, exactly resembling, on a small scale, the explosions from cannon or the still more attractive ones from Vesuvius.

If phosphoretted hydrogen or any other gas, which takes fire in air, were thus projected upwards, a very singular kind of fire-work would be produced.

It is possible in dark nights or in fogs that by such means signals might be made to communicate news or to warn vessels of danger.

Vesuvius was then in a state of moderate activity. It had a huge cone of ashes on its summit, surrounding an extensive crater of great depth. In one corner of this was a smaller crater, quite on a diminutive scale, which from time to time ejected red-hot fragments of lava occasionally to the height of from a thousand to fifteen hundred feet above the summit of the mountain.

I had taken apartments in the Chiaja, just opposite the volcano, in order that I might watch it with a telescope. In fact, as I lay in my bed I had an excellent view of the mountain. My next step was to consult with Salvatori, the most experienced of the guides, from whom I had purchased a good many minerals, as to the possibility of getting a peep down the volcano's throat.

Salvatori undertook to report to me from time to time the state of the mountain, round the base of which I made frequent excursions. After about a fortnight, the explosions were more regular and uniform, and Salvatori assured me that all the usual known indications led him to think that it was a fit time for my expedition. As I wished to see as much as possible, I made arrangements to economize my strength by using horses or mules to carry me wherever they could go. Where they could not carry me, as for instance, up the steep slope of the cone of ashes, I employed men to convey me in a chair.

By these means, I saw in the afternoon and evening of one day a good deal of the upper part of the mountain, then took a few hours' repose in a hut, and reached the summit of the cone long before sunrise.

It was still almost dark: we stood upon the irregular edge of a vast gulf spread out below at the depth of about five hundred feet. The plain at the bottom would have been invisible but for an irregular network of bright-red cracks spread over the whole of its surface. Now and then the silence was broken by a rush upwards of a flight of red-hot scoria from the diminutive crater within the large one. These missiles, however, although projected high above the summit of the cone, never extended themselves much beyond the small cavity from which they issued.

Those who have seen the blood-vessels of their own eye by the aid of artificial light, will have seen on a small scale a perfect resemblance of the plain which at that time formed the bottom of the great crater of Vesuvius.

As the morning advanced the light increased, and some time before sunrise we had completed the tour of the top of the great crater. Then followed that glorious sight—the sun when seen rising from the top of some lofty mountain.

I now began to speculate upon the means of getting a nearer view of the little miniature volcano in action at one corner of the gulf beneath us. We had brought ropes with us, and I had observed, in our tour round the crater, every dike of congealed lava by which the massive cone was split. These presented buttresses with frequent ledges or huge steps by which I hoped, with the aid of ropes, to descend into the Tartarus below.

Having consulted with our chief guide Salvatori, I found that he was unwilling to accompany us, and proposed remaining with the other guides on the upper edge of the crater. Upon the whole, I was not discontented with the arrangement, because it left a responsible person to keep the other guides in order, and also sufficient force to lift us up bodily by the ropes if that should become necessary.

The abruptness of the rocky buttresses compelled us to use ropes, but the attempt to traverse the steep inclines of light ashes and of fine sand would have been more dangerous from the risk of being engulfed in them.

Having well examined the several disadvantages of these rough-hewn irregular Titanic stairs, I selected one which seemed the most promising for facilitating our descent into the crater. I was encumbered with one of Troughton's heavy barometers, strapped to my back, looking much like Cupid's quiver, though probably rather heavier. In my pocket I had an excellent box sextant, and in a rough kind of basket two or three thermometers, a measuring tape, and a glass bottle enclosed in a leather case, commonly called a pocket-pistol, accompanied by a few biscuits.

We began our descent by the aid of two ropes, each supported above by two guides. I proceeded, trusting to my rope to step wherever I could, and then cautiously holding on by the rope to spring down to the next ledge. In this manner we descended until we arrived at the last projecting ledge of the dike. Nothing then remained for us but to slide down a steep and lengthened incline of fine sand. Fortunately, the sand itself was not very deep, and was supported by some solid material beneath it. I soon found that it was impossible to stand, so I sat down upon this moving mass, which evidently intended to accompany us in our journey. At first, to my great dismay, I was relieved from the care of my barometer, of which the runaway sand immediately took charge. I then found myself getting deeper and deeper in the sand, and still accelerating my downward velocity.

Gravity had at last done its work and became powerless. I soon dug myself out of my sandy couch, and rushed to my faithful barometer lying at some distance from me with its head just unburied. Fortunately, it was uninjured. My companion, with more skill or good fortune, or with less incumbrances, had safely alighted on the burning plain we now stood upon.

The area of this plain, for it was perfectly flat, was in shape somewhat elliptical. The surface consisted of a black scoriacious rock, reticulated with ditches from one to three feet wide, intersecting each other in every direction. From some of these, fumes not of the most agreeable odour were issuing. All those above two feet deep showed that at that depth below us everything was of a dull-red heat. It was these ditches with red-hot bottoms which, in the darkness of the night, had presented the singular spectacle I described as having witnessed on the evening before.

At one extremity of this oval plain there was a small cone, from which the eruptions before described appeared to issue.

My first step, after examining the few instruments I had brought with me, was to select a spot upon which to measure a base for ascertaining the depth of the crater from its upper edge.

Having decided upon my base line, I took with my sextant the angle of elevation of the rim of the crater above a remarkable spot on a level with my eye. Then fixing my walking-stick into a little crack in the scoria, I proceeded to measure with a tape a base line of 340 feet. Arrived at this point, I again took the angle of elevation of the same part of the rim from the same remarkable spot on a level with the eye. Then, by way of verification, I remeasured my base line and found it only differed from the former measure by somewhat less than one foot. But my walking-stick, which had not penetrated the crack more than a few inches, was actually in flames.

Having noted down these facts, including the state of the thermometer and barometer, in my pocket-book, I took first a survey and then a tour about my fiery domain. I afterwards found, from the result of this measurement, that our base line was 570 feet below one of the lowest points of the edge of the crater. Having collected a few mineral specimens, I applied myself to observe and register the eruptions of the little embryo volcano at the further extremity of the elliptical plain.

These periodical eruptions interested me very much. I proceeded to observe and register them, and found they occurred at tolerably regular intervals. At first, I performed this operation at a respectful distance and out of the reach of the projected red-hot scoria. But as I acquired confidence in their general regularity, I approached from time to time more nearly to the little cone of scoria produced by its own eruptions.

I now perceived an opening in this little cone close to the perpendicular rock of the interior of the great crater. I was very anxious to see real fluid lava; so immediately after an eruption, I rushed to the opening and thus got within the subsidiary crater. But my curiosity was not gratified, for I observed, about forty or fifty feet below me, a huge projecting rock, which being somewhat in advance, effectively prevented me from seeing the lava lake, if any such existed. I then retreated to a respectful distance from this infant volcano to wait for the next explosion.

I continued to note the intervals of time between these jets of red-hot matter, and found that from ten to fifteen minutes was the range of the intervals of repose. Having once more reconnoitred the descent into the little volcano, I seized the opportunity of the termination of one of the most considerable of its eruptions to run towards the gap and cautiously to pick my way down to the rock which hid from me, as I supposed, the liquid lava. I was armed with two phials, one of common smelling salts, and the other containing a solution of ammonia. On reaching the rock, I found it projected over a lake which was really filled by liquid fiery lava. I immediately laid myself down, and looking over its edge, saw, with great delight, lava actually in a state of fusion.

Presently I observed a small bubble swelling up on the surface of the fluid lava: it became gradually larger and larger, but did not burst. I had some vague suspicion that this indicated a coming eruption; but on looking at my watch, I was assured that only one minute had elapsed since the termination of the last. I therefore watched its progress; after a time the bubble slowly subsided without breaking.

I now found the heat of the rock on which I was reposing and the radiation from the fluid lava, almost insupportable, whilst the sulphurous effluvium painfully affected my lungs. On looking around, I fortunately observed a spot a few feet above me, from which I could, in a standing position, get a better view of the lake, and perhaps suffer less inconvenience from its vapours. Having reached this spot, I continued to observe the slow formation and absorption of these vesicles of lava. One of them soon appeared. Another soon followed at a different part of the fiery lake, but, like its predecessor, it disappeared as quietly.

Another swelling now arose about half way distant from the centre of the cauldron, which enlarged much beyond its predecessors in point of size. It attained a diameter of about three feet, and then burst, but not with any explosion. The waves it propagated in the fiery fluid passed on to the sides, and were thence reflected back just as would have happened in a lake of water of the same dimensions.

This phenomenon reappeared several times, some of the bubbles being considerably larger in size, and making proportionally greater disturbance in the liquid of this miniature crater. I would gladly have remained a longer time, but the excessive heat, the noxious vapours, and the warning of my chronometer forbade it. I climbed back through the gap by which I had descended, and rushed as fast as I could to a safe distance from the coming eruption.

I was much exhausted by the heat, although I suffered still greater inconvenience from the vapours. From my observations of the eruptions before my descent into this little crater, I had estimated that I might safely allow myself six minutes, but not more than eight, if I descended into the crater immediately after an eruption.

If my memory does not fail me, I passed about six minutes in examining it, and the next explosion occurred ten minutes after the former one. On my return to Naples I found that a pair of thick boots I had worn on this expedition were entirely destroyed by the heat, and fell to pieces in my attempt to take them off.

On my return from the pit of burning fire, I sat down with my companion to refresh myself with a few biscuits contained in our basket. Cold water would have been the most refreshing fluid we could have desired, but we had none, and my impatient friend cried out, "I wish I had a glass of whisky!" It immediately occurred to me to feel in my own basket for a certain glass bottle preserved in a tight leather case, which fortunately being found, I presented to my astonished friend, with the remark that it contained half a pint of the finest Irish whisky. This piece of good luck for my fellow-traveller arose not from my love but from my dislike of whisky. Shortly before my Italian tour I had been travelling in the north of Ireland, and having exhausted my brandy, was unable to replace it by anything but whisky, a drink which I can only tolerate under very exceptional circumstances.

Hot Springs

During my residence at Naples in 1828, the government appointed a commission of members of the Royal Academy of Naples to visit Ischia and make a report upon the hot springs in that island. Being a foreign member of the Academy, they did me the honour of placing my name upon that commission. The weather was very favourable, the party was most agreeable, and during three or four days I enjoyed the society of my colleagues, the delightful scenery, and the highly interesting natural phenomena of that singular island.

None of the hot springs were deep: in several we made excavations which, in all cases, gave increased heat to the water. In one or two, I believe if we had excavated to a small depth or bored a few feet, we might have met with boiling water.

I took the opportunity of this visit to view the devastations made by the recent earthquake in the small town which had been destroyed.

The greater part of the town consisted of narrow streets formed by small houses built of squared stone. In some of these streets the houses on one side were thrown down, whilst those a few feet distant, on the opposite side, although severely damaged, had their walls left standing.

The landlord of the hotel at which we took up our quarters assured me the effects of the recent earthquake were entirely confined to a small portion of the island which he pointed out from the front of his hotel, and added that it was scarcely felt in other parts.

Earthquakes

At the commencement of this chapter I mentioned that I had never been *consciously* sensible of the occurrence of an earthquake. I think it may perhaps be useful to state that on a recent occasion I really perceived the effects of an earthquake, although at the time I assigned them to a different cause.

On the 6th of last October, about half-past three, a.m., most of the inhabitants of London who were awake at that hour perceived several shocks of an earthquake. I also was awake, although not conscious of the shocks of an earthquake.

As soon as I read of the event in the morning papers, I was forcibly struck by its coincidence with my own observations, although I had attributed to them an entirely different cause. In order to explain this, it is necessary to premise that I had on a former occasion instituted some experiments for the purpose of ascertaining how far off the passing of a cart or carriage would affect the steadiness of a star observed by reflection. Amongst other methods, I had fixed a looking-glass of about 12 by 16 inches, by a pair of hinges, to the front wall of my bedroom. It was usually so placed that, as I lay in bed, at the distance of about 10 or 12 feet, I could see by reflection a small gas-light burner, which was placed on my left hand.

By this arrangement any tremors propagated through the earth from passing carriages would be communicated to the looking-glass by means of the front wall of the house, which rose about 40 feet from the surface. The image of the small gas-burner reflected in the looking-glass would be proportionally disturbed. In this state of things, at about half-past three o'clock of the morning in question, I observed the reflected image of the gas-light move downwards and upwards two or three times. I then listened attentively, expecting to hear the sound of a distant carriage or cart. Hearing nothing of the kind, I concluded that the earth wave had travelled beyond the limit of the sound wave, arising from the carriage which produced it. Presently the image of the gaslight again vibrated up and down, and then suddenly fell about four or five inches lower down in the glass,

where it remained fixed for a time.   Still thinking the observation of
no consequence, I shut my eyes, and after perhaps another minute,
again saw the image in its lower position.   It then rose to its former
position, vibrated, and shortly again descended: it remained down for
some time and then resumed its first position.

Fire Damp

An opportunity presented itself several years after my examination
of Vesuvius of witnessing another form under which fire occasionally
exerts its formidable power.

I was visiting a friend* at Merthyr Tydfil, who possessed very
extensive coal-mines.   I inquired of my host whether any fire-damp
existed in them.   On receiving an affirmative answer, I expressed a
wish to become personally acquainted with the miner's invisible but
most dangerous enemy.   Arrangements were therefore made for my
visit to the subterranean world on the following day.   Professor Moll
of Utrecht, who was also a guest, expressed a wish to accompany me.

The entrance to the mine is situated in the side of a mountain.   Its
chief manager conducted our expedition to visit the "fire-king."

We found a coal-waggon drawn by a horse, and filled with clean
straw, standing on the railway which led into the workings.

The manager, Professor Moll, and myself, together with two or
three assistants, with candles, lanterns, and Davy-lamps, got into this
vehicle, which immediately entered the adit of the mine.   We
advanced at a good pace, passing at intervals doors which opened
on our approach and then instantly closed.   Each door had an
attendant boy, whose duty was confined to the regulation of his own
door.

Many were the doors we passed before we arrived at the termination
of the tram-road.   After travelling about a mile and a half, our
carriage stopped and we alighted.   We now proceeded on foot, each
carrying his own candle, until we reached a kind of chamber where
one of our attendants was left with the candles.

We, each holding a Davy-lamp in our hand, advanced towards a
small opening in the side of this chamber, which was so low that we
were compelled to crawl, one after another, on our hands and knees.
A powerful current of air rushed through this small passage.   On
reaching the end of it, we found ourselves in a much larger chamber
from which the coal had been excavated.   At a little distance opposite
to the path by which we entered was a continuation of the same narrow
hole which had led us to the waste in which we now stood.   From

* The late Sir John J. Guest, Bart.

this opening issued the powerful stream of air which seemed to pass in a direct course from one opening to the other.

On our right hand the large chamber we had entered appeared to spread to a very considerable distance, its termination being lost in darkness. The floor was covered with fragments which had fallen from the roof; so that, besides the risk from explosion, there was also a minor one arising from the possible fall of some huge mass of slate from the roof of the excavation beneath which we stood: an accident which I had already witnessed in the waste of another coal-mine. As we advanced over this flaky flooring it was evident that we were making a considerable ascent. We, in fact, now occupied a vast cavern, which had been originally formed by the extraction of the coal, and then partially filled up by the falling in from time to time of portions of the slaty roof.

As we advanced cautiously with our Davy-lamps beyond the current of air which had hitherto accompanied us, it was evident that a change had taken place in their light: for the flames became much enlarged. Professor Moll and myself mounted a huge heap of these fragments, and thus came into contact with air highly charged with carburetted hydrogen. At this point there was a very sensible difference in the atmosphere, even by a change of three feet in the elevation of the lamp.

Holding up the lamp at the level of my head, I could not see the wick of the lamp, but a general flame seemed to fill the inside of its wire-covering. On lowering it to the height of my knee, the wick resumed its large nebulous appearance.

My companion, Professor Moll, was very much delighted with this experiment. He told me he had often at his lectures explained these effects to his pupils, but that this was the first exhibition of them he had ever witnessed in their natural home.

Although well acquainted with the miniature explosions of the experimentalist, I found it very difficult to realize in my own mind the effects which might result from an explosion under the circumstances in which we were then placed. I inquired of the manager, who stood by my side, what would probably be the effect, if an explosion were to take place? Pointing to the vast heap of shale from which I had just descended, he said the whole of that would be blown through the narrow channel by which we entered, and every door we had passed through would be blown down.

We now retraced our steps, and crawling back through the narrow passage, rejoined our carriage, and were rapidly conveyed to the light of day.

CHAPTER XVIII

PICKING LOCKS AND DECIPHERING

Interview with Vidocq—Remarkable Power of altering his Height—A Bungler in picking Locks—Mr. Hobb's Lock and the Duke of Wellington—Strong belief that certain Ciphers are inscrutable—Davies Gilbert's Cipher—The Author's Cipher both deciphered—Classified Dictionaries of the English Language—Anagrams—Squaring Words—Bishop not easily squared—Lesser Dignitaries easier to work upon.

THESE two subjects are in truth much more nearly allied than might appear upon a superficial view of them. They are in fact closely connected with each other as small branches of the same vast subject of *combinations*.

Several years ago, the celebrated thief-taker, Vidocq, paid a short visit to London. I had an interview of some duration with this celebrity, who obligingly conveyed to me much information, which, though highly interesting, was not of a nature to become personally useful to me.

He possessed a very remarkable power, which he was so good as to exhibit to me. It consisted in altering his height to about an inch and a half less than his ordinary height. He threw over his shoulders a cloak, in which he walked round the room. It did not touch the floor in any part, and was, I should say, about a inch and a half above it. He then altered his height and took the same walk. The cloak then touched the floor and lay upon it in some part or other during the whole walk. He then stood still and altered his height alternately, several times to about the same amount.

I inquired whether the altered height, if sustained for several hours, produced fatigue. He replied that it did not, and that he had often used it during a whole day without any additional fatigue. He remarked that he had found this gift very useful as a disguise. I asked whether any medical man had examined the question; but it did not appear that any satisfactory explanation had been arrived at.

I now entered upon a favourite subject of my own—the art of picking locks—but, to my great disappointment, I found him not at all strong upon that question. I had myself bestowed some attention upon it, and had written a paper, 'On the Art of Opening all Locks,' at the conclusion of which I had proposed a plan of partially defeating my own method. My paper on that subject is not yet published.

Several years after Vidocq's appearance in London, the Exhibition of 1851 occurred. On one of my earliest visits, I observed a very curious lock of large dimensions with its internal mechanism fully exposed to view. I found, on inquiry, that it belonged to the American department. Having discovered the exhibitor, I asked for an explanation of the lock. I listened with great interest to a very profound disquisition upon locks and the means of picking them, conveyed to me with the most unaffected simplicity.

I felt that the maker of that lock surpassed me in knowledge of the subject as much as I had thought I excelled Vidocq. Having mentioned it to the late Duke of Wellington, he proposed that we should pay a visit to the lock the next time I accompanied him to the Exhibition. We did so a few days after, when the Duke was equally pleased with the lock and its inventor. Mr. Hobbs, the gentleman of whom I am speaking, and whose locks have now become so celebrated, was good enough to explain to me from time to time many difficult questions in the science of constructing and of picking locks. He informed me that he had devised a system for defeating all these methods of picking locks, for which he proposed taking out a patent. I was, however, much gratified when I found that it was precisely the plan I had previously described in my own unpublished pamphlet.

Deciphering

Deciphering is, in my opinion, one of the most fascinating of arts, and I fear I have wasted upon it more time than it deserves. I practised it in its simplest form when I was at school. The bigger boys made ciphers, but if I got hold of a few words, I usually found out the key. The consequence of this ingenuity was occasionally painful: the owners of the detected ciphers sometimes thrashed me, though the fault really lay in their own stupidity.

There is a kind of maxim amongst the craft of decipherers (similar to one amongst the locksmiths), that every cipher can be deciphered.

I am myself inclined to think that deciphering is an affair of time, ingenuity, and patience; and that very few ciphers are worth the trouble of unravelling them.

One of the most singular characteristics of the art of deciphering is the strong conviction possessed by every person, even moderately acquainted with it, that he is able to construct a cipher which nobody else can decipher. I have also observed that the cleverer the person, the more intimate is his conviction. In my earliest study of the subject I shared in this belief, and maintained it for many years.

In a conversation on that subject which I had with the late Mr. Davies Gilbert, President of the Royal Society, each maintained that he possessed a cipher which was absolutely inscrutable. On comparison, it appeared that we had both imagined the same law, and we were thus confirmed in our conviction of the security of our cipher.

Many years after, the late Dr. Fitton, having asked my opinion of the possibility of making an inscrutable cipher, I mentioned the conversation I had had with Davies Gilbert, and explained the law of the cipher, which we both thought would baffle the greatest adept in that science. Dr. Fitton fully agreed in my view of the subject; but even whilst I was explaining the law, an indistinct glimpse of defeating it presented itself vaguely to my imagination. Having mentioned my newly-conceived doubt, it was entirely rejected by my friend. I then proposed that Dr. Fitton should write a few sentences in a cipher constructed according to this law, and that I should make some attempts to unravel it. I offered to give a few hours to the subject; and if I could see my way to a solution, to continue my researches; but if not on the road to success, to tell him I had given up the task.

Late in the evening of that day I commenced a preparatory inquiry into the means of unravelling this new cipher, and I soon arrived at a tolerable certainty that I should succeed. The next night, on my return from a party, I found Dr. Fitton's cipher on my table. I immediately commenced my attempt. After some time I found that it would not yield to my means of treating it; and on further examination I succeeded in proving that it was not written according to the law agreed upon. At first my friend was very positive that I was mistaken; and having taken it to his sister, by whose aid it was composed, he returned and told me that it *was* constructed upon the very law I had proposed. I then assured him that they *must* have made some mistake, and that my evidence was so irresistible, that if my life depended upon the result I should have no hesitation in making my election.

Dr. Fitton again retired to consult his sister; and after the lapse of a considerable interval of time again returned, and informed me that I was right—that his sister had inadvertently mistaken the enunciation of the law. I now remarked that I possessed an absolute demonstration of the fact I had communicated to him; and added that, having conjectured the origin of the mistake, I would decipher the cipher with the erroneous law before he could send me the new cipher to be made according to the law originally proposed.   Before the evening of the next day both ciphers had been translated.

This cipher was arranged upon the following principle:—Two concentric circles of cardboard were formed, each divided into twenty-six or more divisions.

On the outer were written in regular order the letters of the alphabet. On the inner circle were written the same twenty-six letters, but in any irregular manner.

In order to use this cipher, look for the first letter of the word to be ciphered on the outside circle.   Opposite to it, on the inner circle, will be another letter, which is to be written as the cipher for the former.

Now turn round the inner circle until the cipher just written is opposite the letter *a* on the *outer* circle.   Proceed in the same manner for the next, and so on for all succeeding letters.

Many varieties of this cipher may be made by inserting other characters to represent the divisions between words, the various stops, or even blanks.   Although Davies Gilbert, I believe, and myself, both arrived at it from our own efforts, I have reason to think that it is of very much older date.   I am not sure that it may not be found in the "Steganographia" of Schott, or even of Trithemius.

One great aid in deciphering is, a complete analysis of the language in which the cipher is written.   For this purpose I took a good English dictionary, and had it copied out into a series of twenty-four other dictionaries.   They comprised all words of

One letter,
Two letters,
Three letters,
&c.   &c.:
Twenty-six letters.

Each dictionary was then carefully examined, and all the modifications of each word, as, for instance, the plurals of substantives, the comparatives and superlatives of adjectives, the tenses and participles of verbs, &c., were carefully indicated.   A second edition of these twenty-six dictionaries was then made, including these new derivatives.

Each of these dictionaries was then examined, and every word which contained any two or more letters of the same kind was carefully marked.   Thus, against the word *tell* the numbers 3 and 4 were placed to indicate that the third and fourth letters are identical.   Similarly, the word *better* was followed by the numbers 25, 34.   Each of these dictionaries was then re-arranged thus:—In the first or original one each word was arranged according to the alphabetical order of its *initial* letter.

In the next the words were arranged alphabetically according to the *second* letter of each word, and so in the other dictionaries on to the last letter.

Again, each dictionary was divided into several others, according to the numerical characteristics placed at the end of each word. Many words appeared repeatedly in several of these subdivisions.

The work is yet unfinished, although the classification already amounts, I believe, to nearly half a million words.

From some of these, dictionaries were made of those words only which by transposition of their letters formed anagrams. A few of these are curious:—

Opposite		*Similarity*		*Satirical*	
vote	veto	fuel	flue	odes	dose
acre	care	taps	pats	bard	drab
evil	veil	tubs	buts	poem	mope
ever	veer	vast	vats	poet	tope
lips	slip	note	tone	trio	riot
cask	sack	cold	clod	star	rats
fowl	wolf	evil	vile	wive	view
gods	dogs	arms	mars	nabs	bans
tory	tyro	rove	over	tame	mate
tars	rats	lips	lisp	acts	cats

There are some verbal puzzles costing much time to solve which may be readily detected by these dictionaries. Such, for instance, is the sentence,

I tore ten Persian MSS.,

which it is required to form into one word of eighteen letters.

The first process is to put opposite each letter the number of times it occurs, thus:—

i	2	p	1	It contains—	
t	2	s	3	2 triplets.	
o	1	a	1	4 pairs.	
r	2	m	1	4 single letters.	
e	3	—		—	
n	2		6	18	
	—		12		
	12		—		
			18		

Now, on examining the dictionary of all words of eighteen letters, it will be observed that they amount to twenty-seven, and that they may be arranged in six classes:—

7	having five letters of the same kind.	
5	„ four	„
3	„ three triplets.	
7	„ two triplets.	
3	„ one triplet.	
2	„ seven pairs.	

27

Hence it appears that the word sought must be one of those seven having two triplets, and also that it must have four pairs; this reduces the question to the two words—

misinterpretations,
misrepresentations.

The latter is the one sought, because its triplets are e and s, whilst those of the former are i and t.

The reader who has leisure may try to find out the word of eighteen letters formed by the following sentence:—

Art is not in, but Satan.

Another amusing puzzle may be greatly assisted by these dictionaries. It is called squaring words, and is thus practised:—Let the given word to be squared be Dean.   It is to be written horizontally, and also vertically, thus:—

D e a n.
e . . .
a . . .
n . . .

And it is required to fill up the blanks with such letters that each vertical column shall be the same as its corresponding horizontal column, thus:—

D e a n
e a s e
a s k s
n e s t

The various ranks of the church are easily squared; but it is stated, I know not on what authority, that no one has yet succeeded in squaring the word bishop.

Having obtained one squared word, as in the case of Dean, it will be observed that any of the letters in the two diagonals, d, a, k, t,— n, s, s, n, may be changed into any other letter which will make an English word.

Thus Dean may be changed into such words as

dear	peas	weak	beam
fear	seas	lead	seal
deaf	bear	real	team

In fact there are upwards of sixty substitutes: possibly some of these might render the two diagonals, d, a, k, t, and n, s, s, n, also English words.

CHAPTER XXV

RAILWAYS

AT the commencement of the railway system I naturally took a great interest in the subject, from its bearings upon mechanism as well as upon political economy.

I accompanied Mr. Woolryche Whitmore, the member for Bridge-north, to Liverpool, at the opening of the Manchester and Liverpool Railway. The morning previous to the opening, we met Mr. Huskisson at the Exchange, and my friend introduced me to him. The next day the numerous trains started with their heavy load of travellers. All went on pleasantly until we reached Parkside, near Newton. During the time the engines which drew us were taking in their water and their fuel, many of the passengers got out and recognized their friends in other trains.

At a certain signal all resumed their seats; but we had not proceeded a mile before the whole of our trains came to a stand-still without any ostensible cause. After some time spent in various conjectures, a single engine almost flew past us on the other line of rail, drawing with it the ornamental car which the Duke of Wellington and other officials had so recently occupied. Instead of its former numerous company it appeared to convey only two, or at most three, persons; but the rapidity of its flight prevented any close observation of the passengers.

A certain amount of alarm now began to pervade the trains, and various conjectures were afloat of some serious accident. After a while Mr. Whitmore and myself got out of our carriage and hastened back towards the halting place. At a little distance before us, in the middle of the railway, stood the Duke of Wellington, Sir Robert Peel, and the Boroughreeve of Manchester, discussing the course to be pursued in consequence of the dreadful accident which had befallen Mr. Huskisson, whom I had seen but a few minutes before standing at the door of the carriage conversing with the Duke of Wellington. The Duke was anxious that the whole party should return to Liverpool; but the chief officer of Manchester pressed upon them the necessity of continuing the journey, stating that if it were given up he could not be answerable for the safety of the town.

It was at last mournfully resolved to continue our course to Manchester, where a luncheon had been prepared for us; but to give up all the ceremonial, and to return as soon as we could to Liverpool.

For several miles before we reached our destination the sides of the railroad were crowded by a highly-excited populace shouting and yelling. I feared each moment that some still greater sacrifice of life might occur from the people madly attempting to stop by their feeble arms the momentum of our enormous trains.

Having rapidly taken what refreshment was necessary, we waited with anxiety for our trains; but hour after hour passed away before they were able to start. The cause of this delay arose thus. The Duke of Wellington was the guest of the Earl of Wilton, the nearest station to whose residence was almost half way between Manchester and Liverpool. A train therefore was ordered to convey the party to Heaton House. Unfortunately, our engines had necessarily gone a considerable distance upon that line to get their supply of water, and were thus cut off by the train conveying the Duke, from returning direct to Manchester.

There were not yet at this early period of railway history any sidings to allow of a passage, or any crossing to enable the engines to get upon the other line of rails. Under these circumstances the drivers took the shortest course open to them. Having taken in their water, they pushed on as fast as they could to a crossing at a short distance from Liverpool. They backed into the other line of rail, and thus returned to Manchester to pick up their trains.

In the meantime the vague rumour of some great disaster had reached Liverpool. Thousands of persons, many of whom had friends and relatives in the excursion trains, were congregated on the bridges and at the railway station, anxious to learn news of their friends and relatives.

About five o'clock in the evening they perceived at a distance half-a-dozen engines without any carriages, rushing furiously towards them—suddenly checking their speed—then backing into the other line of rail—again flying away towards Manchester, without giving any signs or explanation of the mystery in which many of them were so deeply interested.

It is difficult to estimate the amount of anxiety and misery which was thus unwillingly but inevitably caused amongst all those who had friends, connections, or relatives in the missing trains.

When these engines returned to Manchester, our trains were unfortunately connected together, and three engines were attached to the front of each group of three trains.

This arrangement considerably diminished their joint power of traction. But another source of delay arose: the couplings which were strong enough when connecting an engine and its train were not sufficiently strong when three engines were coupled together. The consequence was that there were frequent fractures of our couplings and thus great delays arose.

About half-past eight in the evening I reached the great building in which we were to have dined. Its tables were half filled with separate groups of three or four people each, who being strangers in Liverpool, had no other resource than to use it as a kind of coffee-room in which to get a hasty meal, and retire.

The next morning I went over to see the plate-glass manufactory at about ten miles from Liverpool.

On my arrival I found, to my great disappointment, that there were orders that nobody should be admitted on that day, as the Duke of Wellington and a large party were coming over from Lord Wilton's. This was the only day at my disposal, and it wanted nearly an hour to the time appointed: so I asked to be permitted to see the works, promising to retire as soon as the Earl of Wilton's party arrived. I added incidentally that I was not entirely unknown to the Duke of Wellington.

On the arrival of the party I quietly made my retreat unobserved, and had just entered the carriage which had conveyed me from Liverpool, when a messenger arrived with the Duke's compliments, hoping that I would join his party. I willingly accepted the invitation; the Duke presented me to each of his friends, and I had the advantage of having another survey of the works. This was my first acquaintance with the late Lady Wilton, who afterwards called on me with the Duke of Wellington, and put that sagacious question relative to the Difference Engine which I have mentioned in another part of this volume.* Amongst the party were Mr. and Mrs. Arbuthnot, with the former of whom I afterwards had several interesting discussions relative to subjects connected with the ninth "Bridgewater Treatise."

* See page xxiv.

A few days after, I met at dinner a large party at the house of one of the great Liverpool merchants. Amongst them were several officers of the new railway, and almost all the party were more or less interested in its success.

In these circumstances the conversation very naturally turned upon the new mode of locomotion. Its various difficulties and dangers were suggested and discussed. Amongst others, it was observed that obstacles might be placed upon the rail, either accidentally or by design, which might produce expensive and fatal effects.

To prevent the occurrence of these evils, I suggested two remedies.

1st. That every engine should have just in advance of each of its front wheels a powerful framing, supporting a strong piece of plate-iron, descending within an inch or two of the upper face of the rail. These iron plates should be fixed at an angle of 45° with the line of rail, and also at the same angle with respect to the horizon. Their shape would be something like that of ploughshares, and their effect would be to pitch any obstacle obliquely off the rail unless its heavier portion were between the rails.

Some time after, a strong vertical bar of iron was placed in front of the wheels of every engine. The objection to this is, that it has a tendency to throw the obstacle straight forward upon another part of the rail.

2nd. The second suggestion I made, was to place in front of each engine a strong leather apron attached to a powerful iron bar, projecting five or six feet in front of the engine and about a foot above the ballast. The effect of this would be, that any animal straying over the railway would be pitched into this apron, probably having its legs broken, but forming no impediment to the progress of the train.

I have been informed that this contrivance has been adopted in America, where the railroads, being unenclosed, are subject to frequent obstruction from cattle. If used on enclosed roads, it still might occasionally save the lives of incautious persons, although possibly at the expense of broken limbs.

Another question discussed at this party was, whether, if an engine went off the rail, it would be possible to separate it from the train before it had dragged the latter after it. I took out my pencil and sketched upon a card a simple method of accomplishing that object. It passed round the table, and one of the party suggested that I should communicate the plan to the Directors of the railway.

My answer was, that having a great wish to diminish the dangers of this new mode of travelling, I declined making any such communication to them; for, I added, unless these Directors are quite unlike all of whom I have had any experience, I can foresee the inevitable result of such a communication.

It might take me some time and trouble to consider the best way of carrying out the principle and to make the necessary drawings. Some time after I have placed these in the hands of the Company, I shall receive a very pretty letter from the secretary, thanking me in the most flattering terms for the highly ingenious plan I have placed in their hands, but regretting that their engineer finds certain practical difficulties in the way.

Now, if the same Company had taken the advice of some eminent engineer, to whom they would have to pay a large fee, no practical difficulties would ever be found to prevent its trial.

It was evident from the remarks of several of the party that I had pointed out the most probable result of any such communication.

It is possible that some report of this plan subsequently reached the Directors; for about six months after, I received from an officer of the railway Company a letter, asking my assistance upon this identical point. I sent them my sketch and all the information I had subsequently acquired on the subject. I received the stereotype reply I had anticipated, couched in the most courteous language; in short, quite a model letter for a young secretary to study.

Several better contrivances than mine were subsequently proposed; but experience seems to show that the whole train ought to be connected together as firmly as possible.

Not long after my return from Liverpool I found myself seated at dinner next to an elderly gentleman, an eminent London banker. The new system of railroads, of course, was the ordinary topic of conversation. Much had been said in its favour, but my neighbour did not appear to concur with the majority. At last I had an opportunity of asking his opinion. "Ah," said the banker, "I don't approve of this new mode of travelling. It will enable our clerks to plunder us, and then be off to Liverpool on their way to America at the rate of *twenty* miles an hour." I suggested that science might perhaps remedy this evil, and that possibly we might send lightning to outstrip the culprit's arrival at Liverpool, and thus render the railroad a sure means of arresting the thief. I had at the time I uttered those words no idea how soon they would be realized.

In 1838 and 1839 a discussion of considerable public importance had arisen respecting the Great Western Railway. Having an interest in that undertaking, it was the wish of Mr. Brunel and the Directors that I should state my own opinion upon the question. I felt that I could not speak with confidence without making certain experiments.

The Directors therefore lent me steam-power, and a second-class carriage to fit up with machinery of my own contrivance, and appointed one of their officers to accompany me, through whom I might give such directions as I deemed necessary during my experiments.

I removed the whole of the internal parts of the carriage. Through its bottom firm supports, fixed upon the framework below, passed up into the body of the carriage, and supported a long table entirely independent of its motions.

On this table slowly rolled sheets of paper, each a thousand feet long. Several inking pens traced curves on this paper, which expressed the following measures:—

1. Force of traction.
2. Vertical shake of carriage at its middle.
3. Lateral ditto.
4. End ditto.
5, 6, and 7. The same shakes at the end of the carriage.
8. The curve described upon the earth by the centre of the frame of the carriage.
9. A chronometer marked half seconds on the paper.

Above two miles of paper were thus covered. These experiments cost me about 300*l.*, and took up my own time, and that of all the people I was then employing, during five months.

I had previously travelled over most of the railways then existing in this country, in order to make notes of such facts as I could observe during my journeys.

The result of my experiments convinced me that the broad gauge was most convenient and safest for the public. It also enabled me fearlessly to assert that an immense array of experiments which were exhibited round the walls of the meeting-room by those who opposed the Directors were made with an instrument which could not possibly measure the quantities proposed, and that the whole of them were worthless for the present argument. The production of the work of such an instrument could not fail to damage even a good cause.

On the discussion at the general meeting at the London Tavern, I made a statement of my own views, which was admitted at the time to have had considerable influence on the decision of the proprietors. Many years after I met a gentleman who told me he and a few other proprietors holding several thousand proxies came up from Liverpool intending to vote according to the weight of the arguments adduced. He informed me that he and all his friends decided their votes on hearing my statement. He then added, "But for that speech, the broad gauge would not now exist in England."

These experiments were not unaccompanied with danger. I sometimes attached my carriage to a public train to convey me to the point where my experiments commenced, and I had frequently to interrupt their course, in order to run on to a siding to avoid a coming train.

I then asked to be allowed to make such experiments during the night when there were no trains; but Brunel told me it was too dangerous to be permitted, and that ballast-waggons, and others, carrying machinery and materials for the construction and completion of the railroad itself, were continually traversing various parts of the line at uncertain hours.

The soundness of this advice became evident a very short time after it was given. On arriving one morning at the terminus, the engine which had been promised for my experimental train was not ready, but another was provided instead. On further inquiry, I found that the "North Star," the finest engine the Company then possessed, had been placed at the end of the great polygonal building devoted to engines, in order that it might be ready for my service in the morning; but that, during the night, a train of twenty-five empty ballast-waggons, each containing two men, driven by an engine, both the driver and stoker of which were asleep, had passed right through the engine-house and damaged the "North Star."

Most fortunately, no accident happened to the men beyond a severe shaking. It ought, however, in extenuation of such neglect, to be observed that engine-drivers were at that period so few, and so thoroughly overworked, that such an occurrence was not surprising.

It then occurred to me, that being engaged on a work which was anything but profitable to myself, but which contributed to the safety of all travellers, I might, without impropriety, avail myself of the repose of Sunday for advancing my measures. I therefore desired Brunel to ask for the Directors' permission. The next time I saw Brunel, he told me the Directors did not like to give an official permission, but it was remarked that having put one of their own officers under my orders, I had already the power of travelling on whatever day I preferred.

I accordingly availed myself of the day on which, at that time, scarcely a single train or engine would be in motion upon it.

Upon one of these Sundays, which were, in fact, the only really safe days, I had proposed to investigate the effect of considerable additional weight. With this object, I had ordered three waggons laden with thirty tons of iron to be attached to my experimental carriage.

On my arrival at the terminus a few minutes before the time appointed, my aide-de-camp informed me that we were to travel on the north line. As this was an invasion of the usual regulations, I inquired very minutely into the authority on which it rested. Being satisfied on this point, I desired him to order my train out immediately. He returned shortly with the news that the fireman had neglected his duty, but that the engine would be ready in less than a quarter of an hour.

A messenger arrived soon after to inform me that the obstructions had been removed, and that I could now pass upon the south, which was the proper line.

I was looking at the departure of the only Sunday train, and conversing with the officer, who took much pains to assure me that there was no danger on whichever line we might travel; because, he observed, when that train had departed, there can be no engine except our own on either line until five o'clock in the evening.

Whilst we were conversing together, my ear, which had become peculiarly sensitive to the distant sound of an engine, told me that one was approaching. I mentioned it to my railway official: he did not hear it, and said, "Sir, it is impossible."—"Whether it is possible or impossible," I said, "an engine *is* coming, and in a few minutes we shall see its steam." The sound soon became evident to both, and our eyes were anxiously directed to the expected quarter. The white cloud of steam now faintly appeared in the distance; I soon perceived the line it occupied, and then turned to watch my companion's countenance. In a few moments more I saw it slightly change, and he said, "It *is*, indeed, on the north line."

Knowing that it would stop at the engine-house, I ran as fast I could to that spot. I found a single engine, from which Brunel, covered with smoke and blacks, had just descended. We shook hands, and I inquired what brought my friend here in such a plight. Brunel told me that he had posted from Bristol, to meet the only train at the furthest point of the rail then open, but had missed it. "Fortunately," he said, "I found this engine with its fire up, so I ordered it out, and have driven it the whole way up at the rate of fifty miles an hour."

I then told him that but for the merest accident I should have met him on the *same* line at the rate of forty miles, and that I had attached to my engine my experimental carriage, and three waggons with thirty tons of iron. I then inquired what course he would have pursued if he had perceived another engine meeting him upon his own line.

Brunel said, in such a case he should have put on all the steam he could command, with a view of driving off the opposite engine by the superior velocity of his own.

If the concussion had occurred, the probability is, that Brunel's engine would have been knocked off the rail by the superior momentum of my train, and that my experimental carriage would have been buried under the iron contained in the waggons behind.

These rates of travelling were then unusual, but have now become common. The greatest speed which I have personally witnessed, occurred on the return of a train from Bristol, on the occasion of the floating of the "Great Britain." I was in a compartment, in conversation with three eminent engineers, when one of them remarked the unusual speed of the train: my neighbour on my left took out his watch, and noted the time of passage of the distance posts, whence it appeared that we were then travelling at the rate of seventy-eight miles an hour. The train was evidently on an incline, and we did not long sustain that dangerous velocity.

One very cold day I found Dr. Lardner making experiments on the Great Western Railway. He was drawing a series of trucks with an engine travelling at known velocities. At certain intervals, a truck was detached from his train. The time occupied by this truck before it came to rest was the object to be noted. As Dr. Lardner was short of assistants, I and my son offered to get into one of his trucks and note for him the time of coming to rest.

Our truck having been detached, it came to rest, and I had noted the time. After waiting a few minutes, I thought I perceived a slight motion, which continued, though slowly. It then occurred to me that this must arise from the effect of the wind, which was blowing strongly. On my way to the station, feeling very cold, I had purchased three yards of coarse blue woollen cloth, which I wound round my person. This I now unwound; we held it up as a sail, and gradually acquiring greater velocity, finally reached and sailed across the whole of the Hanwell viaduct at a very fair pace.

The question of the best gauge for a system of railways is yet undecided. The present gauge of 4.8½ was the result of the accident that certain tram-roads adjacent to mines were of that width. When the wide gauge of the Great Western was suggested and carried out, there arose violent party movements for and against it. At the meeting of the British Association at Newcastle, in 1838, there were two sources of anxiety to the Council—the discussion of the question of Steam Navigation to America, and what was called "The battles of the Gauges." Both these questions bore very strongly upon pecuniary interests, and were expected to be fiercely contested.

On the Council of the British Association, of course, the duty of nominating the Presidents and Vice-Presidents of its various sections devolves. During the period in which I took an active part in that body, it was always a principle, of which I was ever the warm advocate, that we should select those officers from amongst the persons most distinguished for their eminence in their respective subjects, who were born in or connected with the district we visited.

In pursuance of this principle, I was deputed by the Council to invite Mr. George Stephenson to become the President of the Mechanical Section. In case he should decline it, I was then empowered to offer it to Mr. Buddle, the eminent coal-viewer; and in case of these both declining, I was to propose it to the late Mr. Bryan Donkin, of London, a native of that district, and connected with it by family ties.

On my arrival at Newcastle, I immediately called on George Stephenson, and represented to him the unanimous wish of the Council of the British Association. To my great surprise, and to my still greater regret, I found that he at once declined the offer. All my powers of persuasion were exercised in vain. Knowing that the two great controverted questions to be discussed most probably formed the real obstacle, I mentioned them, and added that, as I should be one of his Vice-Presidents, I would, if he wished it, take the Chair upon either or upon both the discussions of the Gauges and of the Atlantic Steam Voyage, or upon any other occasion that might be agreeable or convenient to himself: I found him immoveable in his decision. I made another attempt the next day, and renewed the expression of my own strong feeling, that we should pay respect and honour to the most distinguished men of the district we visited. I then told him the course I was instructed by the Council to pursue.

My next step was to apply to Mr. Buddle. I need not repeat the arguments I employed: I was equally unsuccessful with each of the eminent men the Council had wished to honour. I therefore now went back to George Stephenson, told him of the failure of my efforts, and asked him, if he still persisted in declining the Chair, would he do me the favour to be one of the Vice-Presidents, as the Council had now no resource but to place me in the Chair, which I had hoped would have been occupied by a more competent person.

To this latter application he kindly acceded; and I felt that, with the assistance of George Stephenson's and Mr. Donkin's professional knowledge, and their presence by my side, I should be able to keep order in these dreaded discussions.

The day before the great discussion upon Atlantic Steam Navigation, I had a short conversation with Dr. Lardner: I told him that in my opinion some of his views were hasty; but that much stronger opinions had been assigned to him than those he had really expressed, and I recommended him to admit as much as he fairly could.

At the appointed hour the room was filled with an expectant and rather angry audience. Dr. Lardner's beautiful apparatus for illustrating his views was before them, and the Doctor commenced his statement. He was listened to with the greatest attention, and was really most judicious as well as very instructive. At the very moment which seemed to me the most favourable for it, he turned to the explanation of the instruments he proposed to employ, and having concluded his statement, it became my duty to invite discussion upon the question.

I did so in very few words, merely observing that several opinions had been attributed to Dr. Lardner which he had never maintained, and that additional information had induced him candidly to admit that some of those doctrines which he had supported were erroneous. I added, that nothing was more injurious to the progress of truth than to reproach any man who honestly admitted that he had been in error.

The discussion then commenced: it was continued with considerable energy, but with great temper; and after a long and instructive debate the assembled multitude separated. Some few who attended in expectation of a scene were sorely disappointed. As I was passing out, one of my acquaintance remarked, "You have saved that —— —— Lardner": to which I replied, "I have saved the British Association from a scandal."

Before I terminate this Chapter on Railways, it will perhaps be expected by some of my readers that I should point out such measures as occur to me for rendering this universal system more safe. Since the long series of experiments I made in 1839, I have had no experience either official or professional upon the subject. My opinions, therefore, must be taken only at what they are worth, and will probably be regarded as the dreams of an amateur. I have indeed formed very decided opinions upon certain measures relative to railroads; but my hesitation to make them public arises from the circumstance, that by publishing them I may possibly delay their adoption. It may happen, as is now happening to my system of distinguishing lighthouses from each other, and of night telegraphic communication between ships at sea—that although officially communicated to all the great maritime governments, and even publicly exhibited for months during the Exhibition of 1851, it will be allowed to go to sleep for years, until some official person, casually hearing of it, or perhaps re-inventing it, shall have *interest* with the higher powers to get it quietly adopted as his own invention. I have given, in a former page, a list of the self-registering apparatus I employed in my own experiments.

In studying the evidence given upon the inquiries into the various lamentable accidents which have occurred upon railways, I have been much struck by the discordance of that evidence as to the speed with which the engines were travelling when they took place.

Even the best and most unbiassed judgment ought not to be trusted when mechanical evidence can be produced. The first rule I propose is, that—

Every engine should have mechanical self-registering means of recording its own velocity at every instant during the whole course of its journey.

In my own experiments this was the first point I attended to. I took a powerful spring clock, with a chronometer movement, which every half second lifted a peculiar pen, and left a small dot of ink upon the paper, which was moving over a table with the velocity given to it by the wheels of the carriage.

Thus the comparative frequency of these dots indicated the rate of travelling at the time. But the instrument was susceptible of giving different scales of measurement. Thus it might be that only three inches of paper passed under the pen in every mile, or any greater length of paper, up to sixty feet per mile, might be ordered to pass under the paper during an equal space. Again, the number of dots per second could, if required, be altered.

The clock was broken four or five times during the earliest experiments. This arose from its being fixed upon the platform carrying the axles of the wheels. I then contrived a kind of parallel motion, by which I was enabled to support the clock upon the carriage-springs, and yet allow it to impress its dots upon the paper, which did not require that advantage. After this, the clock was never injured.

The power of regulating the length of paper for each mile was of great importance; it enabled me to examine, almost microscopically, the junctions of the rails. When a large scale of paper was allowed, every joining was marked upon the paper.

I find, on referring to my paper records, that on 3rd March, 1839, the "Atlas" engine drew my experimental carriage, with two other carriages attached behind it, from Maidenhead to Drayton, with its paper travelling only eleven feet for each mile of journey; whilst from Drayton to Slough, forty-four feet of paper passed under the pen during each mile of progress.

The inking pens at first gave me some trouble, but after successively discovering their various defects, and remedying them at an expense of nearly £20, they performed their work satisfactorily. The information they gave might be fully relied upon.

We had an excellent illustration of this on one occasion when we were returning, late in the evening, from Maidenhead, after a hard day's work.   The pitchy darkness of the night, which prevented us from seeing any objects external to our carriage, was strongly contrasted with the bright light of four argand lamps within it.   I was accompanied by my eldest son, Mr. Herschel Babbage, and three assistants.   A roll of paper a thousand feet in length was slowly unwinding itself upon the long table extended before us, and winding itself up on a corresponding roller at its other extremity.   About a dozen pens connected with a bridge crossing the middle of the table were each marking its own independent curve gradually or by jumps, as the circumstances attending our railway course was dictating. The self-feeding pens, which the self-acting roller of blotting-paper continually followed, but never overtook, were quietly marking their inevitable courses.   All had gone on well for a considerable time amidst perfect silence, if the steady pace of thirty miles an hour, the dogged automatic action of the material, and the muteness of the living machinery, admitted of such a term.   Being myself entirely ignorant of our position upon the rail, I disturbed this busy repose by inquiring whether any one knew where we were?   To this question there was no reply.   Each continued to watch in silence for the duties which his own department might at any moment require, but no such demands were made.

After some minutes, as I was watching the lengthening curves, I perceived a slight indication of our position on the railroad.   I instantly looked at my son, and saw, by a faint smile on his countenance, that he also perceived our situation on the line.   I had scarcely glanced back at the growing curves upon the paper, to confirm my interpretation, when each of my three assistants at the same instant called out "Thames Junction."

At the period I speak of the double line of a small railway, called the Thames Junction, crossed the Great Western line on a level at between two and three miles from its terminus.   The interruption caused certain jerks in several of our curves, which, having once noticed, it was impossible to mistake.

I would suggest that every engine should carry a spring clock, marking small equal intervals of time by means of a needle-point impinging upon paper, the speed of whose transit should be regulated by the speed of the engine.   It might, perhaps, be desirable to have a differently-formed mark to indicate each five minutes.   Also, two or more studs on the driving-wheel should mark upon the same paper the number of its revolutions.   Besides this, it might be imperative on the engine-driver to mark upon the paper a dot upon passing each of certain prescribed points upon the railway.   This latter is not absolutely necessary, but may occasionally supply very valuable information.

The second point which I consider of importance is, that—

Between every engine and its train there should be interposed a dynamometer, that is, a powerful spring to measure the force exerted by the engine.

It may, perhaps, be objected that this would require a certain amount of movement between the engine and its train. A very small quantity would be sufficient, say half an inch, or less. The forces in action are so very large, that even a still smaller amount of motion than this might be sufficiently magnified. Its indications should be marked by self-acting machinery governing points impinging upon the paper on which the velocity is marked.

Whenever any unusual resistance has opposed the progress of the train, it will thus be marked upon the paper. It will indicate in some measure the state of the road, and it will assuredly furnish valuable information in case an accident happens, and the train or the engine gets off the rails.

The third recommendation I have to make is—

That the curve described by the centre of the engine itself upon the plane of the railway should be laid down upon the paper.

Finding this a very important element, I caused a plate of hardened steel to be pressed by a strong spring against the inner edge of the rail. It was supported by a hinge upon a strong piece of timber descending from the platform supporting the carriage itself. The motion of this piece of steel, arising from the varying position of the wheels themselves upon the rail, was conveyed to a pen which transferred to the paper the curve traversed by the centre of the carriage referred to the plane of the rail itself.

The contrivance and management of this portion of my apparatus was certainly the most difficult part of my task, and probably the most dangerous. I had several friendly cautions, but I knew the danger, and having examined its various causes, adopted means of counteracting its effect.

After a few trials we found out how to manage it, and although it often broke four or five times in the course of the day's work, the fracture inevitably occurred at the place intended for it, and my first notice of the fact often arose from the blow the fragment made when suddenly drawn by a strong rope up to the under side of the floor of our experimental carriage.

I have a very strong opinion that the adoption of such mechanical registrations would add greatly to the security of railway travelling, because they would become the unerring record of facts, the incorruptible witnesses of the immediate antecedents of any catastrophe.

I have, however, little expectation of their adoption, unless Directors can be convinced that the knowledge derived from them would, by pointing out incipient defects, and by acting as a check upon the vigilance of all their officers, considerably diminish the repairs and working expenses both of the engine and of the rail. Nor should I be much surprised even if they were pronounced impracticable, although they existed very nearly a quarter of a century ago.

The question of the gauges has long been settled. A small portion of broad gauge exists, but it is probable that it will ultimately be changed. The vast expense of converting the engines and the rolling stock for use on the narrower gauge presents the greatest obstacle.

It may, however, be interesting to learn the opinion of the father of railways at an early period of their progress. I have already mentioned the circumstances under which my acquaintance with George Stephenson began. They were favourable to that mutual confidence which immediately arose. I was naturally anxious to ascertain the effect of the existing experience upon his own mind, but I waited patiently until a favourable opportunity presented itself.

At a large public dinner, during the meeting of the British Association at Newcastle, I sat next to George Stephenson. It occurred to me that the desired opportunity had now arrived. I said little about railways until after the first glass of champagne. I mentioned several that I had travelled upon, and the conclusions I had drawn relative to the mechanical department. I then referred to the economy of management, and pointed out one railway in which the accounts were so well arranged, that I had been able to arrive at a testing point of an opinion I had formed from my own observations.

One great evil of the narrow gauge was, that when some trifling derangement in the engine occurred, which might be repaired at the expense of two or three shillings, it frequently became necessary to remove uninjured portions of the machine, in order to get at the fault; that the re-making the joints and replacing these parts thus temporarily removed, frequently led to an expense of several pounds.

The second glass of champagne now interrupted a conversation which was, I hope, equally agreeable to both, and was certainly very instructive for me. I felt that the fairest opportunity I could desire of ascertaining my friend's real opinion of the gauge had now arrived. Availing myself of the momentary pause after George Stephenson's glass was empty, I said—

"Now, Mr. Stephenson, will you allow me to ask you to suppose for an instant that no railways whatever existed, and yet that you were in full possession of all that large amount of knowledge which you have derived from your own experience. Under such circumstances, if you were consulted respecting the gauge of a system of railways about to be inaugurated, would you advise the gauge of 4 feet 8½ inches?"

"Not exactly that gauge," replied the creator of railroads; "I would take a few inches more, but a very few."

I was quite satisfied with this admission, though I confess it reminded me of the frail fair one who, when reproached by her immaculate friend with having had a child—an ecclesiastical licence not being first obtained—urged, as an extenuating circumstance, that it was a very small one.

In this age of invention, it is difficult to predict the railroads of the future. Already it has been suggested to give up wheels and put carriages upon sledges. This would lower the centre of gravity considerably, and save the expense of wheels. On the other hand, every carriage must have an apparatus to clean and grease the rails, and the wear and tear of these latter might overbalance the economy arising from abolishing wheels.

Again, short and much-frequented railways might be formed of a broad, continuous strap, always rolling on. At each station means must exist for taking up and putting down the passengers without stopping the rolling strap.

The exhaustion of air in a continuous tunnel was proposed many years ago for the purpose of sucking the trains along. This has recently been applied with success to the transmission of parcels and letters.

Possibly in the next International Exhibition a light railway might be employed within the building.*

1st. A quick train to enable visitors to get rapidly from end to end, avoiding the crowd and saving time, say at the expense of a penny.

2nd. A very slow train passing along the most attractive line, and occasionally stopping, to enable persons not capable of bearing the fatigue of pushing on foot through crowds.

If such railways were considered in the original design of the building, they might be made to interfere but little with the general public, and would bring in a considerable revenue to the concern.

---

* A gallery, elevated about seven feet, in the centre of each division of the new National Gallery, might be used either for a light railway, or for additional means of seeing the pictures on the walls.

CHAPTER XXVIII

HINTS FOR TRAVELLERS

New Inventions—Stomach Pump—Built a Carriage—Description of Thames Tunnel—Barton's Iridescent Buttons—Chinese Orders of Nobility—Manufactory of Gold Chains at Venice—Pulsations and Respirations of Animals—Punching a Hole in Glass without cracking it—Specimen of an Enormous Smash—Proteus Anguineus—Travellers' Hotel at Sheffield—Wentworth House.

In this chapter I propose to throw together a few suggestions, which may assist in rendering a tour successful for its objects and agreeable in its reminiscences.

Money is the fuel of travelling. I can give the traveller a few hints how to get money, although I never had any skill in making it myself.

In one tour, extending over more than a twelvemonth, I took with me two letters of credit, each for half the sum I should probably require. My reasons for this were, that in case one was lost the other might still be available. One of these was generally kept about my person, the other concealed in my writing-case. Another reason was, that if I were unluckily carried off and detained for a ransom, it might thus be mitigated.

It is of great advantage to a traveller to have some acquaintance with the use of tools. It is often valuable for his own comfort, and sometimes renders him able to assist a friend. I met at Frankfort the eldest son of the coachmaker of the Emperor of Russia. He had been travelling over the western part of Europe, and showed me drawings he had made of all the most remarkable carriages he had met with. Some of these were selected for their elegance, others for the reverse; take, as an example, the Lord Mayor's.

We travelled together to Munich, and I took that opportunity of discussing, seriatim, with my very intelligent young friend, every part of the structure of a carriage.

I made notes of certain portions in case I should find occasion to have a carriage built for my own use.

The young Russian was on his way to Moscow, and was very anxious to prevail on me to accompany him thither, for which purpose he offered to wait my own time at Munich. As, however, I wished to reach Italy as soon as possible, I declined his proposition with much regret.

However, in the following year, I profited by the information I then gained. I had built for me at Vienna, from my own design, a strong light four-wheeled calèche in which I could sleep at full length. Amongst its conveniences were a lamp by which I occasionally boiled an egg or cooked my breakfast. A large shallow drawer in which might be placed, without folding, plans, drawings, and dress-coats. Small pockets for the various kinds of money, a larger one for travelling books and telescopes, and many other conveniences. It cost somewhat above sixty pounds. After carrying me during six months, at the expense of only five francs for repair, I sold it at the Hague for thirty pounds.

It is always advantageous for a traveller to carry with him anything of use in science or in art if it is of a portable nature, and still more so if it has also the advantage of novelty. At the time I started on a lengthened tour the stomach-pump had just been invented. It appeared to give promise of great utility. I therefore arranged in a small box the parts of an instrument which could be employed either as a syringe, a stomach-pump, or for cupping. As a stomach-pump, it was in great request from its novelty and utility. I had many applications for permission to make drawings of it, to which I always most willingly acceded. At Munich, Dr. Weisbrod, the king's physician, was greatly interested with it, and at his wish I lent it to the chief surgical instrument-maker who produced for him an exact copy of the whole apparatus.

Having visited the Thames Tunnel a day or two before I started for the Continent, I purchased a dozen copies of the very lucid account of that most interesting work. Six of the copies were in French and the other six in the German language. I frequently lent a copy, and upon some occasions I gave one away; but if I had had twice that number I should have found that I might have distributed them with advantage as acknowledgments of the many attentions I received.

Another most valuable piece of travelling merchandise consisted of a dozen large and a dozen small gold buttons stamped by Barton's steel dies. These buttons displayed the most beautiful iridescence, especially in the light of the sun. They were formed by ruling the steel die in parallel lines in various forms. The lines were from the four to the ten thousandth of an inch apart.

I possessed a die which Mr. Barton had kindly given me. This I kept in my writing-case; but I had had a small piece of steel ruled in the same way, though not with quite the same perfection, which I always kept in my waistcoat pocket; it was also accompanied by a small gold button in a sandalwood case. These were frequently of great service. The mere sight of them procured me many little attentions in diligences and steamboats.

Of course I never appeared to be the possessor of more than one of these treasured buttons; so that if any one had saved my life, its gift would have been thought a handsome acknowledgment. If I had travelled in the East, as I had originally intended until the battle of Navarino prevented me, my buttons might have given me unlimited success in the celestial empire.

The Chinese, like ourselves, have five orders of nobility. They are indicated by spherical buttons. The Chinese nobles, however, wear them on the top of their caps, whilst our nobility wear their pearls and strawberry-leaves in their armorial bearings.

It is a curious circumstance that the most anciently civilized nation should have invented an order of knighthood almost exactly similar to our own—the order of the Peacock's Feather—which, like our own Garter, is confined to certain classes of nobility of the highest rank. Of the two the decoration of the Chinese noble is certainly the more graceful.

One out of many illustrations may show the use I made of a button. During my first visit to Venice I wished to see a manufactory of gold chains for which that city is justly celebrated. I readily got permission, and the proprietor was so good as to accompany me round his factory. I had inquired the price of various chains, and had expressed my wish to purchase a few inches of each kind; but I was informed that they never sold less than a braccia of any one chain. This amount would have made my purchase more costly than I proposed, so I gave it up.

In the meantime we proceeded through several rooms in which various processes were going on. Observing some tools in one of the shops, I took up a file and asked whence it was procured. This led to a conversation on the subject, in which the proprietor gave me some account of files from various countries, but concluded by observing that the Lancashire files, when they could be got, were by far the best. I took this opportunity of asking him whether he had seen any of our latest productions in steel: then pulling out of my waistcoat-pocket the piece of hardened steel, ruled by a diamond, I put it into his hands. The sun was shining brightly, and he was very much interested with it. I remarked that in a darkened room, and with a single lamp, it would be seen with still greater advantage. A room was soon darkened, and a single lamp produced, and the effect was still more perfect. My conductor then observed that his managing man was a very skilful workman, and if I could afford the time, he should much wish to show him this beautiful sight. I said it always gave me pleasure to see and converse with a skilful workman, and that I considered it as time well spent. The master sent for his superintendent, who, being of a judicious turn of mind, was lavish in admiring what his master approved. The master himself, gratified by this happy confirmation, turning to me, said that he would let me have pieces of any or all of his gold chains of any length, however short I might wish them to be.

I thanked him for thus enabling me to make my countrymen appreciate the excellence of Venetian workmanship, and purchased small samples of every kind of chain then manufactured. These, on my return to London, I weighed and measured, and referred to them in the economy of manufactures as illustrations of the different proportions in which skilled labour and price of raw material occur in the same class of manufactured articles.

A friend of mine, then at Venice, again visited that city about five years afterwards. He subsequently informed me that he had purchased, at the manufactory I visited, samples of gold chains about an inch or two long, fixed on black velvet, and that it formed a regular article of trade in some demand.

A man may, without being a proficient in any science, and indeed with only the most limited knowledge of a small portion of it, yet make himself useful to those who are most- instructed. However limited the path he may himself pursue, he will insensibly acquire other information in return for that which he can communicate. I will illustrate this by one of my own pursuits. I possess the slightest possible acquaintance with the vast fields of animal life, but at an early period I was struck by the numerical regularity of the pulsation and of the breathings. It appeared to me that there must exist some relation between these two functions. Accordingly, I took every opportunity of counting the numbers of the pulsations and of the breathings of various animals. The pig fair at Pavia and the book fair at Leipsic equally placed before me menageries in which I could collect such facts. Every zoological collection of living animals which I visited thus gained an additional interest, and occasionally excited the attention of those in charge of it to making a collection of facts relating to that subject. This led me at another period to generalize the subject of inquiry, and to print a skeleton form for the constants of the class mammalia. It was reprinted by the British Association at Cambridge in 1833, and also at Brussels in the 'Travaux du Congrès Général de Statistique,' Brussels, 1853.

I have so frequently been mortified by having the utterly-undeserved reputation of knowing everything that I was led to inquire into the probable ground of the egregious fallacy. The most frequent symptom was an address of this kind:—"Now Mr. Babbage, will you, who know everything, kindly explain to me — — —." Perhaps the thing whose explanation was required might be the metre of some ancient Chinese poem: or whether there were any large rivers in the planet Mercury.

One of the most useful accomplishments for a philosophical traveller with which I am acquainted, I learned from a workman, who taught me how to punch a hole in a sheet of glass without making a crack in it.

The process is very simple. Two centre-punches, a hammer, an ordinary bench-vice, and an old file, are all the tools required. These may be found in any blacksmith's shop. Having decided upon the part of the glass in which you wish to make the hole, scratch a cross (X) upon the desired spot with the point of the old file; then turn the bit of glass over, and scratch on the other side a similar mark exactly opposite the former.

Fix one of the small centre-punches with its point upwards in the vice. Let an assistant gently hold the bit of glass with its scratched point exactly resting upon the point of the centre-punch.

Take the other centre-punch in your own left hand and place its point in the centre of the upper scratch, which is of course nearly, if not exactly, above the fixed centre-punch. Now hit the upper centre-punch a *very* slight blow with the hammer: a mere touch is almost sufficient. This must be carefully repeated two or three times. The result of these blows will be to cause the centre of the cross to be, as it were, gently pounded.

Turn the glass over and let the slight cavity thus formed rest upon the fixed centre-punch. Repeat the light blows upon this side of the glass, and after turning it two or three times, a very small hole will be made through the glass. It not unfrequently happens that a small crack occurs in the glass; but with a little skill this can be cut out with the pane of the hammer.

The next process is to enlarge the hole and cut it into the required shape with the pane of the hammer. This is accomplished by supporting the glass upon the point of the fixed centre-punch, very close to the edge required to be cut. A light blow must then be struck with the pane of the hammer upon the edge to be broken. This must be repeated until the required shape is obtained.

The principles on which it depends are, that glass is a material breaking in every direction with a conchoidal fracture, and that the vibrations which would have caused cracking or fracture are checked by the support of the fixed centre-punch in close contiguity with the part to be broken off.

When by hastily performing this operation I have caused the glass to crack, I have frequently, by using more care, cut an opening all round the cracked part, and so let it drop out without spreading.

This process is rendered still more valuable by the use of the diamond. I usually carried in my travels a diamond mounted on a small circle of wood, so that I could easily cut out circles of glass with small holes in the centre.   The description of this process is sufficient to explain it to an experienced workman; but if the reader should wish to employ it, his readiest plan would be to ask such a person to show him how to do it.

The above technical description will doubtless be rather dry and obscure to the general reader; so I hope to make him amends by one or two of the consequences which have resulted to me from having instructed others in the art.

In the year 1825, during a visit to Devonport, I had apartments in the house of a glazier, of whom one day I inquired whether he was acquainted with the art of punching a hole in glass, to which he answered in the negative, and expressed great curiosity to see it done. Finding that at a short distance there was a blacksmith whom he sometimes employed, we went together to pay him a visit, and having selected from his rough tools the centre-punches and the hammer, I proceeded to explain and execute the whole process, with which my landlord was highly delighted.

On the eve of my departure I asked for the landlord's account, which was duly sent up and quite correct, except the omission of the charge for the apartments which I had agreed for at two guineas a week.   I added the four weeks for my lodgings, and the next morning, having placed the total amount upon the bill, I sent for my host in order to pay him, remarking that he had omitted the principal article of his account, which I had inserted.

He replied that he had intentionally omitted the lodgings, as he could not think of taking payment for them from a gentleman who had done him so great a service.   Quite unconscious of having rendered him any service, I asked him to explain how I had done him any good. He replied that he had the contract for the supply and repair of the whole of the lamps of Devonport, and that the art in which I had instructed him would save him more than twenty pounds a year.   I found some difficulty in prevailing on my grateful landlord to accept what was justly his due.

The second instance I shall mention of the use to which I turned this art of punching a hole in glass occurred in Italy, at Bologna.

I spent some weeks very agreeably in that celebrated university, which is still proud of having had the discoverer of the circulation of the blood amongst its students. One morning an Italian friend accompanied me round the town, to point out the more remarkable shops and manufactories. Passing through a small street, he remarked that there was a very well-informed man who kept a little shop for the sale of needles and tape and a few other such articles, but who also made barometers and thermometers, and had a very respectable knowledge of such subjects. I proposed that we should look in upon him as we were passing through the street. On entering his small shop, I was introduced to its tenant, who conversed very modestly and very sensibly upon various mathematical instruments.

I had invited several of my friends and professors to spend the evening with me at my hotel, for the purpose of examining various instruments I had brought with me. I knew that the sight of them would be quite a treat to the occupier of this little shop, so I mentioned the idea to my friend, and inquired whether my expected guests in the evening would think I had taken a liberty with them in inviting the humble constructor of instruments at the same time.

My friend and conductor immediately replied that he was well known to most of the professors, and much respected by them, and that they would think it very kind of me to give him that opportunity of seeing the instruments I possessed. I therefore took the opportunity of asking him to join the very agreeable party which assembled in my apartments in the evening.

We now made a tour of the city, and reached the factory of the chief philosophical instrument-maker of Bologna. He took great pleasure in showing me the various instruments he manufactured; but still there was a certain air of presumption about him, which seemed to indicate a less amount of knowledge than I should otherwise have assigned to him. I had on the preceding day mentioned to my Italian friend, who now accompanied me, that there existed a very simple method of punching a hole in a piece of glass, which, as he was much interested about it, I promised to show him on the earliest opportunity.

Finding myself in the workshop of the first instrument maker in Bologna, and observing the few tools I wanted, I thought it a good opportunity to explain the process to my friend; but I could only do this by applying to the master for the loan of some tools. I also thought it possible that the method was known to him, and that, having more practice, he would do the work better than myself.

I therefore mentioned the circumstance of my promise, and asked the master whether he was acquainted with the process. His reply was, "Yes; we do it every day." I then handed over to him the punch and the piece of glass, declaring that a mere amateur, who only occasionally practised it could not venture to operate before the first instrument-maker in Bologna, and in his own workshop.

I had observed a certain shade of surprise glance across the face of one of the workmen who heard the assertion of this daily practice of his master's, and, as I had my doubts of it, I contrived to put him in such a position that he must either retract his statement or else attempt to do the trick.

He then called for a flat piece of iron with a small hole in it. Placing the piece of glass upon the top of this bit of iron, and holding the punch upon it directly above the aperture, he gave a strong blow of the hammer, and smashed the glass into a hundred pieces.

I immediately began to console him, remarking that I did not myself always succeed, and that unaccountable circumstances sometimes defeated the skill even of the most accomplished workman. I then advised him to try a larger* piece of glass. Just after the crash I had put my hand upon a heavier hammer, which I immediately withdrew on his perceiving it. Thus encouraged, he called for a larger piece of glass, and a bit of iron with a smaller hole in it. In the meantime all the men in the shop rested from their work to witness this feat of every-day occurrence. Their master now seized the heavier hammer, which I had previously just touched. Finding him preparing for a strong and decided blow, I turned aside my head, in order to avoid seeing him blush—and also to save my own face from the coming cloud of splinters.

I just saw the last triumphant flourish of the heavy hammer waving over his head, and then heard, on its thundering fall, the crash made by the thousand fragments of glass which it scattered over the workshop.

I still, however, felt it my duty to administer what consolation I could to a fellow-creature in distress; so I repeated to him (which was the truth) that I, too, occasionally failed. Then looking at my watch, and observing to my companion that these tools were not adapted to my mode of work, I reminded him that we had a pressing engagement. I then took leave of this celebrated instrument-maker, with many thanks for àll he had shown me.

After such a misadventure, I thought it would be cruel to invite him to meet the learned professors who would be assembled at my evening

* The larger the piece of glass to be punched the more certainly the process succeeds.

party, especially as I knew that I should be asked to show my friends a process with which he had assured me he was so familiar. The unpretending maker of thermometers and barometers did however join the party; and the kind and considerate manner in which my guests of the university and of the city treated him raised both parties in my estimation.

I will here mention another mode of treating glass, which may occasionally be found worth communicating.

Ground glass is frequently employed for transmitting light into an apartment, whilst it effectually prevents persons on the outside from seeing into the room. Rough plate-glass is now in very common use for the same purpose. In both these circumstances there is a reciprocity, for those who are within such rooms cannot see external forms.

It may in some cases be desirable partially to remedy this difficulty. In my own case, I cut with my diamond a small disc of window-glass, about two inches in diameter, and cemented it with Canada balsam to the rough side of my rough plate-glass. I then suspended a circular piece of card by a thread, so as to cover the circular disc. When the Canada balsam is dry, it fills up all the little inequalities of the rough glass with a transparent substance, of nearly the same refracting power; consequently, on drawing aside the suspended card, the forms of external objects become tolerably well defined.

The smooth surface of the rough plate-glass, not being perfectly flat, produces a slight distortion, which might, if it were worth while, be cured by cementing another disc of glass upon that side. In case the ground glass itself happens to be plate-glass, the image of external objects is perfect.

Occasionally I met, in the course of my travels, with various things which, though not connected with my own pursuits, might yet be highly interesting to others. If the cost suited my purse, and the subject was easily carried, or the specimen of importance, I have in many instances purchased them. Such was the case with respect to that curious creature the *proteus anguineus*, a creature living only in the waters of dark caverns, which has eyes, but the eylids cannot open.

When I visited the caves of Adelsburg, in Styria, I inquired whether any of these singular creatures could be procured. I purchased all I could get, being six in number. I conveyed them in large bottles full of river water, which I changed every night. During the greater part of their journey the bottles were placed in large leathern bags lashed to the barouche seat of my calash.

The first of these pets died at Vienna, and another at Prague. After three months, two only survived, and reached Berlin, where they also died—I fear from my servant having supplied them with water from a well instead of from a river.

At night they were usually placed in a large wash-hand basin of water, covered over with a napkin.

They were very excitable under the action of light. On several occasions when I have visited them at night with a candle, one or more have jumped out of their watery home.

These rare animals were matters of great interest to many naturalists whom I visited in my rambles, and procured for me several very agreeable acquaintances. When their gloomy lives terminated I preserved them in spirits, and sent the specimens to the collections of our own universities, to India, and some of our colonies.

When I was preparing materials for the 'Economy of Manufactures,' I had occasion to travel frequently through our manufacturing and mining districts. On these occasions I found the travellers' inn or the travellers' room was usually the best adapted to my purpose, both in regard to economy and to information. As my inquiries had a wide range, I found ample assistance in carrying them on. Nobody doubted that I was one of the craft; but opinions were widely different as to the department in which I practised my vocation.

In one of my tours I passed a very agreeable week at the Commercial Hotel at Sheffield. The society of the travellers' room is very fluctuating. Many of its frequenters arrive at night, have supper, breakfast early the next morning, and are off soon after: others make rather a longer stay. One evening we sat up after supper much later than is usual, discussing a variety of commercial subjects.

When I came down rather late to breakfast, I found only one of my acquaintance of the previous evening remaining. He remarked that we had had a very agreeable party last night, in which I cordially concurred. He referred to the intelligent remarks of some of the party in our discussion, and then added, that when I left them they began to talk about me. I merely observed that I felt myself quite safe in their hands, but should be glad to profit by their remarks. It appeared, when I retired for the night, they debated about what trade I travelled for. "The tall gentleman in the corner," said my informant, "maintained that you were in the hardware line; whilst the fat gentleman who sat next to you at supper was quite sure that you were in the spirit trade." Another of the party declared that they were both mistaken: he said he had met you before, and that you were travelling for a great iron-master. "Well," said I, "you, I presume, knew my vocation better than our friends."—"Yes," said my informant, "I knew perfectly well that you were in the Nottingham lace trade." The waiter now appeared with his bill, and announced that my friend's trap was at the door.

I had passed nearly a week at the Commercial Inn without having broken the eleventh commandment; but the next day I was doomed to be found out. A groom, in the gay livery of the Fitzwilliams, having fruitlessly searched for me at all the great hotels, at last in despair thought of inquiring for me at the Commercial Hotel. The landlady was sure I was not staying in her house; but, in deference to the groom's urgent request, went to make inquiries amongst her guests. I was the first person she questioned, and was, of course, obliged to admit the impeachment. The groom brought a very kind note from the late Lord Fitzwilliam, who had heard of my being in Sheffield, to invite me to spend a week at Wentworth.

I gladly availed myself of this invitation, and passed it very agreeably. During the few first days the party in the house consisted of the family only. Then followed three days of open house, when their friends came from great distances, even as far as sixty or eighty miles, and that at a period when railroads were unknown.

On the great day upwards of a hundred persons sat down to dinner, a large number of whom slept in the house. This was the first time the ancient custom of open house had been kept up at Wentworth since the death of the former Earl, the celebrated Whig Lord Lieutenant of Yorkshire.

CHAPTER XXXIV

THE AUTHOR'S FURTHER CONTRIBUTIONS TO HUMAN KNOWLEDGE

Glaciers—Uniform Postage—Weight of the Bristol Bags—Parcel Post—Plan for transmitting Letters along Aërial Wires—Cost of Verification is part of Price—Sir Rowland Hill—Submarine Navigation—Difference Engine—Analytical Engine—Cause of Magnetic and Electric Rotations—Mechanical Notation—Occulting Lights—Semi-occultation may determine Distances—Distinction of Lighthouses numerically—Application from the United States—Proposed Voyage—Loss of the Ship and Mr. Reed—Congress of Naval Officers at Brussels in 1853—My Portable Occulting Light exhibited—Night Signals—Sun Signals—Solar Occulting Lights—Afterwards used at Sebastopol—Numerical Signals applicable to all Dictionaries—Zenith-Light Signals—Telegraph for Ships on Shore—Greenwich Time Signals—Theory of Isothermal Surfaces to account for the Geological Facts of the successive Uprising and Depression of various parts of the Earth's Surface—Games of Skill—Tit-tat-to—Exhibitions—Problem of the Three Magnetic Bodies.

Of Glaciers

MUCH has been written upon the subject of glaciers. The view which I took of the question on my first acquaintance with them still seems to me to afford a sufficient explanation of the phenomena. It is probable that I may have been anticipated in it by Saussure and others; but, having no time to inquire into its history, I shall give a very brief statement of those views.

The greater part of the material which ultimately constitutes a glacier arises from the rain falling and the snow deposited in the higher portions of mountain ranges, which naturally first fill up the ravines and valleys, and rests on the tops of the mountains, covering them to various depths.

The chief facts to be explained are—first, the causes of the descent of these glaciers into the plains; second, the causes of the transformation of the opaque consolidated snow at the sources of the glacier into pure transparent ice at its termination.

The glaciers usually lying in valleys having a steep descent, gravity must obviously have a powerful influence; but its action is considerably increased by another cause.

The heat of the earth and that derived from the friction of the glacier and its broken fragments against the rock on which it rests, as well as from the friction of its own fragments, slowly melts the ice, and thus diminishing the amount of its support, the ice above cracks and falls down upon the earth, again to be melted and again to be broken.

But as the ice is upon an inclined plane, the pressure from above, on the upper side of the fragment, will be greater than that on the lower; consequently, at every fall the fallen mass will descend by a very small quantity further into the valley. Another consequence of the melting of the lower part of the centre of the glacier will be that the centre will advance faster than the sides, and its termination will form a curve convex towards the valley.

The above was, I believe, the common explanation of the formation of glaciers. The following part explains my own views:—

Of the Causes of the Transformation of Condensed Snow into Transparent Ice

It is a well-known fact that water rapidly frozen retains all the air it held in solution, and is opaque.

It is also known that water freezing very slowly is transparent.

Whenever, by the melting of the lower portion of any part of a glacier, a piece of it cracks and falls to a lower level, the friction of the broken sides will produce heat, and melt a small portion of water. This water, trickling down very slowly, will form a thin layer on the broken surface, and a portion will be retained in the narrowest part of the crack. But, since the temperature of a glacier is very near the freezing point, that water will freeze very slowly. It will, therefore, become transparent ice, and will, as it were, solder together the two adjacent surfaces by a thin layer of transparent ice.

But the transparent ice is much stronger and more difficult to break than opaque ice; consequently, the next time the soldered fragments are again broken, they will not break in the strongest part, which is the transparent ice: but the next fracture will occur in the opaque ice, as it was at first.

Thus, by the continued breaking and falling downward of the fragments of the glacier, as it proceeds down the valley, a series of vertical, rudely parallel veins of transparent ice will be formed. As these masses descend the valley, fresh vertical layers of transparent ice will be interposed between those already existing until the whole takes that beautiful transparent cerulean tint which we so frequently see at the lower termination of a glacier. Another effect of this vertical fracture at the surfaces of least resistance will be alternate vertical layers of opaque and transparent ice shading into each other. This would, in some of its stages, give a kind of ribboned appearance to the ice. Probably traces of it would still be exhibited even in the most transparent ice. Speaking roughly, this ribboned structure ought to be closer together the nearer the piece examined is to the end of the glacier. It ought also to be more apparent towards the centre of the glacier than towards the sides. The effect of this progress downward is to produce a very powerful friction between the masses of ice and the earth over which they are pushed, and, consequently, a continual accession to that stream of water which is found issuing from all glaciers.

The result of this continual breaking up is to cause all the water melted by the friction of the blocks of ice which is not retained in the interstices to fall towards the lowest part of the descending valley, and thus increase the stream, and so take away more and more of the support of the central part of the glacier. Hence the advance of the surface of the glacier will be much quicker towards its middle than near the sides.

The consequence of these actions is, that cracks in the ice will occur generally in planes perpendicular to its surface. The rain which falls upon the glacier, the water produced from its surface by the sun's rays and by the effect of the temperature of the atmosphere, as well as the water produced by the friction of its descending fragments, will penetrate through these cracks, and be retained by capillary action on the surfaces, and still more where the distance of the adjacent surfaces is very small. The rest of this unfrozen water will reach the rocky bottom of the glacier, and give up some of its heat to the bed over which it passes, to be again employed in melting away the lowest support of the glacier ice. Although the temperature of the glacier should differ but by a very small quantity from that of the freezing point of water, yet these films will only freeze the more slowly, and therefore become more solid and transparent ice. Their very thinness will enable all the air to be more readily extricated by freezing.

The question of the *regelation* of pounded ice, if by that term is meant anything more than welding ice by heat, or of joining its parts by a process analogous to that which is called *burning together* two separate portions of a bronze statue, has always appeared to me unsatisfactory.

The process of "burning together" is as follows:—Two portions of a large statue, which have been cast separately, are placed in a trough of sand, with their corresponding ends near to each other. A channel is made in the sand, leading through the junction of the parts to be united.

A stream of melted bronze is now allowed to run out from the furnace through the channel between the contiguous ends which it is proposed to unite. The first effect of this is to heat the ends of the two fragments. After the stream of melted metal has continued some time, the ends of those fragments themselves begin to melt. When a small quantity of each end is completely melted, the further flow of the melted metal is stopped, and as soon as the pool of melted metal connects, the two ends of the pieces to be united begin to consolidate: the whole is covered up with sand and allowed to cool gradually. When cold, the unnecessary metal is cut away, and the fragments are as perfectly united as if they had been originally cast in one piece.

The sudden consolidation, by physical force, of pounded ice or snow appears to me to arise from the first effect of the pressure producing heat, which melts a small portion into water, and brings the particles of ice or snow nearer to each other. The portion of water thus produced then, having its heat abstracted by the ice, connects the particles of the latter more firmly together by freezing.

If two flat surfaces of clear ice had a heated plate of metal put between them, two very thin layers of water would be formed between the ice and the heated plate. If the hot plate were suddenly withdrawn, and the two plates of ice pressed together, they would then be frozen together. This would be equivalent to welding. In all these cases the temperature of the ice must be a very little lower than the freezing-point. The more nearly it approached that point the slower the process of freezing would be, and therefore the more transparent the ice thus formed.

In the Exhibition of 1862 there were two different processes by which ice was produced in abundance, even in the heat of the Machinery Annex, in which they were placed.

In both the water was quickly converted into ice, and in both cases the ice was opaque.

In one of them the ice was produced in the shape of long hollow cylinders. These were quite opaque, and were piled up in stacks. The temperature of the place caused the ice to melt slowly; consequently, the interstices where the cylinders rested upon each other, received and retained a small portion of the water, which, trickling down, was detained by capillary attraction. Here it was very slowly frozen, and formed at the junction of the cylinders a thin film of transparent ice. This gradually increased as the upper cylinders of the ice melted away, and, after several hours' exposure, I have seen clear transparent ice a quarter of an inch thick, where at the commencement, there had not been even a trace of translucency.

On inquiring of the operator why the original cylinders were opaque, he told me, because they were frozen quickly. I then pointed out to him the small portions of transparent ice, which I have described, and asked him the cause. He immediately said, because they had been frozen slowly.

It appeared to be an axiom, derived from his own experience, that water quickly frozen is always opaque, and water slowly frozen always transparent. I pointed out this practical illustration to many of the friends I accompanied in their examination of the machinery of the Annex.

It would follow from this explanation, that glaciers on lofty mountains and in high latitudes may, by their own action, keep the surface of the earth on which they rest at a higher temperature than it would otherwise attain.

Book and Parcel Post

When my friend, the late General Colby, was preparing the materials and instruments for the intended Irish survey, he generally visited me about once a week to discuss and talk over with me his various plans. We had both of us turned our attention to the Post-office, and had both considered and advocated the question of a uniform rate of postage. The ground of that opinion was, that the actual *transport* of a letter formed but a small item in the expense of transmitting it to its destination; whilst the heaviest part of the cost arose from the *collection* and *distribution*, and was, therefore, almost independent of the length of its journey. I got some returns of the weight of the Bristol mail-bag for each night during one week, with a view to ascertain the possibility of a more rapid transmission. General Colby arrived at the conclusion that, supposing every letter paid sixpence, and that the same number of letters were posted, then the revenue would remain the same. I believe, when an official comparison was subsequently made, it was found that the equivalent sum was fivepence halfpenny. I then devised means for transmitting letters enclosed in small cylinders, along wires suspended from posts, and from towers, or from church steeples. I made a little model of such an apparatus, and thus transmitted notes from my front drawing-room, through the house, into my workshop, which was in a room above my stables. The date of these experiments I do not exactly recollect, but it was certainly earlier than 1827.

I had also, at a still earlier period, arrived at the remarkable economical principle, *that one element in the price of every article is the cost of its verification.* It arose thus:—

In 1815 I became possessed of a house in London, and commenced my residence in Devonshire Street, Portland Place, in which I resided until 1827. A kind relative of mine sent up a constant supply of game. But although the game cost nothing, the expense charged for its carriage was so great that it really was more expensive than butchers' meat. I endeavoured to get redress for the constant overcharges, but as the game was transferred from one coach to another I found it practically impossible to discover where the overcharge arose, and thus to remedy the evil. These efforts, however, led me to the fact that *verification*, which in this instance constituted a considerable part of the *price of the article, must form a portion of its price in every case.*

Acting upon this, I suggested that if the Government were to become, through the means of the Post-office, parcel carriers, they would derive a greater profit from it than any private trader, because the whole price of verification would be saved by the public. I therefore recommended the enlargement of the duties of the Post-office by employing it for the conveyance of books and parcels.

I mention these facts with no wish to disparage the *subsequent* exertions of Sir Rowland Hill. His devotion to the subject, his unwearied industry, and his long and at last successful efforts to overcome the notorious official friction of that department, required all the enduring energy he so constantly bestowed upon the subject. The benefit conferred upon the country by the improvements he introduced is as yet scarcely sufficiently estimated.

These principles were published afterwards in the "Economy of Manufactures."—See First Edition, 8th June, 1832; Second Edition, 22nd November, 1832. See chap. on the "Influence of Verification on Price," p. 134, and "Conveyance of Letters," p. 273.

Submarine Navigation

Of this it is not necessary to do more than mention the title and refer for the detail to the chapter on Experience by Water: and also to the article Diving Bell in the "Encyclopedia Metropolitana."

I have only to add my opinion that in open inverted vessels it may probably be found, under certain circumstances, of important use.

Difference Engine

Enough has already been said about that unfortunate discovery in the previous part of this volume. The first and great cause of its discontinuance was the inordinately extravagant demands of the person whom I had employed to construct it for the Government. Even this might, perhaps, by great exertions and sacrifices, have been surmounted. There is, however, a limit beyond which human endurance cannot go. If I survive some few years longer, the Analytical Engine will exist, and its works will afterwards be spread over the world. If it is the will of that Being, who gave me the endowments which led to that discovery, that I should not survive to complete my work, I bow to that decision with intense gratitude for those gifts: conscious that through life I have never hesitated to make the severest sacrifices of fortune, and even of feelings, in order to accomplish my imagined mission.

The great principles on which the Analytical Engine rests have been examined, admitted, recorded, and demonstrated. The mechanism itself has now been reduced to unexpected simplicity. Half a century may probably elapse before any one without those aids which I leave behind me, will attempt so unpromising a task. If, unwarned by my example, any man shall undertake and shall succeed in really constructing an engine embodying in itself the whole of the executive department of mathematical analysis upon different principles or by simpler mechanical means, I have no fear of leaving my reputation in his charge, for he alone will be fully able to appreciate the nature of my efforts and the value of their results.

Explanation of the Cause of Magnetic and Electric Rotations

In 1824 Arago published his experiments on the magnetism manifested by various substances during rotation. I was much struck with the announcement, and immediately set up some apparatus in my own workshop in order to witness the facts thus announced.

My friend Herschel, who assisted at some of the earliest experiments, joined with me in repeating and varying those of Arago. The results were given in a joint paper on that subject, published in the "Transactions of the Royal Society" in 1825.

I had previously made some magnetic experiments on a large magnet which would, under peculiar management, sustain about $32\frac{1}{2}$ lbs. It was necessary to commence with a weight of about 28 lbs., and then to add at successive intervals additional weights, but each less and less than the former. This led me to an explanation of the cause of those rotations, which I still venture to think is the true cause, although it is not so recognized by English philosophers.

The history is a curious one, and whether the cause which I assigned is right or wrong, the train of thought by which I was led to it is valuable as an illustration of the mode in which the human mind works in its progress towards new discoveries.

The first experiment, showing that the weight suspended might be increased at successive intervals of time, was stated in most treatises on magnetism. But the visible fact impressed strongly on my mind the conclusion that the production and discharge of magnetism is not instantaneous, but requires time for its complete action. It appeared, therefore, to me that this principle was sufficient for the explanation of the rotations observed by Arago.

In the following year it occurred to me that electricity possessed the same property, namely, that of requiring time for its communication. I then instituted a new series of experiments, and succeeded, as I had anticipated, in producing electric rotations. But a new fact now presented itself: in certain cases the electric needle moved back in the contrary direction to that indicated by the influences to which it was subjected. Whenever this occurred the retrograde motion was always very slow. After eliminating successively by experiment every cause which I could imagine, the fact which remained was, that in certain cases there occurred a motion in the direction opposite to that which was expected. But whenever such a motion occurred it was always very slow. Upon further reflection, I conjectured that it might arise from the screen, interposed between the electric and the needle itself, becoming electrified possibly in the opposite direction. New experiments confirmed this view and proved that the original cause was sufficient for the production of all the observed effects.

These experiments and their explanation were printed in the "Phil. Trans." 1826. But they met with so little acceptance in England that I had ceased to contend for them against more popular doctrines, and was too deeply occupied with other inquiries to enter on their defence. Several years after, during a visit to Berlin, taking a morning walk with Mitscherlich, I asked what explanation he adopted of the magnetic rotations of Arago. He instantly replied, "There can be no doubt that yours is the true one."

It will be a curious circumstance in the history of science, if an erroneous explanation of new and singular experiments in one department should have led to the prevision of another similar set of facts in a different department, and even to the explanation of new facts at first apparently contradicting it.

Mechanical Notation

This also has been described in a former chapter. I look upon it as one of the most important additions I have made to human knowledge. It has placed the construction of machinery in the rank of a demonstrative science. The day will arrive when no school of mechanical drawing will be thought complete without teaching it.

Occulting Lights

The great object of all my inquiries has ever been to endeavour to ascertain those laws of thought by which man makes discoveries. It was by following out one of the principles which I had arrived at that I was led to the system of occulting numerical lights for distinguishing lighthouses and for night signals at sea, which I published about twelve years ago. The principle I allude to is this:—

Whenever we meet with any defect in the means we are contriving for the accomplishing a given object, that defect should be noted and reserved for future consideration, and inquiry should be made—

Whether that which is a defect as regards the object in view may not become a source of advantage in some totally different subject.

I had for a long series of years been watching the progress of electric, magnetic, and other lights of that order, with the view of using them for domestic purposes; but their want of uniformity seemed to render them hopeless for that object. Returning from a brilliant exhibition of voltaic light, I thought of applying the above rule. The accidental interruptions might, by breaking the circuit, be made to recur at any required intervals. This remark suggested their adaptation to a system of signals. But it was immediately followed by another, namely: that the interruptions were equally applicable to all lights, and might be effected by simple mechanism.

I then, by means of a small piece of clock-work and an argand lamp, made a *numerical* system of occultation, by which any number might be transmitted to all those within sight of the source of light. Having placed this in a window of my house, I walked down the street to the distance of about 250 yards. On turning round I perceived the number 32 clearly indicated by its occultations. There was, however, a small defect in the apparatus. After each occultation there was a kind of semi-occultation. This arose from the arm which carried the shade rebounding from the stop on which it fell. Aware that this defect could be easily remedied, I continued my onward course for about 250 yards more, with my back towards the light. On turning round I was much surprised to observe that the signal 32 was repeated distinctly without the slightest trace of any semi-occultation or blink.

I was very much astonished at this change; and on returning towards my house had the light constantly in view. After advancing a short distance I thought I perceived a very faint trace of the blink. At thirty or forty paces nearer it was clearly visible, and at the half-way point it was again perfectly distinct. I knew that the remedy was easy, but I was puzzled as to the cause.

After a little reflection I concluded that it arose from the circumstance that the small hole through which the light passed was just large enough to be visible at five hundred yards, yet that when the same hole was partially covered by the rebound there did not remain sufficient light to be seen at the full distance of five hundred yards.

Thus prepared, I again applied the principle I had commenced with and proceeded to examine whether this defect might not be converted into an advantage.

I soon perceived that a lighthouse, whose number was continually repeated with a blink, obscuring just half its light, would be seen *without any blink* at all distances beyond half its range; but that at all distances within its half range that fact would be indicated by a blink. Thus with two blinks, properly adjusted, the distance of a vessel from a first-class light would be distinguished at from twenty to thirty miles by occultations indicating its number without any blink; between ten and twenty miles by an occultation with one blink, and within ten miles by an occultation with two blinks.

But another advantage was also suggested by this defect. If the opaque cylinder which intercepts the light consists of two cylinders, A and B, connected together by rods: thus—

If the compound cylinder descend to *a*, and then rise again, there will be a single
occultation.

,,   ,,   ,,   *b*   ,,   double occultation.
,,   ,,   ,,   *c*   ,,   triple occultation.

Such occultations are very distinct, and are specially applicable to lighthouses.

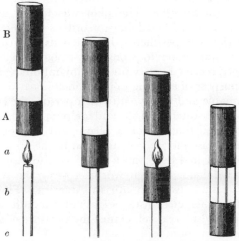

In the year 1851, during the Great Exhibition, the light I have described was exhibited from an upper window of my house in Dorset Street during many weeks.   It had not passed unnoticed by foreigners, who frequently reminded me that they had passed my door when I was asleep by writing upon their card the number exhibited by the occulting light and dropping it into my letter-box.

About five or six weeks after its first appearance I received a letter from a friend of mine in the United States, expressing great interest about it, and inquiring whether its construction was a secret.   My answer was, that I made no secret of it, and would prepare and send him a short description of it.

I then prepared a description, of which I had a very few copies printed.   I sent twelve of these to the proper authorities of the great maritime countries.   Most of them were accompanied by a private note of my own to some person of influence with whom I happened to be acquainted.

One of these was addressed to the present Emperor of the French, then a member of their Representative Chamber.   It was dated the 30th November, 1852.   Three days after I read in the newspapers the account of the *coup* of December 2, and smiled at the inopportune time at which my letter had accidentally been forwarded.   However, three days after I received from M. Mocquard the prettiest note, saying that he was commanded by the Prince President to thank me for the communication, and to assure me that the Prince was as much attached as ever to science, and should always continue to promote its cultivation.

The letter which was sent to the United States was placed in the hands of the Coast Survey.   The plan was highly approved, and Congress made a grant of 5,000 dollars, in order to try it experimentally.   After a long series of experiments, in which its merits were severely tested, a report was made to Congress strongly recommending its adoption.   I then received a very pressing invitation to visit the United States, for the purpose of assisting to put it in action.   It was conveyed to me by an amiable and highly cultivated person, the late Mr. Reed, Professor of English Literature at Philadelphia, who, on his arrival in London, proposed that I should accompany him on his return in October, the best season for the voyage, and in the finest vessel of their mercantile navy.   I had long had a great wish to visit the American continent, but I did not think it worth crossing the Atlantic, unless I could have spent a twelvemonth in America.   Finding this impossible under the then circumstances, about a month before the time arrived I resigned with great reluctance the pleasure of accompanying my friend to his own country.

It was most fortunate that I was thus prevented from embarking on board the Arctic, a steamer of the largest class.

Steaming at the rate of thirteen knots an hour over the banks of Newfoundland during a dense fog, the Arctic was run into by a steamer of about half its size, moving at the rate of seven knots. The concussion was in this instance fatal to the larger vessel.

This sad catastrophe was thus described by the brother of my lost friend:—

On the 20th of September, 1854, Mr. Reed, with his sister, embarked at Liverpool for New York, in the United States steamship Arctic. Seven days afterwards, at noon, on the 27th, when almost in sight of his native land, a fatal collision occurred, and before sundown every human being left upon the ship had sunk under the waves of the ocean. The only survivor who was personally acquainted with my brother, saw him about two o'clock, P.M., after the collision, and not very long before the ship sank, sitting with his sister in the small passage aft of the dining-saloon. They were tranquil and silent, though their faces wore the look of painful anxiety. They probably afterwards left this position, and repaired to the promenade deck. For a selfish struggle for life, with a helpless companion dependent upon him, with a physical frame unsuited for such a strife, and above all, with a sentiment of religious resignation which taught him in that hour of agony, even with the memory of his wife and children thronging in his mind, to bow his head in submission to the will of God,—for such a struggle he was wholly unsuited; and his is the praise, that he perished with the women and children.

In 1853 I spent some weeks at Brussels. During my residence in that city a Congress of naval officers from all the maritime nations assembled to discuss and agree upon certain rules and observations to be arranged for the common benefit of all. One evening I had the great pleasure of receiving the whole party at my house for the purpose of witnessing my occulting lights.

The portable occulting light which I had brought with me was placed in the verandah on the first floor, and we then went along the Boulevards to see its effect at different distances and with various numerical signals. On our return several papers relating to the subject were lying upon the table. The Russian representative, M. ——, took up one of the original printed descriptions and was much interested in it. On taking leave he asked, with some hesitation, whether I would lend it to him for a few hours. I told him at once that if I possessed another copy I would willingly give it to him; but that not

being the case I could only offer to lend it. M. ——— therefore took
it home with him, and when I sat down to breakfast the next morning
I found it upon my table.   In the course of the day I met my Russian
friend in the Park.   I expressed my hope that he had been interested
by the little tract he had so speedily returned.   He replied that it had
interested him so much that he had sat up all night, had copied the
whole of it, and that his transcript and a despatch upon the subject
was now on its way by the post to his own Government.

Several years after I was informed that *occulting solar lights* were used
by the Russians during the siege of Sebastopol.

Night Signals

The system of occulting light applies with remarkable facility to
night signals, either on shore or at sea.   If it is used numerically, it
applies to all the great dictionaries of the various maritime nations.
I may here remark, that there exist means by which all such signals
may, if necessary, be communicated in cipher.

Sun Signals

The distance at which such signals can be rendered visible exceeds
that of any other class of signals by means of light.   During the Irish
Trigonometrical Survey, a mountain in Scotland was observed, with
an angular instrument from a station in Ireland, at the distance of
108 miles.   This was accomplished by stationing a party on the summit of the mountain in Scotland with a looking-glass of about a foot
square, directing the sun's image to the opposite station.   No occultations were used; but if the mirror had been larger, and occultation
employed, messages might have been sent, and the time of residence
upon the mountain considerably diminished.   When I was occupied
with occulting signals, I made this widely known.   I afterwards
communicated the plan, during a visit to Paris, to many of my friends
in that capital, and, by request, to the Minister of Marine.

I have observed in the "Comptes Rendus" that the system has to
a certain extent been since used in the south of Algeria, where, during
eight months of the year, the sun is generally unobscured by clouds as
long as it is above the horizon.   I have not, however, noticed in those
communications to the Institute any reference to my own previous
publication.

Zenith-light Signals

Another form of signal, although not capable of use at very great
distances, may, however, be employed with considerable advantage,

under certain circumstances. Universality and economy are its great advantages. It consists of a looking-glass, making an angle of 45° with the horizon, placed just behind an opening in a vertical board. This being stuck into the earth, the light of the sky in the zenith, which is usually the brightest, will be projected horizontally through the opening, in whatever direction the person to be communicated with may be placed. The person who makes the signals must stand on one side in front of the instrument; and, by passing his hat slowly before the aperture any number of times, may thus express each unit's figure of his signal.

He must then, leaving the light visible, pause whilst he deliberately counts to himself ten.

He must then with his hat make a number of occultations equal to the tens figure he wishes to express.

This must be continued for each figure in the number of the signal, always pausing between each during the time of counting ten.

When the end of the signal is terminated, he must count sixty in the same manner; and if the signal he gave has not been acknowledged, he should repeat it until it has been observed.

The same simple telegraph may be used in a dark night, by substituting a lantern for the looking-glass. The whole apparatus is simple and cheap, and can be easily carried even by a small boy.

I was led to this contrivance many years ago by reading an account of a vessel stranded within thirty yards of the shore. Its crew consisted of thirteen people, ten of whom got into the boat, leaving the master, who thought himself safer in the ship, with two others of the crew.

The boat put off from the ship, keeping as much out of the breakers as it could, and looking out for a favourable place for landing. The people on shore followed the boat for several miles, urging them not to attempt landing. But not a single word was audible by the boat's crew, who, after rowing several miles, resolved to take advantage of the first favourable lull. They did so—the boat was knocked to pieces, and the whole crew were drowned. If the people on the shore could at that moment have communicated with the boat's crew, they could have informed them that, by continuing their course for half a mile further, they might turn into a cove, and land almost dry.

I was much impressed by the want of easy communication between stranded vessels and those on shore who might rescue them.

I can even now scarcely believe it credible that the very simple means I am about to mention has not been adopted years ago. A list of about a hundred questions, relating to directions and inquiries required to be communicated between the crew of a stranded ship and those on shore who wish to aid it, would, I am told, be amply sufficient for such purposes.   Now, if such a list of inquiries were prepared and printed by competent authority, any system of signals by which a number of two places of figures can be expressed might be used.   This list of inquiries and answers ought to be printed on cards, and nailed up on several parts of every vessel.   It would be still better, by conference with other maritime nations, to adopt the same system of signs, and to have them printed in each language.   A looking-glass, a board with a hole in it, and a lantern would be all the apparatus required.   The lantern might be used for night, and the looking-glass for day signals.

These simple and inexpensive signals might be occasionally found useful for various social purposes.

Two neighbours in the country whose houses, though reciprocally visible, are separated by an interval of several miles, might occasionally telegraph to each other.

If the looking-glass were of large size, its light and its occultation might be seen perhaps from six to ten miles, and thus become by daylight a cheap guiding light through channels and into harbours.

It may also become a question whether it might not in some cases save the expense of buoying certain channels.

For railway signals during daylight it might in some cases be of great advantage, by saving the erection of very lofty poles carrying dark frames through which the light of the sky is admitted.

Amongst my early experiments, I made an occulting hand-lantern, with a shade for occulting by the pressure of the thumb, and with two other shades of red and of green glass.   This might be made available for military purposes, or for the police.

Greenwich Time Signals

It has been thought very desirable that a signal to indicate Greenwich time should be placed on the Start Point, the last spot which ships going down the Channel on distant voyages usually sight.

The advantage of such an arrangement arises from this—that chronometers having had their rates ascertained on shore, may have them somewhat altered by the motions to which they are submitted at sea.   If, therefore, after a run of above two hundred miles, they can be informed of the exact Greenwich time, the sea rate of their chronometers will be obtained.

Of course no other difficulty than that of expense occurs in transmitting Greenwich time by electricity to any points on our coast.   The real difficulty is to convey it to the passing vessels. The firing of a cannon at certain fixed hours has been proposed, but this plan is encumbered by requiring the knowledge of the distance of the vessel from the gun, and also from the variation of the velocity of the transmission of sound under various circumstances.

During the night the flash arising from ignited gunpowder might be employed. But this, in case of rain or other atmospheric circumstances, might be impeded. The best plan for night-signals would be to have an occulting light, which might be that of the lighthouse itself, or another specially reserved for the purpose.

During the day, and when the sun is shining, the time might be transmitted by the occultations of reflected solar light, which would be seen at any distance the curvature of the earth admitted.

The application of my Zenith Light might perhaps fulfil all the required conditions during daylight.

I have found that, even in the atmosphere of London, an opening only five inches square can be distinctly seen, and its occultations counted by the naked eye at the distance of a quarter of a mile. If the side of the opening were double the former, then the light transmitted to the eye would be four times as great, and the occultations might be observed at the distance of one mile.

The looking-glass employed must have its side nearly in the proportion of three to two, so that one of five feet by seven and a half ought to be seen at the distance of about eight or nine miles.

Geological Theory of Isothermal Surfaces

During one portion of my residence at Naples my attention was concentrated upon what in my opinion is the most remarkable building upon the face of the earth, the Temple of Serapis, at Puzzuoli.*

It was obviously built at or above the level of the Mediterranean in order to profit by a hot spring which supplied its numerous baths. There is unmistakable evidence that it has subsided below the present level of the sea, at least twenty-five feet; that it must have remained there during many years; that it then rose gradually up, probably to its former level, and that during the last twenty years it has been again slowly subsiding.

The results of this survey led me in the following year to explain the

* In this inquiry I profited by the assistance of Mr. Head, now the Right Hon. Sir Edmund Head, Bart., K.C.B., late Governor-General of Canada. An abstract of my own observations was printed in the "Abstracts of Proceedings" of the Geological Society, vol. ii. p. 72. My friend's historical views were printed in the "Transactions" of the Antiquarian Society.

various elevations and depressions of portions of the earth's surface, at different periods of time, by a theory which I have called the theory of the earth's isothermal surfaces.

I do not think the importance of that theory has been well understood by geologists, who are not always sufficiently acquainted with physical science. The late Sir Henry De la Beche perceived at an early period the great light those sciences might throw upon his own favourite pursuit, and was himself always anxious to bring them to bear upon geology.

I am still more confirmed in my opinion of the importance of the "Theory of Isothermal Surfaces in Geology" from the fact that a few years afterwards my friend Sir John Herschel arrived independently at precisely the same theory. I have stated this at length in the notes to the "Ninth Bridgewater Treatise."

Games of Skill

A considerable time after the translation of Menabrea's memoir had been published, and after I had made many drawings of the Analytical Engine and all its parts, I began to meditate upon the intellectual means by which I had reached to such advanced and even to such unexpected results. I reviewed in my mind the various principles which I had touched upon in my published and unpublished papers, and dwelt with satisfaction upon the power which I possessed over mechanism through the aid of the Mechanical Notation. I felt, however, that it would be more satisfactory to the minds of others, and even in some measure to my own, that I should try the power of such principles as I had laid down, by assuming some question of an entirely new kind, and endeavouring to solve it by the aid of those principles which had so successfully guided me in other cases.

After much consideration I selected for my test the contrivance of a machine that should be able to play a game of purely intellectual skill successfully; such as tit-tat-to, drafts, chess, &c.

I endeavoured to ascertain the opinions of persons in every class of life and of all ages, whether they thought it required human reason to play games of skill. The almost constant answer was in the affirmative. Some supported this view of the case by observing, that if it were otherwise, then an automaton could play such games. A few of those who had considerable acquaintance with mathematical science allowed the possibility of machinery being capable of such work; but they most stoutly denied the possibility of contriving such

machinery on account of the myriads of combinations which even the
simplest games included.

On the first part of my inquiry I soon arrived at a demonstration
that every game of skill is susceptible of being played by an automaton.

Further consideration showed that if *any position* of the men upon
the board were assumed (whether that position were possible or
impossible), then if the automaton could make the first move rightly,
he must be able to win the game, always supposing that, under the
given position of the men, that conclusion were possible.

Whatever move the automaton made, another move would be made
by his adversary. Now this altered state of the board is *one* amongst
the *many positions* of the men in which, by the previous paragraph, the
automaton was supposed capable of acting.

Hence the question is reduced to that of making the best move under
any possible combinations of positions of the men.

Now the several questions the automaton has to consider are of this
nature:—

1. Is the position of the men, as placed before him on the board, a
   possible position? that is, one which is consistent with the
   rules of the game?
2. If so, has Automaton himself already lost the game?
3. If not, then has Automaton won the game?
4. If not, can he win it at the next move? If so, make that move.
5. If not, could his adversary, if he had the move, win the game.
6. If so, Automaton must prevent him if possible.
7. If his adversary cannot win the game at his next move, Automaton must examine whether he can make such a move that,
   if he were allowed to have two moves in succession, he could at
   the second move have *two* different ways of winning the game;

and each of these cases failing, Automaton must look forward to three
or more successive moves.

Now I have already stated that in the Analytical Engine I had
devised mechanical means equivalent to memory, also that I had
provided other means equivalent to foresight, and that the Engine
itself could act on this foresight.

In consequence of this the whole question of making an automaton
play any game depended upon the possibility of the machine being
able to represent all the myriads of combinations relating to it.
Allowing one hundred moves on each side for the longest game at
chess, I found that the combinations involved in the Analytical Engine
enormously surpassed any required, even by the game of chess.

As soon as I had arrived at this conclusion I commenced an examination of a game called "tit-tat-to," usually played by little children. It is the simplest game with which I am acquainted. Each player has five counters, one set marked with a +, the other set with an O. The board consists of a square divided into nine smaller squares, and the object of each player is to get three of his own men in a straight line. One man is put on the board by each player alternately. In practice no board is used, but the children draw upon a bit of paper, or on their slate, a figure like any of the following.

The successive moves of the two players may be represented as follow:—

Moves 1   2   3   4   5   6   7

```
 |  |       |  |       |  | +      |  | +     + |  | +    + | O | +    + | O | +
-------    -------    -------    -------    -------    -------    -------
 |  |       | O |       | O |      | O |      | O |      | O |      + | O |
-------    -------    -------    -------    -------    -------    -------
+ |  |     + |  |      + |  |      + |  | O    + |  | O    + |  | O    + |  | O
```

In this case + wins at the seventh move.

The next step I made was to ascertain what number of combinations were required for all the possible variety of moves and situations.   I found this to be comparatively insignificant.

I therefore easily sketched out mechanism by which such an automaton might be guided.   Hitherto I had considered only the philosophical view of the subject, but a new idea now entered my head which seemed to offer some chance of enabling me to acquire the funds necessary to complete the Analytical Engine.

It occurred to me that if an automaton were made to play this game, it might be surrounded with such attractive circumstances that a very popular and profitable exhibition might be produced.   I imagined that the machine might consist of the figures of two children playing against each other, accompanied by a lamb and a cock. That the child who won the game might clap his hands whilst the cock was crowing, after which, that the child who was beaten might cry and wring his hands whilst the lamb began bleating.

I then proceeded to sketch various mechanical means by which every action could be produced.   These, when compared with those I had employed for the Analytical Engine, were remarkably simple. A difficulty, however, arose of a novel kind.   It will have been observed, in the explanation I gave of the Analytical Engine, that cases arose in which it became necessary, on the occurrence of certain conditions, that the machine itself should select one out of two or more distinct modes of calculation.   The particular one to be adopted could only be known when those calculations on which the selection depended had been already made.

The new difficulty consisted in this, that when the automaton had to move, it might occur that there were two different moves, each equally conducive to his winning the game. In this case no reason existed within the machine to direct his choice: unless, also, some provision were made, the machine would attempt two contradictory motions.

The first remedy I devised for this defect was to make the machine keep a record of the number of games it had won from the commencement of its existence. Whenever two moves, which we may call A and B, were equally conducive to winning the game, the automaton was made to consult the record of the number of the games he had won. If that number happened to be even, he was directed to take the course A; if it were odd, he was to take the course B.

If there were three moves equally possible, the automaton was directed to divide the number of games he had won by three. In this case the numbers 0, 1, or 2 might be the remainder, and the machine was directed to take the course A, B, or C accordingly.

It is obvious that any number of conditions might be thus provided for. An inquiring spectator, who observed the games played by the automaton, might watch a long time before he discovered the principle upon which it acted. It is also worthy of remark how admirably this illustrates the best definitions of chance by the philosopher and the poet:—

> "Chance is but the expression of man's ignorance."—LAPLACE
> "All chance, design ill understood."—POPE

Having fully satisfied myself of the power of making such an automaton, the next step was to ascertain whether there was any probability, if it were exhibited to the public, of its producing, in a moderate time, such a sum of money as would enable me to construct the Analytical Engine. A friend, to whom I had at an early period communicated the idea, entertained great hopes of its pecuniary success. When it became known that an automaton could beat not merely children but even papa and mamma at a child's game, it seemed not unreasonable to expect that every child who heard of it would ask mamma to see it. On the other hand, every mamma, and some few papas, who heard of it would doubtless take their children to so singular and interesting a sight. I resolved, on my return to London, to make inquiries as to the relative productiveness of the various exhibitions of recent years, and also to obtain some rough estimate of the probable time it would take to construct the automaton, as well as some approximation to the expense.

It occurred to me that if half a dozen were made, they might be exhibited in three different places at the same time. Each exhibitor might then have an automaton in reserve in case of accidental injury. On my return to town I made the inquiries I alluded to, and found that the English machine for making Latin verses, the German talking-machine, as well as several others, were entire failures in a pecuniary point of view. I also found that the most profitable exhibition which had occurred for many years was that of the little dwarf, General Tom Thumb.

On considering the whole question, I arrived at the conclusion, that to conduct the affair to a successful issue it would occupy so much of my own time to contrive and execute the machinery, and then to superintend the working out of the plan, that even if successful in point of pecuniary profit, it would be too late to avail myself of the money thus acquired to complete the Analytical Engine.

Problem of the Three Magnetic Bodies

The problem of the three bodies, which has cost such unwearied labour to so many of the highest intellects of this and the past age, is simple compared with another which is opening upon us. We now possess a very extensive series of well-recorded observations of the positions of the magnetic needle, in various parts of our globe, during about thirty years.

Certain periods of changes of about ten or eleven years are said to be indicated as connected with changes in the amount of solar spots; but the inductive evidence scarcely rests upon three periods, and it seems more probable that these effects arise from some common cause.

(1.) It has been long known that the earth has at least two if not more magnetic poles.

(2.) It is probable, therefore, that the sun and moon also have several magnetic poles.

(3.) In 1826 I proved that when a magnet is brought into proximity to a piece of matter capable of becoming magnetic, the magnetism communicated by it requires *time* for its full development in the body magnetized. Also that when the influence of the magnet is removed, the magnetized body requires *time* to regain its former state.

This being the case, it is required, having assumed certain positions for the poles of these various magnetic bodies, to calculate their reciprocal influences in changing the positions of those poles on the other bodies. The development of the equations representing these forces will indicate cycles which really belong to the nature of the subject. The comparisons of a long series of observations with recorded facts will ultimately enable us to determine both the number and position of those poles upon each body.

Electricity possesses an analogous property with respect to time being required for its full action. If the bodies of our system influence each other electrically, other developments will be required and other cycles discovered.

When the equations resulting from the actions of these causes are formed, and means of developing them arranged, the whole of the rest of the work comes under the domain of machinery.

Charles Babbage, 1792–1871.   Portrait from the *Illustrated London News*, Nov. 4, 1871.

PART II

Selections from

BABBAGE'S CALCULATING
ENGINES

BABBAGE'S

CALCULATING ENGINES.

BEING

A COLLECTION OF PAPERS RELATING TO THEM;
THEIR HISTORY, AND CONSTRUCTION.

LONDON:
E. AND F. N. SPON, 125, STRAND.
1889.

Facsimile of title page from *Babbage's Calculating Engines* published in 1889.

CONTENTS

I

BABBAGE'S CALCULATING ENGINE

By Dr. DIONYSIUS LARDNER

From the *Edinburgh Review*, July, 1834, No. CXX

ART. I.—1. *Letter to Sir Humphry Davy, Bart., P.R.S., on the application of Machinery to Calculate and Print Mathematical Tables.* By CHARLES BABBAGE, Esq., F.R.S., 4to. Printed by order of the House of Commons. No. 370. May 22, 1823.

2. *On the Application of Machinery to the Calculation of Astronomical and Mathematical Tables.* By CHARLES BABBAGE, Esq. Memoirs Astron. Soc. Vol. I. Part 2. London, 1822.

3. *Address to the Astronomical Society, by Henry Thomas Colebrooke, Esq., F.R.S. President, on presenting the first gold medal of the Society to Charles Babbage, Esq., for the invention of the Calculating Engine.* Memoirs Astron. Soc. Vol. I. Part 2. London, 1822.

4. *On the determination of the General Term of a new Class of Infinite Series.* By CHARLES BABBAGE, Esq. Transactions Camb. Phil. Soc. Cambridge, 1824.

5. *On Errors common to many Tables of Logarithms.* By CHARLES BABBAGE, Esq. Memoirs Astron. Soc. London, 1827.

6. *On a Method of Expressing by Signs the Action of Machinery.* By CHARLES BABBAGE, Esq. Phil. Trans. London, 1826.

7. *Report by the Committee appointed by the Council of the Royal Society to consider the subject referred to in a communication received by them from the Treasury, respecting Mr. Babbage's Calculating Engine, and to report thereupon.* London, 1829.

THERE IS no position in society more enviable than that of the few who unite a moderate independence with high intellectual qualities. Liberated from the necessity of seeking their support by a profession, they are unfettered by its restraints, and are enabled to direct the powers of their minds, and to concentrate their intellectual energies on those objects exclusively to which they feel that their powers may be applied with the greatest advantage to the community, and with the most lasting reputation to themselves. On the other hand, their middle station and limited income rescue them from those allurements to frivolity and dissipation, to which rank and wealth ever expose their possessors. Placed in such favourable circumstances, Mr. Babbage selected science as the field of his ambition; and his mathematical researches have conferred on him a high reputation, wherever the exact sciences are studied and appreciated. The suffrages of the mathematical world have been ratified in his own country, where he has been elected to the Lucasian Professorship in his own University—a chair, which, though of inconsiderable emolument, is one on which Newton has conferred everlasting celebrity. But it has been the fortune of this mathematician to surround himself with fame of another and more popular kind, and which rarely falls to the lot of those who devote their lives to the cultivation of the abstract sciences. This distinction he owes to the announcement, some years since, of his celebrated project of a Calculating Engine. A proposition to reduce arithmetic to the dominion of mechanism—to substitute an automaton for a compositor—to throw the powers of thought into wheel-work could not fail to awaken the attention of the world. To bring the practicability of such a project within the compass of popular belief was not easy: to do so by bringing it within the compass of popular comprehension was not possible. It transcended the imagination of the public in general to conceive its possibility; and the sentiments of wonder with which it was received, were only prevented from merging into those of incredulity, by the faith reposed in the high attainments of its projector. This extraordinary undertaking was, however, viewed in a very different light by the small section of the community, who, being sufficiently versed in mathematics, were acquainted with the principle upon which it was founded. By reference to that principle, they perceived at a glance the practicability of the project; and being enabled by the nature of their attainments and pursuits to appreciate the immeasurable importance of its results, they regarded the invention with a proportionately profound interest. The production of numerical tables, unlimited in quantity and variety, restricted to no particular species, and limited by no particular law; extending not merely to the boundaries of existing knowledge, but spreading their powers over the undefined regions of future discovery—were results, the magnitude and the value of which the community in general could neither comprehend nor appreciate. In such a case, the judgment of the world could only rest upon the authority of the philosophical part of it; and the fiat of the scientific community swayed for once political councils. The British Government, advised by the Royal Society, and a committee formed of the most eminent mechanicians and practical engineers, determined on constructing the projected mechanism at the expense of the nation, to be held as national property.

Notwithstanding the interest with which this invention has been regarded in every part of the world, it has never yet been embodied in a written, much less in a published form. We trust, therefore, that some credit will be conceded to us for having been the first to

make the public acquainted with the object, principle, and structure
of a piece of machinery, which, though at present unknown (except as
to a few of its probable results), must, when completed, produce
important effects, not only on the progress of science, but on that of
civilization.

The calculating machinery thus undertaken for the public gratui-
tously (so far as Mr. Babbage is concerned), has now attained a very
advanced stage towards completion; and a portion of it has been put
together, and performs various calculations:—affording a practical
demonstration that the anticipations of those, under whose advice
Government has acted, have been well founded.

There are, nevertheless, many persons who, admitting the great
ingenuity of the contrivance, have, notwithstanding, been accustomed
to regard it more in the light of a philosophical curiosity, than an
instrument for purposes practically useful.   This mistake (than which
it is not possible to imagine a greater) has arisen mainly from the ignorance which prevails of the extensive utility of those numerical tables
which it is the purpose of the engine in question to produce.   There
are also some persons who, not considering the time requisite to bring
any invention of this magnitude to perfection in all its details, incline
to consider the delays which have taken place in its progress as presumptions against its practicability.   These persons should, however,
before they arrive at such a conclusion, reflect upon the time which
was necessary to bring to perfection engines infinitely inferior in
complexity and mechanical difficulty.   Let them remember that—
not to mention the *invention* of that machine—the *improvements* alone
introduced into the steam-engine by the celebrated Watt, occupied a
period of not less than twenty years of the life of that distinguished
person, and involved an expenditure of capital amounting to £50,000.*
The calculating machinery is a contrivance new even in its details.
Its inventor did not take it up already imperfectly formed, after having
received the contributions of human ingenuity exercised upon it for
a century or more.   It has not, like almost all other great mechanical
inventions, been gradually advanced to its present state through a
series of failures, through difficulties encountered and overcome by a
succession of projectors.   It is not an object on which the light of
various minds has thus been shed.   It is, on the contrary, the production of solitary and individual thought—begun, advanced through

* Watt commenced his investigations respecting the steam-engine in 1763, between
which time and the year 1782 inclusive, he took out several patents for improvements
in details.   Bolton and Watt had expended the above sum on their improvements
before they began to receive any return.

each successive stage of improvement, and brought to perfection by one mind.   Yet this creation of genius, from its first rude conception to its present state, has cost little more than half the time, and not one-third of the expense, consumed in bringing the steam-engine (previously far advanced in the course of improvement) to that state of comparative perfection in which it was left by Watt.   Short as the period of time has been which the inventor has devoted to this enterprise, it has, nevertheless, been demonstrated, to the satisfaction of many scientific men of the first eminence, that the design in all its details, reduced, as it is, to a system of mechanical drawings, is complete; and requires only to be constructed in conformity with those plans, to realize all that its inventor has promised.

With a view to remove and correct erroneous impressions, and at the same time to convert the vague sense of wonder at what seems incomprehensible, with which this project is contemplated by the public in general, into a more rational and edifying sentiment, it is our purpose in the present article,

First, To show the immense importance of any method by which numerical tables, absolutely accurate in every individual copy, may be produced with facility and cheapness.   This we shall establish by conveying to the reader some notion of the number and variety of tables published in every country of the world to which civilization has extended, a large portion of which have been produced at the public expense; by showing also, that they are nevertheless rendered inefficient, to a greater or less extent, by the prevalence of errors in them; that these errors pervade not merely tables produced by in-dividual labour and enterprise, but that they vitiate even those on which national resources have been prodigally expended, and to which the highest mathematical ability, which the most enlightened nations of the world could command, has been unsparingly and systematically directed.

Secondly, To attempt to convey to the reader a general notion of the mathematical principle on which the calculating machinery is founded, and of the manner in which this principle is brought into practical operation, both in the process of calculating and printing. It would be incompatible with the nature of this review, and indeed impossible without the aid of numerous plans, sections, and elevations, to convey clear and precise notions of the details of the means by which the process of reasoning is performed by inanimate matter, and the arbitrary and capricious evolutions of the fingers of typographical compositors are reduced to a system of wheel-work.   We are, nevertheless, not without hopes of conveying, even to readers unskilled in mathematics, some satisfactory notions of a general nature on this subject.

Thirdly, To explain the actual state of the machinery at the present time; what progress has been made towards its completion; and what are the probable causes of those delays in its progress, which must be a subject of regret to all friends of science. We shall indicate what appears to us the best and most practicable course to prevent the unnecessary recurrence of such obstructions for the future, and to bring this noble project to a speedy and successful issue.

Viewing the infinite extent and variety of the tables which have been calculated and printed, from the earliest periods of human civilization to the present time, we feel embarrassed with the difficulties of the task which we have imposed on ourselves:—that of attempting to convey to readers unaccustomed to such speculations, anything approaching to an adequate idea of them. These tables are connected with the various sciences, with almost every department of the useful arts, with commerce in all its relations; but, above all, with Astronomy and Navigation. So important have they been considered, that in many instances large sums have been appropriated by the most enlightened nations in the production of them; and yet so numerous and insurmountable have been the difficulties attending the attainment of this end, that after all, even navigators, putting aside every other department of art and science, have, until very recently, been scantily and imperfectly supplied with the tables indispensably necessary to determine their position at sea.

The first class of tables which naturally present themselves, are those of Multiplication. A great variety of extensive multiplication tables have been published from an early period in different countries; and especially tables of *Powers*, in which a number is multiplied by itself successively. In Dodson's *Calculator* we find a table of multiplication extending as far as 10 times 1000.* In 1775, a still more extensive table was published to 10 times 10,000. The Board of Longitude subsequently employed the late Dr. Hutton to calculate and print various numerical tables, and among others, a multiplication table extending as far as 100 times 1000; tables of the squares of numbers, as far as 25,400; tables of cubes, and of the first ten powers of numbers, as far as 100.† In 1814, Professor Barlow, of Woolwich, published, in an octavo volume, the squares, cubes, square roots, cube roots, and reciprocals of all numbers from 1 to 10,000; a table of the

* Dodson's "Calculator." 4to. London, 1747.
† Hutton's "Tables of Products and Powers." Folio. London, 1781.

first ten powers of all numbers from 1 to 100, and of the fourth and fifth powers of all numbers from 100 to 1000.

Tables of Multiplication to a still greater extent have been published in France. In 1785, was published an octavo volume of tables of the squares, cubes, square roots, and cube roots of all numbers from 1 to 10,000; and similar tables were again published in 1801. In 1817, multiplication tables were published in Paris by Voisin; and similar tables, in two quarto volumes, in 1824, by the French Board of Longitude, extending as far as a thousand times a thousand. A table of squares was published in 1810, in Hanover; in 1812, at Leipzig; in 1825, at Berlin; and in 1827, at Ghent. A table of cubes was published in 1827, at Eisenach; in the same year a similar table at Ghent; and one of the squares of all numbers as far as 10,000, was published in that year, in quarto, at Bonn. The Prussian Government has caused a multiplication table to be calculated and printed, extending as far as 1000 times 1000. Such are a few of the tables of this class which have been published in different countries.

This class of tables may be considered as purely arithmetical, since the results which they express involve no other relations than the arithmetical dependence of abstract numbers upon each other. When numbers, however, are taken in a concrete sense, and are applied to express peculiar modes of quantity,—such as angular, linear, superficial, and solid magnitudes,— a new set of numerical relations arise, and a large number of computations are required.

To express angular magnitude, and the various relations of linear magnitude with which it is connected, involves the consideration of a vast variety of Geometrical and Trigonometrical tables; such as tables of the natural sines, co-sines, tangents, secants, co-tangents, &c. &c.; tables of arcs and angles in terms of the radius; tables for the immediate solution of various cases of triangles, &c. Volumes without number of such tables have been from time to time computed and published. It is not sufficient, however, for the purposes of computation to tabulate these immediate trigonometrical functions. Their squares* and higher powers, their square roots, and other roots, occur so frequently, that it has been found expedient to compute tables for them, as well as for the same functions of abstract numbers.

---

* The squares of the sines of angles are extensively used in the calculations connected with the theory of the tides. Not aware that tables of these squares existed, Bouvard, who calculated the tides for Laplace, underwent the labour of calculating the square of each individual sine in every case in which it occurred.

The measurement of linear, superficial, and solid magnitudes, in the various forms and modifications in which they are required in the arts, demands another extensive catalogue of numerical tables. The surveyor, the architect, the builder, the carpenter, the miner, the gauger, the naval architect, the engineer, civil and military, all require the aid of peculiar numerical tables, and such have been published in all countries.

The increased expedition and accuracy which was introduced into the art of computation by the invention of Logarithms, greatly enlarged the number of tables previously necessary. To apply the logarithmic method, it was not merely necessary to place in the hands of the computist extensive tables of the logarithms of the natural numbers, but likewise to supply him with tables in which he might find already calculated the logarithms of those arithmetical, trigonometrical, and geometrical functions of numbers, which he has most frequent occasion to use. It would be a circuitous process, when the logarithm of a sine or co-sine of an angle is required, to refer, first to the table of sines, or co-sines, and thence to the table of the logarithms of natural numbers. It was therefore found expedient to compute distinct tables of the logarithms of the sines, co-sines, tangents, &c., as well as of various other functions frequently required, such as sums, differences, &c.

Great as is the extent of the tables we have just enumerated, they bear a very insignificant proportion to those which remain to be mentioned. The above are, for the most part, general in their nature, not belonging particularly to any science or art. There is a much greater variety of tables, whose importance is no way inferior, which are, however, of a more special nature: Such are, for example, tables of interest, discount, and exchange, tables of annuities, and other tables necessary in life insurances; tables of rates of various kinds necessary in general commerce. But the science in which, above all others, the most extensive and accurate tables are indispensable, is Astronomy; with the improvement and perfection of which is inseparably connected that of the kindred art of Navigation. We scarcely dare hope to convey to the general reader anything approaching to an adequate notion of the multiplicity and complexity of the tables necessary for the purposes of the astronomer and navigator. We feel, nevertheless, that the truly national importance which must attach to any perfect and easy means of producing those tables cannot be at all estimated, unless we state some of the previous calculations necessary in order to enable the mariner to determine with the requisite certainty and precision the place of his ship.

In a word, then, all the purely arithmetical, trigonometrical, and logarithmic tables already mentioned, are necessary, either immediately or remotely, for this purpose.   But in addition to these, a great number of tables, exclusively astronomical, are likewise indispensable.   The predictions of the astronomer, with respect to the positions and motions of the bodies of the firmament, are the means, and the only means, which enable the mariner to prosecute his art.   By these he is enabled to discover the distance of his ship from the Line, and the extent of his departure from the meridian of Greenwich, or from any other meridian to which the astronomical predictions refer.   The more numerous, minute and accurate these predictions can be made, the greater will be the facilities which can be furnished to the mariner. But the computation of those tables, in which the future position of celestial objects are registered, depend themselves upon an infinite variety of other tables which never reach the hands of the mariner. It cannot be said that there is any table whatever, necessary for the astronomer, which is unnecessary for the navigator.

The purposes of the marine of a country whose interests are so inseparably connected as ours are with the improvement of the art of navigation, would be very inadequately fulfilled, if our navigators were merely supplied with the means of determining by *Nautical Astronomy* the position of a ship at sea.   It has been well observed by the Committee of the Astronomical Society, to whom the recent improvement of the Nautical Almanac was confided, that it is not by those means merely by which the seaman is enabled to determine the position of his vessel at sea, that the full intent and purpose of what is usually called *Nautical Astronomy* are answered.   This object is merely a part of that comprehensive and important subject; and might be attained by a very cheap publication, and without the aid of expensive instruments.   A not less important and much more difficult part of nautical science has for its object to determine the precise position of various interesting and important points of the surface of the earth,— such as remarkable headlands, ports, and islands; together with the general trending of the coast between well-known harbours.   It is not necessary to point out here how important such knowledge is to the mariner.   This knowledge, which may be called *Nautical Geography*, cannot be obtained by the methods of observation used on board ship, but requires much more delicate and accurate instruments, firmly placed upon the solid ground, besides all the astronomical aid which can be afforded by the best tables, arranged in the most convenient form for immediate use.   This was Dr. Maskelyne's view of the subject, and his opinion has been confirmed by the repeated wants and demands of those distinguished navigators who have been employed in several recent scientific expeditions.*

Among the tables *directly* necessary for navigation, are those which predict the position of the centre of the sun from hour to hour. These tables include the sun's right ascension and declination, daily, at noon, with the hourly change in these quantities. They also include the equation of time, together with its hourly variation.

Tables of the moon's place for every hour, are likewise necessary, together with the change of declination for every ten minutes. The lunar method of determining the longitude depends upon tables containing the predicted distances of the moon from the sun, the principal planets, and from certain conspicuous fixed stars; which distances being observed by the mariner, he is enabled thence to discover the *time* at the meridian from which the longitude is measured; and, by comparing that time with the time known or discoverable in his actual situation, he infers his longitude. But not only does the prediction of the position of the moon, with respect to these celestial objects, require a vast number of numerical tables, but likewise the observations necessary to be made by the mariner, in order to determine the lunar distances, also require several tables. To predict the exact position of any fixed star, requires not less than ten numerical tables peculiar to that star; and if the mariner be furnished (as is actually the case) with tables of the predicted distances of the moon from one hundred such stars, such predictions must require not less than a thousand numerical tables. Regarding the range of the moon through the firmament, however, it will be readily conceived that a hundred stars form but a scanty supply; especially when it is considered that an accurate method of determining the longitude, consists in observing the extinction of a star by the dark edge of the moon. Within the limits of the lunar orbit there are not less than one thousand stars, which are so situated as to be in the moon's path, and therefore to exhibit, at some period or other, those desirable occultations. These stars are also of such magnitudes, that their occultations may be distinctly observed from the deck, even when subject to all the unsteadiness produced by an agitated sea. To predict the occultations of such stars, would require not less than ten thousand tables. The stars from which lunar distances might be taken are still more numerous; and we may safely pronounce, that, great as has been the improvement effected recently in our Nautical Almanac, it does not yet furnish more than a small fraction of that aid to navigation (in the

* Report of the Committee of the Astronomical Society, prefixed to the "Nautical Almanac" for 1834.

large sense of that term), which, with greater facility, expedition, and economy in the calculation and printing of tables, it might be made to supply.

Tables necessary to determine the places of the planets are not less necessary than those for the sun, moon and stars.   Some notion of the number and complexity of these tables may be formed, when we state that the positions of the two principal planets (and these the most necessary for the navigator), Jupiter and Saturn, require each not less than one hundred and sixteen tables.   Yet it is not only necessary to predict the position of these bodies, but it is likewise expedient to tabulate the motions of the four satellites of Jupiter, to predict the exact times at which they enter his shadow, and at which their shadows cross his disc, as well as the times at which they are interposed between him and the Earth, and he between them and the Earth.

Among the extensive classes of tables here enumerated, there are several which are in their nature permanent and unalterable, and would never require to be recomputed, if they once could be computed with perfect accuracy on accurate data; but the data on which such computations are conducted, can only be regarded as approximations to truth, within limits the extent of which must necessarily vary with our knowledge of astronomical science.   It has accordingly happened, that one set of tables after another has been superseded with each advance of astronomical science.   Some striking examples of this may not be uninstructive.   In 1765, the Board of Longitude paid to the celebrated Euler the sum of 300*l*., for furnishing general formulæ for the computation of lunar tables.   Professor Mayer was employed to calculate the tables upon these formulæ, and the sum of 3000*l*. was voted for them by the British Parliament, to his widow, after his decease.   These tables had been used for ten years, from 1766 to 1776, in computing the Nautical Almanac, when they were superseded by new and improved tables, composed by Mr. Charles Mason, under the direction of Dr. Maskelyne, from calculations made by order of the Board of Longitude, on the observations of Dr. Bradley.   A farther improvement was made by Mason in 1780; but a much more extensive improvement took place in the lunar calculations by the publication of the tables of the Moon, by M. Bürg, deduced from Laplace's theory, in 1806.   Perfect, however, as Bürg's tables were considered, at the time of their publication, they were, within the short period of six years, superseded by a more accurate set of tables published by Burckhardt in 1812; and these also have since been followed by the tables of Damoiseau.   Professor Schumacher has calculated by the latter tables his ephemeris of the Planetary Lunar Distances, and astronomers will hence be enabled to put to the strict test of observation the merits of the tables of Burckhardt and Damoiseau.*

The solar tables have undergone, from time to time, similar changes. The solar tables of Mayer were used in the computation of the Nautical Almanac, from its commencement in 1767, to 1804, inclusive. Within the six years immediately succeeding 1804, not less than three successive sets of solar tables appeared, each improving on the other; the first by Baron de Zach, the second by Delambre, under the direction of the French Board of Longitude, and the third by Carlini. The last, however, differ only in arrangement from those of Delambre.

Similar observations will be applicable to the tables of the principal planets. Bouvard published, in 1808, tables of Jupiter and Saturn; but from the improved state of astronomy, he found it necessary to recompute these tables in 1821.

Although it is now about thirty years since the discovery of the four new planets, Ceres, Pallas, Juno, and Vesta, it was not till recently that tables of their motions were published. They have lately appeared in Encke's Ephemeris.

We have thus attempted to convey some notion (though necessarily a very inadequate one) of the immense extent of numerical tables which it has been found necessary to calculate and print for the purposes of the arts and sciences. We have before us a catalogue of the tables contained in the library of one private individual, consisting of not less than one hundred and forty volumes. Among these there are no duplicate copies; and we observe that many of the most celebrated voluminous tabular works are not contained among them. They are confined exclusively to arithmetical and trigonometrical tables; and, consequently, the myriad of astronomical and nautical tables are totally excluded from them. Nevertheless, they contain an extent of printed surface covered with figures amounting to above sixteen thousand square feet. We have taken at random forty of these tables, and have found that the number of errors *acknowledged* in the respective errata, amounts to above *three thousand seven hundred*.

To be convinced of the necessity which has existed for accurate numerical tables, it will only be necessary to consider at what an immense expenditure of labour and of money even the imperfect ones which we possess have been produced.

To enable the reader to estimate the difficulties which attend the attainment even of a limited degree of accuracy, we shall now explain some of the expedients which have been from time to time resorted to

* A comparison of the results for 1834, will be found in the "Nautical Almanac" for 1835.

for the attainment of numerical correctness in calculating and printing them.

Among the scientific enterprises which the ambition of the French nation aspired to during the Republic, was the construction of a magnificent system of numerical tables. Their most distinguished mathematicians were called upon to contribute to the attainment of this important object; and the superintendence of the undertaking was confided to the celebrated Prony, who co-operated with the government in the adoption of such means as might be expected to ensure the production of a system of logarithmic and trigonometric tables, constructed with such accuracy that they should form a monument of calculation the most vast and imposing that had ever been executed, or even conceived. To accomplish this gigantic task, the principle of the division of labour, found to be so powerful in manufactures, was resorted to with singular success. The persons employed in the work were divided into three sections: the first consisted of half a dozen of the most eminent analysts. Their duty was to investigate the most convenient mathematical formulæ, which should enable the computors to proceed with the greatest expedition and accuracy by the method of Differences, of which we shall speak more fully hereafter. These formulæ, when decided upon by this first section, were handed over to the second section, which consisted of eight or ten properly qualified mathematicians. It was the duty of this second section to convert into numbers certain general or algebraical expressions which occurred in the formulæ, so as to prepare them for the hands of the computers. Thus prepared, these formulæ were handed over to the third section, who formed a body of nearly one hundred computers. The duty of this numerous section was to compute the numbers finally intended for the tables. Every possible precaution was, of course, taken to ensure the numerical accuracy of the results. Each number was calculated by two or more distinct and independent computers, and its truth and accuracy determined by the coincidence of the results thus obtained.

The body of tables thus calculated occupied in manuscript *seventeen* folio volumes.*

---

* These tables were never published. The printing of them was commenced by Didot, and a small portion was actually stereotyped, but never published. Soon after the commencement of the undertaking, the sudden fall of the assignats rendered it impossible for Didot to fulfil his contract with the government. The work was accordingly abandoned, and has never since been resumed. We have before us a copy of 100 pages folio of the portion which was printed at the time the work was stopped, given to a friend on a late occasion by Didot himself. It was remarked in this, as in other similar cases, that the computers who committed fewest errors were those who understood nothing beyond the process of addition.

As an example of the precautions which have been considered necessary to guard against errors in the calculation of numerical tables, we shall further state those which were adopted by Mr. Babbage, previously to the publication of his tables of logarithms. In order to render the terminal figure of tables in which one or more decimal places are omitted as accurate as it can be, it has been the practice to compute one or more of the succeeding figures; and if the first omitted figure be greater than 4, then the terminal figure is always increased by 1, since the value of the tabulated number is by such means brought nearer to the truth.* The tables of Callet, which were among the most accurate published logarithms, and which extended to seven places of decimals, were first carefully compared with the tables of Vega, which extended to ten places, in order to discover whether Callet had made the above correction of the final figure in every case where it was necessary. This previous precaution being taken, and the corrections which appeared to be necessary being made in a copy of Callet's tables, the proofs of Mr. Babbage's tables were submitted to the following test: They were first compared, number by number, with the corrected copy of Callet's logarithms; secondly with Hutton's logarithms; and thirdly, with Vega's logarithms. The corrections thus suggested being marked in the proofs, corrected revises were received back. These revises were then again compared, number by number, first with Vega's logarithms; secondly, with the logarithms of Callet; and thirdly, as far as the first 20,000 numbers, with the corresponding ones in Brigg's logarithms. They were now returned to the printer, and were stereotyped; proofs were taken from the stereotyped plates, which were put through the following ordeal: They were first compared once more with the logarithms of Vega as far as 47,500; they were then compared with the whole of the logarithms of Gardner; and next with the whole of Taylor's logarithms; and as a last test, they were transferred to the hands of a different set of readers, and were once more compared with Taylor. That these precautions were by no means superfluous may be collected from the following circumstances mentioned by Mr. Babbage: In the sheets read immediately previous to stereotyping, thirty-two errors were detected; after stereotyping, eight more were found, and corrected in the plates.

By such elaborate and expensive precautions many of the errors of

---

* Thus, suppose the number expressed at full length were 3.1415927. If the table extend to no more than four places of decimals, we should tabulate the number 3.1416 and not 3.1415. The former would be evidently nearer to the true number, 3.1415927.

computation and printing may certainly be removed; but it is too much to expect that in general such measures can be adopted; and we accordingly find by far the greater number of tables disfigured by errors, the extent of which is rather to be conjectured than determined. When the nature of a numerical table is considered—page after page densely covered with figures, and with nothing else—the chances against the detection of any single error will be easily comprehended; and it may therefore be fairly presumed, that for one error which may happen to be detected, there must be a great number which escape detection. Notwithstanding this difficulty, it is truly surprising how great a number of numerical errors have been detected by individuals no otherwise concerned in the tables than in their use. Mr. Baily states that he has himself detected in the solar and lunar tables, from which our Nautical Almanac was for a long period computed, more than five hundred errors. In the multiplication table already mentioned, computed by Dr. Hutton for the Board of Longitude, a single page was examined and recomputed: it was found to contain about forty errors.

In order to make the calculations upon the numbers found in the Ephemeral Tables published in the Nautical Almanac, it is necessary that the mariner should be supplied with certain permanent tables. A volume of these, to the number of about thirty, was accordingly computed, and published at national expense, by order of the Board of Longitude, entitled "Tables requisite to be used with the Nautical Ephemeris for finding the latitude and longitude at sea." In the first edition of these requisite tables, there were detected, by one individual, above a thousand errors.

The tables published by the Board of Longitude for the correction of the observed distances of the moon from certain fixed stars, are followed by a table of acknowledged errata, extending to seven folio pages, and containing more than eleven hundred errors. Even this table of errata itself is not correct: a considerable number of errors have been detected in it, so that errata upon errata have become necessary.

One of the tests most frequently resorted to for the detection of errors in numerical tables, has been the comparison of tables of the same kind, published by different authors. It has been generally considered that those numbers in which they are found to agree must be correct; inasmuch as the chances are supposed to be very considerable against two or more independent computers falling into precisely the same errors. How far this coincidence may be safely assumed as a test of accuracy we shall presently see.

A few years ago, it was found desirable to compute some very accurate logarithmic tables for the use of the great national survey of Ireland, which was then, and still is in progress; and on that occasion a careful comparison of various logarithmic tables was made.   Six remarkable errors were detected, which were found to be common to several apparently independent sets of tables.   This singular coincidence led to an unusually extensive examination of the logarithmic tables published both in England and in other countries; by which it appeared that thirteen sets of tables, published in London between the years 1633 and 1822, all agreed in these six errors.   Upon extending the inquiry to foreign tables, it appeared that two sets of tables published at Paris, one at Gouda, one at Avignon, one at Berlin, and one at Florence, were infected by exactly the same six errors.   The only tables which were found free from them were those of Vega, and the more recent impressions of Callet.   It happened that the Royal Society possessed a set of tables of logarithms printed in the Chinese character, and on Chinese paper, consisting of two volumes: these volumes contained no indication or acknowledgment of being copied from any other work.   They were examined; and the result was the detection in them of the same six errors.*

It is quite apparent that this remarkable coincidence of error must have arisen from the various tables being copied successively one from another.   The earliest work in which they appeared was Vlacq's Logarithms (folio, Gouda, 1628); and from it, doubtless, those which immediately succeeded it in point of time were copied; from which the errors were subsequently transcribed into all the other, including the Chinese logarithms.

The most certain and effectual check upon errors which arise in the process of computation, is to cause the same computations to be made by separate and independent computers; and this check is rendered still more decisive if they make their computations by different methods.   It is, nevertheless, a remarkable fact, that several computers, working separately and independently, do frequently commit precisely the same error; so that falsehood in this case assumes that character of consistency, which is regarded as the exclusive attribute of truth. Instances of this are familiar to most persons who have had the management of the computation of tables.   We have reason to know, that M. Prony experienced it on many occasions in the management of the great French tables, when he found three, and even a greater number of computers, working separately and independently, to return him the same numerical result, and *that result wrong*.   Mr. Stratford,

* Memoirs Ast. Soc., vol. iii. p. 65.

the conductor of the Nautical Almanac, to whose talents and zeal that work owes the execution of its recent improvements, has more than once observed a similar occurrence. But one of the most signal examples of this kind, of which we are aware, is related by Mr. Baily. The catalogue of stars published by the Astronomical Society was computed by two separate and independent persons, and was afterwards compared and examined with great care and attention by Mr. Stratford. On examining this catalogue, and recalculating a portion of it, Mr. Baily discovered an error in the case of the star, κ Cephei. Its right ascension was calculated *wrongly*, and yet *consistently*, by two computers working separately. Their numerical results agreed precisely in every figure; but Mr. Stratford, on examining the catalogue, failed to detect the error. Mr. Baily having reason, from some discordancy which he observed, to suspect an error, recomputed the place of the star with a view to discover it; and he himself, in the first instance, obtained precisely *the same erroneous numerical result*. It was only on going over the operation a second time that he *accidentally* discovered that all had inadvertently committed the same error.*

It appears, therefore, that the coincidence of different tables, even when it is certain that they could not have been copied one from another, but must have been computed independently, is not a decisive test of their correctness, neither is it possible to ensure accuracy by the device of separate and independent computation.

Besides the errors incidental to the process of computation, there are further liabilities in the process of *transcribing* the final results of each calculation into the fair copy of the table designed for the printer. The next source of error lies with the compositor, in transferring this copy into type. But the liabilities to error do not stop even here; for it frequently happens, that after the press has been fully corrected, errors will be produced in the process of printing. A remarkable instance of this occurs in one of the six errors detected in so many different tables already mentioned. In one of these cases, the last five figures of two successive numbers of a logarithmic table were the following:—

35875

10436.

Now, both of these are erroneous; the figure 8 in the first line should be 4, and the figure 4 in the second should be 8. It is evident that

_____

* Ibid., vol. iv. p. 290.

the types, as first composed, were correct; but in the course of printing the two types 4 and 8 being loose, adhered to the inking-balls, and were drawn out; the pressman in replacing them transposed them, putting the 8 *above* and the 4 *below*, instead of *vice versa*. It would be a curious inquiry, were it possible to obtain all the copies of the original edition of Vlacq's Logarithms, published at Gouda in 1628, from which this error appears to have been copied in all the subsequent tables, to ascertain whether it extends through the entire edition. It would probably, nay almost certainly, be discovered that some of the copies of that edition are correct in this number, while others are incorrect; the former having been worked off before the transposition of the types.

It is a circumstance worthy of notice, that this error in Vlacq's tables has produced a corresponding error in a variety of other tables deduced from them, *in which nevertheless the erroneous figures in Vlacq are omitted*. In no less than sixteen sets of tables published at various times since the publication of Vlacq, in which the logarithms extend only to seven places of figures, the error just mentioned in the *eighth place* in Vlacq causes a corresponding error in the *seventh* place. When the last three figures are omitted in the first of the above numbers, the seventh figure should be 5, inasmuch as the first of the omitted figures is under 5: the erroneous insertion, however, of the figure 8 in Vlacq has caused the figure 6 to be substituted for 5 in the various tables just alluded to. For the same reason, the erroneous occurrence of 4 in the second number has caused the adoption of a 0 instead of a 1 in the seventh place in the other tables. The only tables in which this error does not occur are those of Vega, the more recent editions of Callet, and the still later Logarithms of Mr. Babbage.

The *Opus Palatinum*, a work published in 1596, containing an extensive collection of trigonometrical tables, affords a remarkable instance of a tabular error; which, as it is not generally known, it may not be uninteresting to mention here. After that work had been for several years in circulation in every part of Europe, it was discovered that the commencement of the table of co-tangents and co-secants was vitiated by an error of considerable magnitude. In the first co-tangent the last nine places of figures were incorrect; but from the manner in which the numbers of the table were computed, the error was gradually, though slowly, diminished, until at length it became extinguished in the eighty-sixth page. After the detection of this extensive error, Pitiscus undertook the recomputation of the eighty-six erroneous pages. His corrected calculation was printed; and the erroneous part of the remaining copies of the *Opus Palatinum* was cancelled.   But as the corrected table of Pitiscus was not published until 1607,—thirteen years after the original work,—the erroneous part of the volume was cancelled in comparatively few copies, and consequently correct copies of the work are now exceedingly rare. Thus, in the collection of tables published by M. Schulze,* the whole of the erroneous part of the *Opus Palatinum* has been adopted; he having used the copy of that work which exists in the library of the Academy of Berlin, and which is one of those copies in which the incorrect part was not cancelled.   The corrected copies of this work may be very easily distinguished at present from the erroneous ones; it happened that the former were printed with a very bad and worn-out type, and upon a paper of a quality inferior to that of the original work. On comparing the first eighty-six pages of the volume with the succeeding ones, they are, therefore, immediately distinguishable in the corrected copies.   Besides this test, there is another, which it may not be uninteresting to point out:—At the bottom of page 7 in the corrected copies, there is an error in the position of the words *basis* and *hypothenusa*, their places being interchanged.   In the original uncorrected work this error does not exist.

At the time when the calculation and publication of Taylor's Logarithms were undertaken, it so happened that a similar work was in progress in France; and it was not until the calculation of the French work was completed, that its author was informed of the publication of the English work.   This circumstance caused the French calculator to relinquish the publication of his tables.   The manuscript subsequently passed into the library of Delambre, and, after his death, was purchased at the sale of his books, by Mr. Babbage, in whose possession it now is.   Some years ago it was thought advisable to compare these manuscript tables with Taylor's Logarithms, with a view to ascertain the errors in each, but especially in Taylor.   The two works were peculiarly well suited for the attainment of this end; as the circumstances under which they were produced rendered it quite certain that they were computed independently of each other. The comparison was conducted under the direction of the late Dr. Young, and the result was the detection of the following nineteen errors in Taylor's Logarithms.   To enable those who used Taylor's Logarithms to make the necessary corrections in them, the corrections of the detected errors appeared as follows in the "Nautical Almanac" for 1832.

---

* "Recueil des Tables Logarithmiques et Trigonométriques."   Par J. C. Schulze. 2 vols.   Berlin, 1778.

ERRATA, *detected in* TAYLOR's *Logarithms.   London: 4to,* 1792

				° ′ ″				
1	. E	.	. Co-tangent of .	1.35.55	.	. *for* 43671 *read* 42671
2	. M	.	. Co-tangent of .	4. 4.49	.	. — 66976 — 66979
3	.	.	. Sine of.	4.23.38	.	. — 43107 — 43007
4	.	.	. Sine of .	4.23.39	.	. — 43381 — 43281
5	. S.	.	. Sine of .	6.45.52	.	. — 10001 — 11001
6	. Kk	.	. Co-sine of	14.18. 3	.	. — 3398 — 3298
7	. Ss	.	. Tangent of	18. 1.56	.	. — 5064 — 6064
8	. Aaa	.	. Co-tangent of .	21.11.14	.	. — 6062 — 5962
9	. Ggg	.	. Tangent of	23.48.19	.	. — 6087 — 5987
10	.	.	. Co-tangent of .	23.48.19	.	. — 3913 — 4013
11	. Iii	.	. Sine of .	25. 5. 4	.	. — 3173 — 3183
12	.	.	. Sine of .	25. 5. 5	.	. — 3218 — 3228
13	.	.	. Sine of .	25. 5. 6	.	. — 3263 — 3273
14	.	.	. Sine of .	25. 5. 7	.	. — 3308 — 3318
15	.	.	. Sine of .	25. 5. 8	.	. — 3353 — 3363
16	.	.	. Sine of .	25. 5. 9	.	. — 3398 — 3408
17	. Qqq	.	. Tangent of	28.19.39	.	. — 6302 — 6402
18	. 4H	.	. Tangent of	35.55.51	.	. — 1681 — 1581
19	. 4K	.	. Co-sine of .	37.29. 2	.	. — 5503 — 5603

An error being detected in this list of ERRATA, we find, in the Nautical Almanac for the year 1833, the following ERRATUM of the ERRATA of Taylor's Logarithms:

'In the list of ERRATA detected in Taylor's Logarithms, for cos. 4° 18′ 3″ *read* cos. 14° 18′ 2″.'

Here, however, confusion is worse confounded; for a new error, not before existing, and of much greater magnitude, is introduced! It will be necessary, in the Nautical Almanac for 1836 (that for 1835 is already published), to introduce the following

ERRATUM of the ERRATUM of the ERRATA of TAYLOR's *Logarithms.* For cos. 4° 18′ 3″, *read* cos. 14° 18′ 3″.

If proof were wanted to establish incontrovertibly the utter impracticability of precluding numerical errors in works of this nature, we should find it in this succession of error upon error, produced, in spite of the universally acknowledged accuracy and assiduity of the persons at present employed in the construction and management of the Nautical Almanac. It is only by the *mechanical fabrication of tables* that such errors can be rendered impossible.

On examining this list with attention, we have been particularly struck with the circumstances in which these errors appear to have originated. It is a remarkable fact, that of the above nineteen errors, eighteen have arisen from mistakes in *carrying*. Errors 5, 7, 10, 11, 12, 13, 14, 15, 16, 17, 19, have arisen from a carriage being neglected; and errors 1, 3, 4, 6, 8, 9, and 18, from a carriage being made where none should take place. In four cases, namely, errors 8, 9, 10, and 16, this has caused *two* figures to be wrong. The only error of the nineteen which appears to have been a press error is the second; which has evidently arisen from the type 9 being accidentally inverted, and thus becoming a 6. This may have originated with the compositor, but more probably it took place in the press-work; the type 9 being accidentally drawn out of the form by the inking-ball, as mentioned in a former case, and on being restored to its place, inverted by the pressman.

There are two cases among the above errata, in which an error, committed in the calculation of one number, has evidently been the cause of other errors. In the third erratum, a wrong carriage was made, in computing the sine of 4° 23′ 38″. The next number of the table was vitiated by this error; for we find the next erratum to be in the sine of 4° 23′ 39″, in which the figure similarly placed is 1 in excess. A still more extensive effect of this kind appears in errata 11, 12, 13, 14, 15, 16. A carriage was neglected in computing the sine of 25° 5′ 4″, and this produced a corresponding error in the five following numbers of the table, which are those corrected in the five following errata.

This frequency of errors arising in the process of carrying, would afford a curious subject of metaphysical speculation respecting the operation of the faculty of memory. In the arithmetical process, the memory is employed in a twofold way;—in ascertaining each successive figure of the calculated result by the recollection of a table committed to memory at an early period of life; and by another act of memory, in which the number *carried* from column to column is retained. It is a curious fact, that this latter circumstance, occurring only the moment before, and being in its nature little complex, is so much more liable to be forgotten or mistaken than the results of rather complicated tables. It appears, that among the above errata, the errors 5, 7, 10, 11, 17, 19, have been produced by the computer forgetting a carriage; while the errors 1, 3, 6, 8, 9, 18, have been produced by his making a carriage improperly. Thus, so far as the above list of errata affords grounds for judging, it would seem, (contrary to what might be expected,) that the error by which improper carriages are made is as frequent as that by which necessary carriages are over-looked.

We trust that we have succeeded in proving, first, the great national and universal utility of numerical tables, by showing the vast number of them, which have been calculated and published; secondly, that more effectual means are necessary to obtain such tables suitable to the present state of the arts, sciences and commerce, by showing that the existing supply of tables, vast as it certainly is, is still scanty, and utterly inadequate to the demands of the community;—that it is rendered inefficient, not only in quantity, but in quality, by its want of numerical correctness; and that such numerical correctness is altogether unattainable until some more perfect method be discovered, not only of calculating the numerical results, but of tabulating these,—of reducing such tables to type, and of printing that type so as to intercept the possibility of error during the press-work. Such are the ends which are proposed to be attained by the calculating machinery invented by Mr. Babbage.

The benefits to be derived from this invention cannot be more strongly expressed than they have been by Mr. Colebrooke, President of the Astronomical Society, on the occasion of presenting the gold medal voted by that body to Mr. Babbage:—

> In no department of science, or of the arts, does this discovery promise to be so eminently useful as in that of astronomy, and its kindred sciences, with the various arts dependent on them. In none are computations more operose than those which astronomy in particular requires;—in none are preparatory facilities more needful;—in none is error more detrimental. The practical astronomer is interrupted in his pursuit, and diverted from his task of observation by the irksome labours of computation, or his diligence in observing becomes ineffectual for want of yet greater industry of calculation. Let the aid which tables previously computed afford, be furnished to the utmost extent which mechanism has made attainable through Mr. Babbage's invention, and the most irksome portion of the astronomer's task is alleviated, and a fresh impulse is given to astronomical research.

The first step in the progress of this singular invention was the discovery of some common principle which pervaded numerical tables of every description; so that by the adoption of such a principle as the basis of the machinery, a corresponding degree of generality would be conferred upon its calculations. Among the properties of numerical functions, several of a general nature exist; and it was a matter of no

ordinary difficulty, and requiring no common skill, to select one which
might, in all respects, be preferable to the others.   Whether or not
that which was selected by Mr. Babbage affords the greatest practical
advantages, would be extremely difficult to decide—perhaps impossible, unless some other projector could be found possessed of
sufficient genius, and sustained by sufficient energy of mind and
character, to attempt the invention of calculating machinery on other
principles.   The principle selected by Mr. Babbage as the basis of
that part of the machinery which calculates, is the Method of Differences; and he has, in fact, literally thrown this mathematical principle
into wheel-work.   In order to form a notion of the nature of the
machinery, it will be necessary, first, to convey to the reader some idea
of the mathematical principle just alluded to.

A numerical table, of whatever kind, is a series of numbers which
possess some common character, and which proceed increasing or
decreasing according to some general law.   Supposing such a series
continually to increase, let us imagine each number in it to be subtracted from that which follows it, and the remainders thus successively
obtained to be ranged beside the first, so as to form another table:
these numbers are called the *first differences*.   If we suppose these
likewise to increase continually, we may obtain a third table from
them by a like process, subtracting each number from the succeeding
one: this series is called the *second differences*.   By adopting a like
method of proceeding, another series may be obtained, called the
third differences; and so on.   By continuing this process, we shall at
length obtain a series of differences, of some order, more or less high,
according to the nature of the original table, in which we shall find
the same number constantly repeated, to whatever extent the original
table may have been continued; so that if the next series of differences
had been obtained in the same manner as the preceding ones, every
term of it would be 0.   In some cases this would continue to whatever
extent the original table might be carried; but in all cases a series of
differences would be obtained, which would continue constant for a
very long succession of terms.

As the successive series of differences are derived from the original
table, and from each other, by *subtraction*, the succession of series may
be reproduced in the other direction by *addition*.   But let us suppose
that the first number of the original table, and of each of the series of
differences, including the last, be given: all the numbers of each of
the series may thence be obtained by the mere process of addition.
The second term of the original table will be obtained by adding to
the first the first term of the first difference series; in like manner, the

second term of the first difference series will be obtained by adding
to the first term, the first term of the third difference series, and so on.
The second terms of all the series being thus obtained, the third terms
may be obtained by a like process of addition; and so the series may
be continued. These observations will perhaps be rendered more
clearly intelligible when illustrated by a numerical example. The
following is the commencement of a series of the fourth powers of the
natural numbers:—

No.					*Table*
1	.	.	.	.	1
2	.	.	.	.	16
3	.	.	.	.	81
4	.	.	.	.	256
5	.	.	.	.	625
6	.	.	.	.	1296
7	.	.	.	.	2401
8	.	.	.	.	4096
9	.	.	.	.	6561
10	.	.	.	.	10,000
11	.	.	.	.	14,641
12	.	.	.	.	20,736
13	.	.	.	.	28,561

By subtracting each number from the succeeding one in this series, we
obtain the following series of first differences:—

                    15
                    65
                    175
                    369
                    671
                    1105
                    1695
                    2465
                    3439
                    4641
                    6095
                    7825

In like manner, subtracting each term of this series from the succeeding one, we obtain the following series of second differences:—

50
110
194
302
434
590
770
974
1202
1454
1730

Proceeding with this series in the same way, we obtain the following series of third differences:—

60
84
108
132
156
180
204
228
252
276

Proceeding in the same way with these, we obtain the following for the series of fourth differences:—

24
24
24
24
24
24
24
24
24

It appears, therefore, that in this case the series of fourth differences consists of a constant repetition of the number 24. Now, a slight consideration of the succession of arithmetical operations by which we have obtained this result, will show, that by reversing the process, we could obtain the table of fourth powers by the mere process of addition. Beginning with the first numbers in each successive series of differences, and designating the table and the successive differences by the letters T, D^1 D^2 D^3 D^4, we have then the following to begin with:—

T	D^1	D^2	D^3	D^4
1	15	50	60	24

Adding each number to the number on its left, and repeating 24, we get the following as the second terms of the several series:—

T	D^1	D^2	D^3	D^4
16	65	110	84	24

And, in the same manner, the third and succeeding terms as follows:—

No.	T	D^1	D^2	D^3	D^4
1	1	15	50	60	24
2	16	65	110	84	24
3	81	175	194	108	24
4	256	369	302	132	24
5	625	671	434	156	24
6	1296	1105	590	180	24
7	2401	1695	770	204	24
8	4096	2465	974	228	24
9	6561	3439	1202	252	24
10	10000	4641	1454	276	
11	14641	6095	1730		
12	20736	7825			
13	28561				

There are numerous tables in which, as already stated, to whatever order of differences we may proceed, we should not obtain a series of rigorously constant differences; but we should always obtain a certain number of differences which to a given number of decimal places would remain constant for a long succession of terms. It is plain that such a table might be calculated by addition in the same manner as those which have a difference rigorously and continuously constant; and if at every point where the last difference requires an increase, that increase be given to it, the same principle of addition may again be applied for a like succession of terms, and so on.

By this principle it appears, that all tables in which each series of differences continually increases, may be produced by the operation of addition alone; provided the first terms of the table, and of each series of differences, be given in the first instance. But it sometimes happens, that while the table continually increases, one or more series of differences may continually diminish. In this case, the series of differences are found by subtracting each term of the series, not from that which follows, but from that which precedes it; and consequently, in the re-production of the several series, when their first terms are given, it will be necessary in some cases to obtain them by *addition*, and in others by *subtraction*. It is possible, however, still to perform all the operations by addition alone; this is effected in performing the operation of subtraction, by substituting for the subtrahend its *arithmetical complement*, and adding that, omitting the unit of the highest order in the result. This process, and its principle, will be readily comprehended by an example. Let it be required to subtract 357 from 768.

The common process would be as follows:—

From	768
Subtract	357
Remainder . . .	411

The *arithmetical complement* of 357, or the number by which it falls short of 1000, is 643. Now, if this number be added to 768, and the first figure on the left be struck out of the sum, the process will be as follows:—

To	768
Add	643
Sum	1411
Remainder sought .	411

The principle on which this process is founded is easily explained. In the latter process we have first added 643, and then subtracted 1000. On the whole, therefore, we have subtracted 357, since the number actually subtracted exceeds the number previously added by that amount.

Since, therefore, subtraction may be effected in this manner by addition, it follows that the calculation of all series, so far as an order of differences can be found in them which continues constant, may be conducted by the process of addition alone.

It also appears, from what has been stated, that each addition consists only of two operations. However numerous the figures may be of which the several pairs of numbers to be thus added may consist, it is obvious that the operation of adding them can only consist of repetitions of the process of adding one digit to another; and of carrying one from the column of inferior units to the column of units next superior when necessary. If we would therefore reduce such a process to machinery, it would only be necessary to discover such a combination of moving parts as are capable of performing these two processes of *adding* and *carrying* on two single figures; for, this being once accomplished, the process of adding two numbers, consisting of any number of digits, will be effected by repeating the same mechanism as often as there are pairs of digits to be added. Such was the simple form to which Mr. Babbage reduced the problem of discovering the calculating machinery; and we shall now proceed to convey some notion of the manner in which he solved it.

For the sake of illustration, we shall suppose that the table to be calculated shall consist of numbers not exceeding six places of figures; and we shall also suppose that the difference of the fifth order is the constant difference. Imagine, then, six rows of wheels, each wheel carrying upon it a dial-plate like that of a common clock, but consisting of *ten* instead of *twelve* divisions; the several divisions being marked 1, 2, 3, 4, 5, 6, 7, 8, 9, 0. Let these dials be supposed to revolve whenever the wheels to which they are attached are put in motion, and to turn in such a direction that the series of increasing numbers shall pass under the index which appears over each dial:— thus, after 0 passes the index, 1 follows, then 2, 3, and so on, as the dial revolves. In fig. 1 are represented six horizontal rows of such dials.*

The method of differences, as already explained, requires, that in proceeding with the calculation, this apparatus should perform continually the addition of the number expressed upon each row of dials, to the number expressed upon the row immediately above it. Now, we shall first explain how this process of addition may be conceived to be performed by the motion of the dials; and in doing so, we shall consider separately the processes of addition and carriage, considering the addition first, and then the carriage.

Let us first suppose the line D^1 to be added to the line T. To accomplish this, let us imagine that while the dials on the line D^1 are quiescent, the dials on the line T are put in motion, in such a manner, that as many divisions on each dial shall pass under its

* [The figure referred to is not reproduced here.]

index, as there are units in the number at the index immediately below it. It is evident that this condition supposes that if 0 be at any index on the line D^1, the dial immediately above it in the line T shall not move. Now the motion here supposed, would bring under the indices on the line T such a number as would be produced by adding the number D^1 to T, neglecting all the carriages; for a carriage should have taken place in every case in which the figure 9 of every dial in the line T had passed under the index during the adding motion. To accomplish this carriage, it would be necessary that the dial immediately on the left of any dial in which 9 passes under the index, should be advanced one division, independently of those divisions which it may have been advanced by the addition of the number immediately below it. This effect may be conceived to take place in either of two ways. It may be either produced at the moment when the division between 9 and 0 of any dial passes under the index; in which case the process of carrying would go on simultaneously with the process of adding; or the process of carrying may be postponed in every instance until the process of addition, without carrying, has been completed; and then by another distinct and independent motion of the machinery, a carriage may be made by advancing one division all those dials on the right of which a dial had, during the previous addition, passed from 9 to 0 under the index. The latter is the method adopted in the calculating machinery, in order to enable its inventor to construct the carrying machinery independent of the adding mechanism.

Having explained the motion of the dials by which the addition, excluding the carriages of the number on the row D^1, may be made to the number on the row T, the same explanation may be applied to the number on the row D^2 to the number on the row D^1; also, of the number D^3 to the number on the row D^4, and so on. Now it is possible to suppose the additions of all the rows, except the first, to be made to all the rows except the last, simultaneously; and after these additions have been made, to conceive all the requisite carriages to be also made by advancing the proper dials one division forward. This would suppose all the dials in the scheme to receive their adding motion together; and, this being accomplished, the requisite dials to receive their carrying motions together. The production of so great a number of simultaneous motions throughout any machinery, would be attended with great mechanical difficulties, if indeed it be practicable. In the calculating machinery it is not attempted. The additions are performed in two successive periods of time, and the carriages in two other periods of time, in the following manner. We shall suppose one complete revolution of the axis which moves the machinery, to make one complete set of additions and carriages; it will then make them in the following order:—

The first quarter of a turn of the axis will add the second, fourth, and sixth rows to the first, third, and fifth, omitting the carriages; this it will do by causing the dials on the first, third, and fifth rows, to turn through as many divisions as are expressed by the numbers at the indices below them, as already mentioned.

The second quarter of a turn will cause the carriages consequent on the previous addition, to be made by moving forward the proper dials one division.

(During these two quarters of a turn, the dials of the first, third, and fifth row alone have been moved; those of the second, fourth, and sixth, have been quiescent.)

The third quarter of a turn will produce the addition of the third and fifth rows to the second and fourth, omitting the carriages; which it will do by causing the dials of the second and fourth rows to turn through as many divisions as are expressed by the numbers at the indices immediately below them.

The fourth and last quarter of a turn will cause the carriages consequent on the previous addition, to be made by moving the proper dials forward one division.

This evidently completes one calculation, since all the rows except the first have been respectively added to all the rows except the last.

To illustrate this: let us suppose the table to be computed to be that of the fifth powers of the natural numbers, and the computation to have already proceeded so far as the fifth power of 6, which is 7776. This number appears, accordingly, in the highest row, being the place appropriated to the number of the table to be calculated. The several differences as far as the fifth, which is in this case constant, are exhibited on the successive rows of dials in such a manner as to be adapted to the process of addition by alternate rows, in the manner already explained. The process of addition will commence by the motion of the dials in the first, third, and fifth rows, in the following manner: The dial A, fig. 1, must turn through one division, which will bring the number 7 to the index; the dial B must turn through three divisions, which will bring 0 to the index; this will render a carriage necessary, but that carriage will not take place during the present motion of the dial. The dial C will remain unmoved, since 0 is at the index below it; the dial D must turn through nine divisions; and as, in doing so, the division between 9 and 0 must pass under the index, a carriage must subsequently take place upon the dial to the left; the remaining dials of the row T, fig. 1, will remain unmoved. In the row D^2 the dial A^2 will remain unmoved, since 0 is at the index below it; the dial B^2 will be moved through five divisions, and will render a subsequent carriage on the dial to the left necessary; the dial C^2 will be moved through five divisions; the dial D^2 will be moved through three divisions, and the remaining dials of this row will remain unmoved.   The dials of the row D^4 will be moved according to the same rules; and the whole scheme will undergo a change exhibited in fig. 2; a mark (*) being introduced on those dials to which a carriage is rendered necessary by the addition which has just taken place.

The second quarter of a turn of the moving axis will move forward through one division all the dials which in fig. 2† are marked (*), and the scheme will be converted into the scheme expressed in fig. 3.

In the third quarter of a turn, the dial A^1, fig. 3,† will remain unmoved, since 0 is at the index below it; the dial B^1 will be moved forward through three divisions; C^1 through nine divisions, and so on; and in like manner the dials of the row D^3 will be moved forward through the number of divisions expressed at the indices in the row D^4.   This change will convert the arrangement into that expressed in fig. 4,† the dials to which a carriage is due, being distinguished as before by (*).

The fourth quarter of a turn of the axis will move forward one division all the dials marked (*); and the arrangement will finally assume the form exhibited in fig. 5,† in which the calculation is completed.   The first row T in this expresses the fifth power of 7; and the second expresses the number which must be added to the first row, in order to produce the fifth power of 8; the numbers in each row being prepared for the change which they must undergo, in order to enable them to continue the computation according to the method of alternate addition here adopted.

Having thus explained what it is that the mechanism is required to do, we shall now attempt to convey at least a general notion of some of the mechanical contrivances by which the desired ends are attained. To simplify the explanation, let us first take one particular instance— the dials B and B^1, fig. 1, for example.   Behind the dial B^1 is a bolt, which, at the commencement of the process, is shot between the teeth of a wheel which drives the dial B: during the first quarter of a turn this bolt is made to revolve, and if it continued to be engaged in the teeth of the said wheel, it would cause the dial B to make a complete revolution; but it is necessary that the dial B should only move through three divisions, and, therefore, when three divisions of this dial have

† [The figure referred to is not reproduced here, but like Fig. 1, described on p. 189, represents six horizontal rows of dials in various appropriate positions.]

passed under its index, the aforesaid bolt must be withdrawn: this is accomplished by a small wedge, which is placed in a fixed position on the wheel behind the dial B^1, and that position is such that this wedge will press upon the bolt in such a manner, that at the moment when three divisions of the dial B have passed under the index, it shall withdraw the bolt from the teeth of the wheel which it drives. The bolt will continue to revolve during the remainder of the first quarter of a turn of the axis, but it will no longer drive the dial B, which will remain quiescent. Had the figure at the index of the dial B^1 been any other, the wedge which withdraws the bolt would have assumed a different position, and would have withdrawn the bolt at a different time, but at a time always corresponding with the number under the index of the dial B^1: thus, if 5 had been under the index of the dial B^1, then the bolt would have been withdrawn from between the teeth of the wheel which it drives, when five divisions of the dial B had passed under the index, and so on. Behind each dial in the row D^1 there is a similar bolt and a similar withdrawing wedge, and the action upon the dial above is transmitted and suspended in precisely the same manner. Like observations will be applicable to all the dials in the scheme here referred to, in reference to their adding actions upon those above them.

There is, however, a particular case which here merits notice: it is the case in which 0 is under the index of the dial from which the addition is to be transmitted upwards. As in that case nothing is to be added, a mechanical provision should be made to prevent the bolt from engaging in the teeth of the wheel which acts upon the dial above: the wedge which causes the bolt to be withdrawn, is thrown into such a position as to render it impossible that the bolt should be shot, or that it should enter between the teeth of the wheel, which in other cases it drives. But inasmuch as the usual means of shooting the bolt would still act, a strain would necessarily take place in the parts of the mechanism, owing to the bolt not yielding to the usual impulse. A small shoulder is therefore provided, which puts aside, in this case, the piece by which the bolt is usually struck, and allows the striking implement to pass without encountering the head of the bolt or any other obstruction. This mechanism is brought into play in the scheme, fig. 1, in the cases of all those dials in which 0 is under the index.

Such is the general description of the nature of the mechanism by which the adding process, apart from the carriages, is effected. During the first quarter of a turn, the bolts which drive the dials in the first, third, and fifth rows, are caused to revolve, and to act upon these dials, so long as they are permitted by the position of the several wedges on the second, fourth, and sixth rows of dials, by which these bolts are respectively withdrawn: and, during the third quarter of a turn, the bolts which drive the dials of the second and fourth rows are made to revolve and act upon these dials so long as the wedges on the dials of the third and fifth rows, which withdraw them, permit. It will hence be perceived, that, during the first and third quarters of a turn, the process of addition is continually passing upwards through the machinery; alternately from the even to the odd rows, and from the odd to the even rows, counting downwards.

We shall now attempt to convey some notion of the mechanism by which the process of carrying is effected during the second and fourth quarters of a turn of the axis. As before, we shall first explain it in reference to a particular instance. During the first quarter of a turn the wheel B^2, fig. 1, is caused by the adding bolt to move through five divisions; and the fifth of these divisions, which passes under the index, is that between 9 and 0. On the axis of the wheel C^2, immediately to the left of B^2, is fixed a wheel, called in mechanics a ratchet wheel, which is driven by a claw which constantly rests in its teeth. This claw is in such a position as to permit the wheel C^2 to move in obedience to the action of the adding bolt, but to resist its motion in the contrary direction. It is drawn back by a spiral spring, but its recoil is prevented by a hook which sustains it; which hook, however, is capable of being withdrawn, and when withdrawn, the aforesaid spiral spring would draw back the claw, and make it fall through one tooth of the ratchet wheel. Now, at the moment that the division between 9 and 0 on the dial B^2 passes under the index, a thumb placed on the axis of this dial touches a trigger which raises out of the notch the hook which sustains the claw just mentioned, and allows it to fall back by the recoil of the spring, and to drop into the next tooth of the ratchet wheel. This process, however, produces no immediate effect upon the position of the wheel C^2, and is merely preparatory to an action intended to take place during the second quarter of a turn of the moving axis. It is in effect a memorandum taken by the machine of a carriage to be made in the next quarter of a turn.

During the second quarter of a turn, a finger placed on the axis of the dial B^2 is made to revolve, and it encounters the heel of the above-mentioned claw. As it moves forward it drives the claw before it; and this claw, resting in the teeth of the ratchet wheel fixed upon the axis of the dial C^2, drives forward that wheel, and with it the dial. But the length and position of the finger which drives the claw limits its action, so as to move the claw forward through such a space only as will cause the dial C^2 to advance through a single division; at which point it is again caught and retained by the hook. This will be added to the number under its index, and the requisite carriage from B² to C² will be accomplished.

In connection with every dial is placed a similar ratchet wheel with a similar claw, drawn by a similar spring, sustained by a similar hook, and acted upon by a similar thumb and trigger; and therefore the necessary carriages, throughout the whole machinery, take place in the same manner and by similar means.

During the second quarter of a turn, such of the carrying claws as have been allowed to recoil in the first, third, and fifth rows, are drawn up by the fingers on the axes of the adjacent dials; and, during the fourth quarter of a turn, such of the carrying claws on the second and fourth rows as have been allowed to recoil during the third quarter of a turn, are in like manner drawn up by the carrying fingers on the axes of the adjacent dials. It appears that the carriages proceed alternately from right to left along the horizontal rows during the second and fourth quarters of a turn; in the one, they pass along the first, third, and fifth rows, and in the other, along the second and fourth.

There are two systems of waves of mechanical action continually flowing from the bottom to the top; and two streams of similar action constantly passing from the right to the left. The crests of the first system of adding waves fall upon the last difference, and upon every alternate one proceeding upwards; while the crests of the other system touch upon the intermediate differences. The first stream of carrying action passes from right to left along the highest row and every alternate row, while the second stream passes along the intermediate rows.

Such is a very rapid and general outline of this machinery. Its wonders, however, are still greater in its details than even in its broader features. Although we despair of doing it justice by any description which can be attempted here, yet we should not fulfil the duty we owe to our readers, if we did not call their attention at least to a few of the instances of consummate skill which are scattered with a prodigality characteristic of the highest order of inventive genius, throughout this astonishing mechanism.

In the general description which we have given of the mechanism for *carrying*, it will be observed, that the preparation for every carriage is stated to be made during the previous addition, by the disengagement of the carrying claw before mentioned, and by its consequent recoil, urged by the spiral spring with which it is connected; but it may, and does, frequently happen, that though the process of addition may not have rendered a carriage necessary, one carriage may itself produce the necessity for another. This is a contingency not provided against in the mechanism as we have described it: the case would occur in the scheme represented in fig. 1, if the figure under the index of C^2 were 4 instead of 3. The addition of the number 5 at the index of C^3 would, in this case, in the first quarter of a turn, bring 9 to the index of C^2; this would obviously render no carriage necessary, and of course no preparation would be made for one by the mechanism— that is to say, the carrying claw of the wheel D^2 would not be detached. Meanwhile a carriage upon C^2 has been rendered necessary by the addition made in the first quarter of a turn to B^2. This carriage takes place in the ordinary way, and would cause the dial C^2, in the second quarter of a turn, to advance from 9 to 0: this would make the necessary preparation for a carriage from C^2 to D^2. But unless some special arrangement were made for the purpose, that carriage would not take place during the second quarter of a turn. This peculiar contingency is provided against by an arrangement of singular mechanical beauty, and which, at the same time, answers another purpose—that of equalizing the resistance opposed to the moving power by the carrying mechanism. The fingers placed on the axes of the several dials in the row D^2, do not act at the same instant on the carrying claws adjacent to them; but they are so placed that their action may be distributed throughout the second quarter of a turn in regular succession. Thus the finger on the axis of the dial A^2 first encounters the claw upon B^2, and drives it through one tooth immediately forwards; the finger on the axis of B^2 encounters the claw upon C^2, and drives it through one tooth; the action of the finger on C^2 on the claw on D^2 next succeeds, and so on. Thus, while the finger on B^2 acts on C^2, and causes the division from 9 to 0 to pass under the index, the thumb on C^2 at the same instant acts on the trigger, and detaches the carrying claw on D^2, which is forthwith encountered by the carrying finger on C^2, and driven forward one tooth. The dial D^2 accordingly moves forward one division, and 5 is brought under the index. This arrangement is beautifully effected by placing the several fingers, which act upon the carrying claws, *spirally* on their axes, so that they come into action in regular succession.

We have stated that, at the commencement of each revolution of the moving axis, the bolts which drive the dials of the first, third, and fifth rows, are shot. The process of shooting these bolts must therefore have taken place during the last quarter of the preceding revolution; but it is during that quarter of a turn that the carriages are effected in the second and fourth rows. Since the bolts which drive the dials at the first, third, and fifth rows, have no mechanical connexion with the dials in the second and fourth rows, there is nothing in the process of shooting those bolts incompatible with that of moving the dials of the second and fourth rows: hence these two processes may both take place during the same quarter of a turn. But in order to equalize the resistance to the moving power, the same expedient is here adopted as that already described in the process of carrying. The arms which shoot the bolts of each row of dials are arranged *spirally*, so as to act successively throughout the quarter of a turn. There is, however, a contingency which, under certain circumstances, would here produce a difficulty which must be provided against. It is possible, and in fact does sometimes happen, that the process of carrying causes a dial to move under the index from 0 to 1. In that case, the bolt, preparatory to the next addition, ought not to be shot until after the carriage takes place; for if the arm which shoots it passes its point of action before the carriage takes place, the bolt will be moved out of its sphere of action, and will not be shot, which, as we have already explained, must always happen when 0 is at the index: therefore no addition would in this case take place during the next quarter of a turn of the axis; whereas, since 1 is brought to the index by the carriage, which immediately succeeds the passage of the arm which ought to bolt, 1 should be added during the next quarter of a turn. It is plain, accordingly, that the mechanism should be so arranged, that the action of the arms, which shoot the bolts successively, should immediately follow the action of those fingers which raise the carrying claws successively; and therefore either a separate quarter of a turn should be appropriated to each of those movements, or if they be executed in the same quarter of a turn, the mechanism must be so constructed, that the arms which shoot the bolts successively, shall severally follow immediately after those which raise the carrying claws successively. The latter object is attained by a mechanical arrangement of singular felicity, and partaking of that elegance which characterises all the details of this mechanism. Both sets of arms are spirally arranged on their respective axes, so as to be carried through their period in the same quarter of a turn; but the one spiral is shifted a few degrees, in angular position, behind the other, so that each pair of corresponding arms succeed each other in the most regular order,—equalizing the resistance, economizing time, harmonizing the mechanism, and giving to the whole mechanical action the utmost practical perfection.

The system of mechanical contrivances by which the results, here attempted to be described, are attained, form only one order of expedients adopted in this machinery; although such is the perfection of their action, that in any ordinary case they would be regarded as having attained the ends in view with an almost superfluous degree of precision. Considering, however, the immense importance of the purposes which the mechanism was destined to fulfil, its inventor determined that a higher order of expedients should be superinduced upon those already described; the purpose of which should be to obliterate all small errors or inequalities which might, even by remote possibility, arise either from defects in the original formation of the mechanism, from inequality of wear, from casual strain or derangement,—or, in short, from any other cause whatever. Thus the movements of the first and principal parts of the mechanism were regarded by him merely as a first, though extremely nice approximation, upon which a system of small corrections was to be subsequently made by suitable and independent mechanism. This supplementary system of mechanism is so contrived, that if one or more of the moving parts of the mechanism of the first order be slightly out of their places, they will be forced to their exact position by the action of the mechanical expedients of the second order to which we now allude. If a more considerable derangement were produced by any accidental disturbance, the consequence would be that the supplementary mechanism would cause the whole system to become locked, so that not a wheel would be capable of moving; the impelling power would necessarily lose all its energy, and the machine would stop. The consequence of this exquisite arrangement is, that the machine will either calculate rightly, or not at all.

The supernumerary contrivances which we now allude to, being in a great degree unconnected with each other, and scattered through the machinery to a certain extent, independent of the mechanical arrangement of the principal parts, we find it difficult to convey any distinct notion of their nature or form.

In some instances they consist of a roller resting between certain curved surfaces, which has but one position of stable equilibrium, and that position the same, however the roller or the curved surfaces may wear. A slight error in the motion of the principal parts would make this roller for the moment rest on one of the curves; but, being constantly urged by a spring, it would press on the curved surface in such a manner as to force the moving piece on which that curved surface is formed, into such a position that the roller may rest between the two surfaces; that position being the one which the mechanism should have. A greater derangement would bring the roller to the crest of the curve, on which it would rest in instable equilibrium; and the machine would either become locked, or the roller would throw it as before into its true position.

In other instances a similar object is attained by a solid cone being pressed into a conical seat; the position of the axis of the cone and that of its seat being necessarily invariable, however the cone may wear; and the action of the cone upon the seat being such, that it cannot rest in any position except that in which the axis of the cone coincides with the axis of its seat.

Having thus attempted to convey a notion, however inadequate, of the calculating section of the machinery, we shall proceed to offer some explanation of the means whereby it is enabled to print its calculations in such a manner as to preclude the possibility of error in any individual printed copy.

On the axle of each of the wheels which express the calculated number of the table T, there is fixed a solid piece of metal, formed into a curve, not unlike the wheel in a common clock, which is called the *snail*. This curved surface acts against the arm of a lever, so as to raise that arm to a higher or lower point according to the position of the dial with which the snail is connected. Without entering into a more minute description, it will be easily understood that the snail may be so formed that the arm of the lever shall be raised to ten different elevations, corresponding to the ten figures of the dial which may be brought under the index. The opposite arm of the lever here described puts in motion a solid arch, or sector, which carries ten punches: each punch bearing on its face a raised character of a figure, and the ten punches bearing the ten characters, 1, 2, 3, 4, 5, 6, 7, 8, 9, 0. It will be apparent from what has been just stated, that this *type sector* (as it is called) will receive ten different attitudes, corresponding to the ten figures which may successively be brought under the index of the dial-plate. At a point over which the type sector is thus moved, and immediately under a point through which it plays, is placed a frame, in which is fixed a plate of copper. Immediately over a certain point through which the type sector moves, is likewise placed a *bent lever*, which, being straightened, is forcibly pressed upon the punch which has been brought under it. If the type sector be moved, so as to bring under the bent lever one of the steel punches above mentioned, and be held in that position for a certain time, the bent lever, being straightened, acts upon the steel punch, and drives it against the face of the copper beneath, and thus causes a sunken impression of the character upon the punch to be left upon the copper. If the copper be now shifted slightly in its position, and the type sector be also shifted so as to bring another punch under the bent lever, another character may be engraved on the copper by straightening the bent lever, and pressing it on the punch as before. It will be evident, that if the copper were shifted from right to left through a space equal to two figures of a number, and, at the same time, the type sector so shifted as to bring the punches corresponding to the figures of the number successively under the bent lever, an engraved impression of the number might thus be obtained upon the copper by the continued action of the bent lever.   If, when one line of figures is thus obtained, a provision be made to shift the copper in a direction at right angles to its former motion, through a space equal to the distance between two lines of figures, and at the same time to shift it through a space in the other direction equal to the length of an entire line, it will be evident that another line of figures might be printed below the first in the same manner.

The motion of the type sector, here described, is accomplished by the action of the snail upon the lever already mentioned.   In the case where the number calculated is that expressed in fig. 1, the process would be as follows:—The snail of the wheel F^1, acting upon the lever, would throw the type sector into such an attitude, that the punch bearing the character 0 would come under the bent lever.   The next turn of the moving axis would cause the bent lever to press on the tail of the punch, and the character 0 would be impressed upon the copper. The bent lever being again drawn up, the punch would recoil from the copper by the action of a spring; the next turn of the moving axis would shift the copper through the interval between two figures, so as to bring the point destined to be impressed with the next figure under the bent lever.   At the same time, the snail of the wheel E would cause the type sector to be thrown into the same attitude as before, and the punch 0 would be brought under the bent lever; the next turn would impress the figure 0 beside the former one, as before described.   The snail upon the wheel D would now come into action, and throw the type sector into that position in which the punch bearing the character 7 would come under the bent lever, and at the same time the copper would be shifted through the interval between two figures; the straightening of the lever would next follow, and the character 7 would be engraved.   In the same manner, the wheels C, B, and A would successively act by means of their snails; and the copper being shifted, and the lever allowed to act, the number 007776 would be finally engraved upon the copper: this being accomplished, the calculating machinery would next be called into action, and another circulation would be made, producing the next number of the Table exhibited in fig. 5.   During this process the machinery would be engaged in shifting the copper both in the direction of its length and its breadth, with a view to commence the printing of another line; and this change of position would be accomplished at the moment when the next calculation would be completed: the printing of the next number would go on like the former, and the operation of the machine would proceed in the same manner, calculating and printing alternately. It is not, however, at all necessary—though we have here supposed it, for the sake of simplifying the explanation—that the calculating part of the mechanism should have its action suspended while the printing part is in operation, or *vice versa*: it is not intended, in fact, to be so suspended in the actual machinery. The same turn of the axis by which one number is printed, executes a part of the movements necessary for the succeeding calculation; so that the whole mechanism will be simultaneously and continuously in action.

Of the mechanism by which the position of the copper is shifted from figure to figure, from line to line, we shall not attempt any description. We feel that it would be quite vain. Complicated and difficult to describe as every other part of this machinery is, the mechanism for moving the copper is such as it would be quite impossible to render at all intelligible, without numerous illustrative drawings.

The engraved plate of copper obtained in the manner above described, is designed to be used as a mould from which a stereotyped plate may be cast; or, if deemed advisable, it may be used as the immediate means of printing. In the one case we should produce a table, printed from type, in the same manner as common letter-press printing; in the other an engraved table. If it be thought most advisable to print from the stereotyped plates, then as many stereotyped plates as may be required may be taken from the copper mould; so that when once a table has been calculated and engraved by the machinery, the whole world may be supplied with stereotyped plates to print it, and may continue to be so supplied for an unlimited period of time. There is no practical limit to the number of stereotyped plates which may be taken from the engraved copper; and there is scarcely any limit to the number of printed copies which may be taken from any single stereotyped plate. Not only, therefore, is the numerical table by these means engraved and stereotyped with infallible accuracy, but such stereotyped plates are producible in unbounded quantity. Each plate, when produced, becomes itself the means of producing printed copies of the table, in accuracy perfect, and in number without limit.

Unlike all other machinery, the calculating mechanism produces, not the object of consumption, but the machinery by which that object may be made. To say that it computes and prints with infallible accuracy, is to understate its merits:—it computes and fabricates *the means* of printing with absolute correctness and an unlimited abundance.

For the sake of clearness, and to render ourselves more easily intelligible to the general reader, we have in the preceding explanation thrown the mechanism into an arrangement somewhat different from that which is really adopted. The dials expressing the numbers of the tables of the successive differences are not placed, as we have supposed them, in horizontal rows, and read from right to left, in the ordinary way; they are, on the contrary, placed vertically, one below the other, and read from top to bottom. The number of the table occupies the first vertical column on the right, the units being expressed on the lowest dial, and the tens on the next above that, and so on. The first difference occupies the next vertical column on the left; and the numbers of the succeeding differences occupy vertical columns, proceeding regularly to the left; the constant difference being on the last vertical column. It is intended in the machine now in progress to introduce six orders of differences, so that there will be seven columns of dials; it is also intended that the calculations shall extend to eighteen places of figures: thus each column will have eighteen dials. We have referred to the dials as if they were inscribed upon the faces of wheels, whose axes are horizontal and planes vertical. In the actual machinery the axes are vertical and the planes horizontal, so that the edges of the *figure wheels*, as they are called, are presented to the eye. The figures are inscribed, not upon the dial-plate, but around the surface of a small cylinder or barrel, placed upon the axis of the figure wheel, which revolves with it; so that as the figure wheel revolves, the figures on the barrel are successively brought to the front, and pass under an index engraved upon a plate of metal immediately above the barrel. This arrangement has the obvious practical advantage, that, instead of each figure wheel having a separate axis, all the figure wheels of the same vertical column revolve on the same axis; and the same observation will apply to all the wheels with which the figure wheels are in mechanical connexion. This arrangement has the further mechanical advantage over that which has been assumed for the purposes of explanation, that the friction of the wheel-work on the axes is less in amount, and more uniformly distributed, than it could be if the axes were placed in the horizontal position.

A notion may therefore be formed of the front elevation of the calculating part of the mechanism, by conceiving seven steel axes erected, one beside another, on each of which shall be placed eighteen wheels,* five inches in diameter, having cylinders or barrels upon them

---

* The wheels, and every other part of the mechanism except the axes, springs, and such parts as are necessarily of steel, are formed of an alloy of copper with a small portion of tin.

an inch and a half in height, and inscribed, as already stated, with the ten arithmetical characters. The entire elevation of the machinery would occupy a space measuring ten feet broad, ten feet high, and five feet deep. The process of calculation would be observed by the alternate motion of the figure wheels on the several axes. During the first quarter of a turn, the wheels on the first, third, and fifth axes would turn, receiving their addition from the second, fourth, and sixth; during the second quarter of a turn, such of the wheels on the first, third, and fifth axes, to which carriages are due, would be moved forward one additional figure: the second, fourth, and sixth columns of wheels being all this time quiescent. During the third quarter of a turn, the second, fourth, and sixth columns would be observed to move, receiving their additions from the third, fifth, and seventh axes: and during the fourth quarter of a turn, such of these wheels to which carriages are due, would be observed to move forward one additional figure; the wheels of the first, third, and fifth columns being quiescent during this time.

It will be observed that the wheels of the seventh column are always quiescent in this process; and it may be asked, of what use they are, and whether some mechanism of a fixed nature would not serve the same purpose? It must, however, be remembered, that for different tables there will be different constant differences; and that when the calculation of a table is about to commence, the wheels on the seventh axis must be moved by the hand, so as to express the constant difference, whatever it may be. In tables, also, which have not a difference rigorously constant, it will be necessary, after a certain number of calculations, to change the constant difference by the hand; and in this case the wheels of the seventh axis must be moved when occasion requires. Such adjustment, however, will only be necessary at very distant intervals, and after a considerable extent of printing and calculation has taken place; and when it is necessary, a provision is made in the machinery by which notice will be given by the sounding of a bell, so that the machine may not run beyond the extent of its powers of calculations.

Immediately behind the seven axes on which the figure wheels revolve, are seven other axes; on which are placed, first, the wheels already described as driven by the figure wheels, and which bear upon them the wedge which withdraws the bolt immediately over these latter wheels, and on the same axis is placed the adding bolt. From the bottom of this bolt there projects downwards the pin, which acts upon the unbolting wedge by which the bolt is withdrawn: from the upper surface of the bolt proceeds a tooth, which, when the bolt is shot, enters between the teeth of the adding wheel, which turns on the same axis, and is placed immediately above the bolt: its teeth, on which the bolt acts, are like the teeth of a crown wheel, and are presented downwards. The bolt is fixed upon this axis, and turns with it; but the adding wheel above the bolt, and the unbolting wheel below it, both turn upon the axis, and independently of it. When the axis is made to revolve by the moving power, the bolt revolves with it; and so long as the tooth of the bolt remains inserted between those of the adding wheel, the latter is likewise moved; but when the lower pin of the bolt encounters the unbolting wedge on the lower wheel, the tooth of the bolt is withdrawn and the motion of the adding wheel is stopped. This adding wheel is furnished with spur teeth, besides the crown teeth just mentioned; and these spur teeth are engaged with those of that unbolting wheel which is in connexion with the adjacent figure wheel to which the addition is to be made. By such an arrangement it is evident that the revolution of the bolt will necessarily add to the adjacent figure wheel the requisite number.

It will be perceived, that upon the same axis are placed an unbolting wheel, a bolt, and an adding wheel, one above the other, for every figure wheel; and as there are eighteen figure wheels there will be eighteen tiers; each tier formed of an unbolting wheel, a bolt, and an adding wheel, placed one above the other; the wheels on this axis all revolving independent of the axis, but the bolts being all fixed upon it. The same observations, of course will apply to each of the seven axes.

At the commencement of every revolution of the adding axes, it is evident that the several bolts placed upon them must be shot in order to perform the various additions. This is accomplished by a third set of seven axes, placed at some distance behind the range of the wheels, which turn upon the adding axes: these are called *bolting axes*. On these bolting axes are fixed, so as to revolve with them, a bolting finger opposite to each bolt: as the bolting axis is made to revolve by the moving power, the bolting finger is turned, and as it passes near the bolt, it encounters the shoulder of a hammer or lever, which strikes the heel of the bolt, and presses it forward so as to shoot its tooth between the crown teeth of the adding wheel. The only exception to this action is the case in which 0 happens to be at the index of the figure wheel; in that case, the lever or hammer, which the bolting finger would encounter, is, as before stated, lifted out of the way of the bolting finger, so that it revolves without encountering it. It is on the bolting axes that the fingers are spirally arranged so as to equalize their action as already explained.

The same axes in front of the machinery on which the figure wheels turn are made to serve the purpose of *carrying*. Each of these bear a series of fingers which turn with them, and which encounter a carrying claw, already described, so as to make the carriage: these carrying fingers are also spirally arranged on their axes, as already described.

Although the absolute accuracy which appears to be ensured by the mechanical arrangements here described is such as to render further precautions nearly superfluous, still it may be right to state, that, supposing it were possible for an error to be produced in calculation, this error could be easily and speedily detected in the printed tables: it would only be necessary to calculate a number of the table taken at intervals, through which the mechanical action of the machine has not been suspended, and during which it has received no adjustment by the hand: if the computed number be found to agree with those printed, it may be taken for granted that all the intermediate numbers are correct; because, from the nature of the mechanism, and the principle of computation, an error occurring in any single number of the table would be unavoidably entailed, in an increasing ratio, upon all the succeeding numbers.

We have hitherto spoken merely of the practicability of executing by the machinery, when completed, that which its inventor originally contemplated—namely, the calculating and printing of all numerical tables, derived by the method of differences from a constant difference. It has, however happened that the actual powers of the machinery greatly transcend those contemplated in its original design:—they not only have exceeded the most sanguine anticipations of its inventor, but they appear to have an extent to which it is utterly impossible, even for the most acute mathematical thinker, to fix a probable limit. Certain subsidiary mechanical inventions have, in the progress of the enterprise, been, by the very nature of the machinery, suggested to the mind of the inventor, which confer upon it capabilities which he had never foreseen. It would be impossible even to enumerate, within the limits of this article, much less to describe in detail, those extraordinary mechanical arrangements, the effects of which have not failed to strike with astonishment every one who has been favoured with an opportunity of witnessing them, and who has been enabled, by sufficient mathematical attainments, in any degree to estimate their probable consequences.

As we have described the mechanism, the axes containing the several differences are successively and regularly added one to another; but there are certain mechanical adjustments, and these of a very simple nature, which being thrown into action, will cause a difference of any order to be added any number of times to a difference of any other order; and that either proceeding backwards or forwards, from a difference of an inferior to one of a superior order, and *vice versa.**

Among other peculiar mechanical provisions in the machinery is one by which, when the table for any order of difference amounts to a certain number, a certain arithmetical change would be made in the constant difference. In this way a series may be tabulated by the machine, in which the constant difference is subject to periodical change; or the very nature of the table itself may be subject to periodical change, and yet to one which has a regular law.

Some of these subsidiary powers are peculiarly applicable to calculations required in astronomy, and are therefore of eminent and immediate practical utility: others there are by which tables are produced, following the most extraordinary, and apparently capricious, but still regular laws. Thus a table will be computed, which, to any required extent, shall coincide with a given table, and which shall deviate from that table for a single term, or for any required number of terms, and then resume its course, or which shall permanently alter the law of its construction. Thus the engine has calculated a table which agreed precisely with a table of square numbers, until it attained the hundred and first term, which was not the square of 101, nor were any of the subsequent numbers squares. Again, it has computed a table which coincided with the series of natural numbers, as far as 100,000,001, but which subsequently followed another law. This result was obtained, not by working the engine through the whole of the first table, for that would have required an enormous length of time; but by showing, from the arrangement of the mechanism, that it must continue to exhibit the succession of natural numbers, until it would reach 100,000,000. To save time, the engine was set by the hand to the number 99999995, and was then put in regular operation. It produced successively the following numbers† :—

---

* The machine was constructed with the intention of tabulating the equation $\Delta^7 u_z = 0$, but, by the means above alluded to, it is capable of tabulating such equations as the following: $\Delta^7 u_z = a \, \Delta u_z$, $\Delta^7 u_z = a \, \Delta^3 u_z$, $\Delta^7 u =$ units figure of Δu.

† Such results as this suggest a train of reflection on the nature and operation of general laws, which would lead to very curious and interesting speculations. The natural philosopher and astronomer will be hardly less struck with them than the metaphysician and theologian.

99,999,996
99,999,997
99,999,998
99,999,999
100,000,000
100,010,002
100,030,003
100,060,004
100,100,005
100,160,006
&c., &c.

Equations have been already tabulated by the portion of the machinery which has been put together, which are so far beyond the reach of
the present power of mathematics, that no distant term of the table
can be predicted, nor any function discovered capable of expressing
its general law. Yet the very fact of the table being produced by
mechanism of an invariable form, and including a distinct principle
of mechanical action, renders it quite manifest that *some* general law
must exist in every table which it produces. But we must dismiss
these speculations: we feel it impossible to stretch the powers of our
own mind, so as to grasp the probable capabilities of this splendid
production of combined mechanical and mathematical genius; much
less can we hope to enable others to appreciate them, without being
furnished with such means of comprehending them as those with
which we have been favoured. Years must in fact elapse, and many
inquirers direct their energies to the cultivation of the vast field of
research thus opened, before we can fully estimate the extent of this
triumph of matter over mind. "Nor is it," says Mr. Colebrooke,
"among the least curious results of this ingenious device, that it affords
a new opening for discovery, since it is applicable, as has been shown
by its inventor, to surmount novel difficulties of analysis. Not
confined to constant differences, it is available in every case of differences that follow a definite law, reducible therefore to an equation.
An engine adjusted to the purpose being set to work will produce
any distant term, or succession of terms, required—thus presenting
the numerical solution of a problem, even though the analytical solution be yet undetermined." That the future path of some important
branches of mathematical inquiry must now in some measure be
directed by the dictates, of mechanism is sufficiently evident; for who
would toil on in any course of analytical inquiry, in which he must

ultimately depend on the expensive and fallible aid of human arithmetic, with an instrument in his hands, in which all the dull monotony of numerical computation is turned over to the untiring action and unerring certainty of mechanical agency.

It is worth notice, that each of the axes in front of the machinery on which the figure wheels revolve, is connected with a bell, the tongue of which is governed by a system of levers, moved by the several figure wheels; an adjustment is provided by which the levers shall be dismissed, so as to allow the hammer to strike against the bell, whenever any proposed number shall be exhibited on the axis. This contrivance enables the machine to give notice to its attendants at any time that an adjustment may be required.

Among a great variety of curious accidental properties (so to speak) which the machine is found to possess, is one by which it is capable of solving numerical equations which have rational roots. Such an equation being reduced (as it always may be) by suitable transformations to that state in which the roots shall be whole numbers, the values 0, 1, 2, 3, &c., are substituted for the unknown quantity, and the corresponding values of the equation ascertained. From these a sufficient number of differences being derived, they are set upon the machine. The machine being then put in motion, the table axis will exhibit the successive values of the formula, corresponding to the substitutions of the successive whole numbers for the unknown quantity: at length the number exhibited on the table axis will be 0, which will evidently correspond to a root of the equation. By previous adjustment, the bell of the table axis will in this case ring and give notice of the exhibition of the value of the root in another part of the machinery.

If the equation have imaginary roots, the formula being necessarily a maximum or minimum on the occurrence of such roots, the first difference will become nothing; and the dials of that axis will under such circumstances present 0 to the respective indices. By previous adjustment, the bell of this axis would here give notice of a pair of imaginary roots.

Mr. Colebrooke speculates on the probable extension of these powers of the machine:

> It may not therefore be deemed too sanguine an anticipation when I express the hope that an instrument which, in its simpler form, attains to the extraction of roots of numbers, and approximates to the roots of equations, may in a more advanced state of improvement, rise to the approximate solution of algebraic equations of elevated degrees. I refer to solutions of such equations proposed by La Grange, and more recently by other annalists, which involve

operations too tedious and intricate for use, and which must remain without efficacy, unless some mode be devised of abridging the labour, or facilitating the means of its performance. In any case this engine tends to lighten the excessive and accumulating burden of arithmetical application of mathematical formulæ, and to relieve the progress of science from what is justly termed by the author of this invention, the overwhelming encumbrance of numerical detail.

Although there are not more than eighteen figure wheels on each axis, and therefore it might be supposed that the machinery was capable of calculating only to the extent of eighteen decimal places; yet there are contrivances connected with it, by which, in two successive calculations, it will be possible to calculate even to the extent of thirty decimal places. Its powers, therefore, in this respect, greatly exceed any which can be required in practical science. It is also remarkable, that the machinery is capable of producing the calculated results *true to the last figure*. We have already explained, that when the figure which would follow the last is greater than 4, then it would be necessary to increase the last figure by 1; since the excess of the calculated number above the true value would in such case be less than its defect from it would be, had the regularly computed final figure been adopted: this is a precaution necessary in all numerical tables, and it is one which would hardly have been expected to be provided for in the calculating machinery.

As might be expected in a mechanical undertaking of such complexity and novelty, many practical difficulties have since its commencement been encountered and surmounted. It might have been foreseen, that many expedients would be adopted and carried into effect, which farther experiments would render it necessary to reject: and thus a large source of additional expense could scarcely fail to be produced. To a certain extent this has taken place; but owing to the admirable system of mechanical drawings, which in every instance Mr. Babbage has caused to be made, and owing to his own profound acquaintance with the practical working of the most complicated mechanism, he has been able to predict in every case what the result of any contrivance would be, as perfectly from the drawing, as if it had been reduced to the form of a working model. The drawings, consequently, form a most extensive and essential part of the enterprise. They are executed with extraordinary ability and precision, and may be considered as perhaps the best specimens of mechanical drawings which have ever been executed. It has been on these, and on these only, that the work of invention has been bestowed. In these, all those progressive modifications suggested by consideration and study have been made; and it was not until the inventor was fully satisfied with the result of any contrivance, that he had it reduced to a working form. The whole of the loss which has been incurred by the necessarily progressive course of invention, has been the expense of rejected drawings. Nothing can perhaps more forcibly illustrate the extent of labour and thought which has been incurred in the production of this machinery, than the contemplation of the working drawings which have been executed previously to its construction: these drawings cover above a thousand square feet of surface, and many of them are of the most elaborate and complicated description.

One of the practical difficulties which presented themselves at a very early stage in the progress of this undertaking, was the impossibility of bearing in mind all the variety of motions propagated simultaneously through so many complicated trains of mechanism. Nothing but the utmost imaginable harmony and order among such a number of movements, could prevent obstructions arising from incompatible motions encountering each other. It was very soon found impossible, by a mere act of memory, to guard against such an occurrence; and Mr. Babbage found, that, without some effective expedient by which he could at a glance see what every moving piece in the machinery was doing at each instant of time, such inconsistencies and obstructions as are here alluded to must continually have occurred. This difficulty was removed by another invention of even a more general nature than the calculating machinery itself, and pregnant with results probably of higher importance. This invention consisted in the contrivance of a scheme of *mechanical notation* which is generally applicable to all machinery whatsoever; and which is exhibited on a table or plan consisting of two distinct sections. In the first is traced, by a peculiar system of signs, the origin of every motion which takes place throughout the machinery; so that the mechanist or inventor is able, by moving his finger along a certain line, to follow out the motion of every piece from effect to cause, until he arrives at the prime mover. The same sign which thus indicates the *source* of motion indicates likewise the *species* of motion, whether it be continuous or reciprocating, circular or progressive, &c. The same system of signs further indicates the nature of the mechanical connexion between the mover and the thing moved, whether it be permanent and invariable (as between the two arms of a lever), or whether the mover and the moved are separate and independent pieces, as is the case when a pinion drives a wheel; also whether the motion of one piece necessarily implies the motion of another; or when such motion in the one is interrupted, and in the other continuous, &c.

The second section of the table divides the time of a complete period of the machinery into any required number of parts; and it exhibits in a map, as it were, that which every part of the machine is doing at each moment of time.   In this way, incompatibility in the motions of different parts is rendered perceptible at a glance.   By such means the contriver of machinery is not merely prevented from introducing into one part of the mechanism any movement inconsistent with the simultaneous action of the other parts; but when he finds that the introduction of any particular movement is necessary for his purpose, he can easily and rapidly examine the whole range of the machinery during one of its periods, and can find by inspection whether there is any, and what portion of time, at which no motion exists incompatible with the desired one, and thus discover a *niche*, as it were, in which to place the required movement.   A further and collateral advantage consists in placing it in the power of the contriver to exercise the utmost possible economy of *time* in the application of his moving power.   For example, without some instrument of mechanical inquiry equally powerful with that now described, it would be scarcely possible, at least in the first instance, so to arrange the various movements that they should be all executed in the least possible number of revolutions of the moving axis.   Additional revolutions would almost inevitably be made for the purpose of producing movements and changes which it would be possible to introduce in some of the phases of previous revolutions; and there is no one acquainted with the history of mechanical invention who must not be aware, that in the progressive contrivance of almost every machine the earliest arrangements are invariably defective in this respect; and that it is only by a succession of improvements, suggested by long experience, that that arrangement is at length arrived at, which accomplishes all the necessary motions in the shortest possible time.   By the application of the mechanical notation, however, absolute perfection may be arrived at in this respect; even before a single part of the machinery is constructed, and before it has any other existence than that which it obtains upon paper.

Examples of this class of advantages derivable from the notation will occur to the mind of every one acquainted with the history of mechanical invention.   In the common suction-pump, for example, the effective agency of the power is suspended during the descent of the piston.   A very simple contrivance, however, will transfer to the descent the work to be accomplished in the next ascent; so that the duty of four strokes of the piston may thus be executed in time of two. In the earlier applications of the steam-engine, that machine was applied almost exclusively to the process of pumping; and the power acted only during the descent of the piston, being suspended during its ascent. When, however, the notion of applying the engine to the general purposes of manufacture occurred to the mind of Watt, he saw that it would be necessary to cause it to produce a continued rotatory motion; and, therefore, that the intervals of intermission must be filled up by the action of the power. He first proposed to accomplish this by a second cylinder working alternately with the first; but it soon became apparent that the blank which existed during the upstroke in the action of the power, might be filled up by introducing the steam at both ends of the cylinder alternately. Had Watt placed before him a scheme of mechanical notation such as we allude to, this expedient would have been so obtruded upon him that he must have adopted it from the first.

One of the circumstances from which the mechanical notation derives a great portion of its power as an instrument of investigation and discovery, is that it enables the inventor to dismiss from his thoughts, and to disencumber his imagination of the arrangement and connexion of the mechanism; which, when it is very complex (and it is in that case that the notation is most useful), can only be kept before the mind by an embarrassing and painful effort. In this respect the powers of the notation may not inaptly be illustrated by the facilities derived in complex and difficult arithmetical questions from the use of the language and notation of algebra. When once the peculiar conditions of the question are translated into algebraical signs, and "reduced to an equation," the computist dismisses from his thoughts all the circumstances of the question, and is relieved from the consideration of the complicated relations of the quantities of various kinds which may have entered it. He deals with the algebraical symbols, which are the representatives of those quantities and relations, according to certain technical rules of a general nature, the truth of which he has previously established; and, by a process almost mechanical, he arrives at the required result. What algebra is to arithmetic, the notation we now allude to is to mechanism. The various parts of the machinery under consideration being once expressed upon paper by proper symbols, the inquirer dismisses altogether from his thoughts the mechanism itself, and attends only to the symbols; the management of which is so extremely simple and obvious, that the most unpractised person, having once acquired an acquaintance with the signs, cannot fail to comprehend their use.

A remarkable instance of the power and utility of this notation occurred in a certain stage of the invention of the calculating machinery. A question arose as to the best method of producing and arranging a certain series of motions necessary to print and calculate a number. The inventor, assisted by a practical engineer of considerable experience and skill, had so arranged these motions, that the whole might be performed by twelve revolutions of the principal moving axis. It seemed, however, desirable, if possible, to execute these motions by a less number of revolutions. To accomplish this, the engineer sat down to study the complicated details of a part of the machinery which had been put together; the inventor at the same time applied himself to the consideration of the arrangement and connection of the symbols in his scheme of notation. After a short time, by some transposition of symbols, he caused the received motions to be completed by eight turns of the axis. This he accomplished by transferring the symbols which occupied the last four divisions of his scheme, into such blank spaces as he could discover in the first eight divisions; due care being taken that no symbols should express actions at once simultaneous and incompatible. Pushing his inquiry, however, still further, he proceeded to ascertain whether his scheme of symbols did not admit of a still more compact arrangement, and whether eight revolutions were not more than enough to accomplish what was required. Here the powers of the practical engineer completely broke down. By no effort could he bring before his mind such a view of the complicated mechanism as would enable him to decide upon any improved arrangement. The inventor, however, without any extraordinary mental exertion, and merely by sliding a bit of ruled pasteboard up and down his plan, in search of a vacancy where the different motions might be placed, at length contrived to *pack* all the motions, which had previously occupied eight turns of the handle, into five turns. The symbolic instrument with which he conducted the investigation, now informed him of the impossibility of reducing the action of the machine to a more condensed form. This appeared by the fulness of every space along the lines of compatible action. It was, however, still possible, by going back to the actual machinery, to ascertain whether movements, which, under existing arrangements, were incompatible, might not be brought into harmony. This he accordingly did, and succeeded in diminishing the number of incompatible conditions, and thereby rendered it possible to make actions simultaneous which were before necessarily successive. The notation was now again called into requisition, and a new disposition of the parts was made. At this point of the investigation, this extraordinary instrument of mechanical analysis put forth one of its most singular exertions of power. It presented to the eye of the engineer two currents of mechanical action, which, from their nature, could not be simultaneous; and each of which occupied a complete revolution of the axis, except about a twentieth; the one occupying the last nineteen-twentieths of a complete revolution of the axis, and the other occupying the first nineteen-twentieths of a complete revolution.   One of these streams of action was, the successive picking up by the carrying fingers of the successive carrying claws; and the other was, the successive shooting of nineteen bolts by the nineteen bolting fingers.   The notation rendered it obvious, that as the bolting action commenced a small space below the commencement of the carrying, and ended an equal space below the termination of the carrying, the two streams of action could be made to flow after one another in one and the same revolution of the axis.   He thus succeeded in reducing the period of completing the action to four turns of the axis; when the notation again informed him that he had again attained a limit of condensed action, which could not be exceeded without a further change in the mechanism. To the mechanism he again recurred, and soon found that it was possible to introduce a change which would cause the action to be completed in three revolutions of the axis.   An odd number of revolutions, however, being attended with certain practical inconveniences, it was considered more advantageous to execute the motions in four turns; and here again the notation put forth its powers, by informing the inventor, *through the eye*, almost independent of his mind, what would be the most elegant, symmetrical, and harmonious disposition of the required motions in four turns.   This application of an almost metaphysical system of abstract signs, by which the motion of the hand performs the office of the mind, and of profound practical skill in mechanics alternately, to the construction of a most complicated engine, forcibly reminds us of a parallel in another science, where the chemist with difficulty succeeds in dissolving a refractory mineral, by the alternate action of the most powerful acids, and the most caustic alkalies, repeated in long-continued succession.

This important discovery was explained by Mr. Babbage, in a short paper read before the Royal Society, and published in the Philosophical Transactions in 1826.*   It is to us more a matter of regret than surprise, that the subject did not receive from scientific men in this country that attention to which its importance in every practical point of view so fully entitled it.   To appreciate it would indeed have been scarcely possible, from the very brief memoir which its inventor presented, unaccompanied by any observations or arguments of a nature to force it upon the attention of minds unprepared for it by the

* Phil. Trans. 1826, Part iii., p. 250, on a method of expressing by signs the action of machinery.

nature of their studies or occupations.   In this country, science has
been generally separated from practical mechanics by a wide chasm.
It will be easily admitted, that an assembly of eminent naturalists and
physicians, with a sprinkling of astronomers, and one or two abstract
mathematicians, were not precisely the persons best qualified to
appreciate such an instrument of mechanical investigation as we have
here described.   We shall not therefore be understood as intending
the slightest disrespect for these distinguished persons, when we express
our regret, that a discovery of such paramount practical value, in a
country pre-eminently conspicuous for the results of its machinery,
should fall still-born and inconsequential through their hands, and
be buried unhonoured and undiscriminated in their miscellaneous
transactions.   We trust that a more auspicious period is at hand; that
the chasm which has separated practical from scientific men will
speedily close; and that that combination of knowledge will be effected,
which can only be obtained when we see the men of science more
frequently extending their observant eye over the wonders of our
factories, and our great practical manufacturers, with a reciprocal
ambition, presenting themselves as active and useful members of our
scientific associations.   When this has taken place, an order of
scientific men will spring up, which will render impossible an oversight
so little creditable to the country as that which has been committed
respecting the mechanical notation.*   This notation has recently
undergone very considerable extension and improvement.   An
additional section has been introduced into it; designed to express
the process of circulation in machines, through which fluids, whether
liquid or gaseous, are moved.   Mr. Babbage, with the assistance of a
friend, who happened to be conversant with the structure and operation
of the steam-engine, has illustrated it with singular felicity and success
in its application to that machine.   An eminent French surgeon, on
seeing the scheme of notation thus applied, immediately suggested the
advantages which must attend it as an instrument for expressing the
structure, operation, and circulation of the animal system; and we
entertain no doubt of its adequacy for that purpose.   Not only the
mechanical connection of the solid members of the bodies of men and
animals, but likewise the structure and operation of the softer parts,
including the muscles, integuments, membranes, &c.; the nature,
motion, and circulation of the various fluids, their reciprocal effects,
the changes through which they pass, the deposits which they leave

---

* This discovery has been more justly appreciated by scientific men abroad.   It
was, almost immediately after its publication, adopted as the topic of lectures, in an
institution on the Continent for the instruction of Civil Engineers.

in various parts of the system; the functions of respiration, digestion, and assimilation—all would find appropriate symbols and representatives in the notation, even as it now stands, without those additions of which, however, it is easily susceptible. Indeed, when we reflect for what a very different purpose this scheme of symbols was contrived, we cannot refrain from expressing our wonder that it should seem, in all respects, as if it had been designed expressly for the purposes of anatomy and physiology.

Another of the uses which the slightest attention to the details of this notation irresistibly forces upon our notice, is to exhibit, in the form of a connected plan or map, the organization of an extensive factory, or any great public institution, in which a vast number of individuals are employed, and their duties regulated (as they generally are or ought to be) by a consistent and well-digested system. The mechanical notation is admirably adapted, not only to express such an organized connection of human agents, but even to suggest the improvements of which such organization is susceptible—to betray its weak and defective points, and to disclose, at a glance, the origin of any fault which may, from time to time, be observed in the working of the system. Our limits, however, preclude us from pursuing this interesting topic to the extent which its importance would justify. We shall be satisfied if the hints here thrown out should direct to the subject the attention of those who, being most interested in such an inquiry, are likely to prosecute it with greatest success.

One of the consequences which have arisen in the prosecution of the invention of the calculating machinery, has been the discovery of a multitude of mechanical contrivances, which have been elicited by the exigencies of the undertaking, and which are as novel in their nature as the purposes were novel which they were designed to attain. In some cases several different contrivances were devised for the attainment of the same end; and that among them which was best suited for the purpose was finally selected: the rejected expedients—those overflowings or waste of the invention—were not, however, always found useless. Like the *waste* in various manufactures, they were soon converted to purposes of utility. These rejected contrivances have found their way, in many cases, into the mills of our manufacturers; and we now find them busily effecting purposes, far different from any which their inventor dreamed of, in the spinning-frames of Manchester.*

* An eminent and wealthy retired manufacturer at Manchester assured us, that on the occasion of a visit to London, when he was favoured with a view of the calculating machinery, he found in it mechanical contrivances, which he subsequently introduced with the greatest advantage into his own spinning-machinery.

Another department of mechanical art, which has been enriched by this invention, has been that of *tools*. The great variety of new forms which it was necessary to produce, created the necessity of contriving and constructing a vast number of novel and most valuable tools, by which, with the aid of the lathe, and that alone, the required forms could be given to the different parts of the machinery with all the requisite accuracy.

The idea of calculation by mechanism is not new. Arithmetical instruments, such as the calculating boards of the ancients, on which they made their computations by the aid of counters—the *Abacus*, an instrument for computing by the aid of balls sliding upon parallel rods—the method of calculation invented by Baron Napier, called by him *Rhabdology*, and since called *Napier's bones*—the Swan Pan of the Chinese—and other similar contrivances, among which more particularly may be mentioned the Sliding Rule, of so much use in practical calculations to modern engineers, will occur to every reader: these may more properly be called *arithmetical instruments*, partaking more or less of a mechanical character. But the earliest piece of mechanism to which the name of "calculating machine" can fairly be given, appears to have been a machine invented by the celebrated Pascal. This philosopher and mathematician, at a very early age, being engaged with his father, who held an official situation in Upper Normandy, the duties of which required frequent numerical calculations, contrived a piece of mechanism to facilitate the performance of them. This mechanism consisted of a series of wheels, carrying cylindrical barrels, on which were engraved the ten arithmetical characters, in a manner not very dissimilar to that already described. The wheel which expressed each order of units was so connected with the wheel which expressed the superior order, that when the former passed from 9 to 0, the latter was necessarily advanced one figure; and thus the process of carrying was executed by mechanism: when one number was to be added to another by this machine, the addition of each figure to the other was performed by the hand; when it was required to add more than two numbers, the additions were performed in the same manner successively; the second was added to the first, the third to their sum, and so on.

Subtraction was reduced to addition by the method of arithmetical complements; multiplication was performed by a succession of additions; and division by a succession of subtractions. In all cases, however, the operations were executed from wheel to wheel by the hand.*

* See a description of this machine by Diderot, in the *Encyc. Method.*; also in the works of Pascal, tom. iv., p. 7; Paris, 1819.

This mechanism, which was invented about the year 1650, does not appear to have been brought into any practical use; and seems to have speedily found its appropriate place in a museum of curiosities. It was capable of performing only particular arithmetical operations, and these subject to all the chances of error in manipulation; attended also with little more expedition (if so much), as would be attained by the pen of an expert computer.

This attempt of Pascal was followed by various others, with very little improvement, and with no additional success. Polenus, a learned and ingenious Italian, invented a machine by which multiplication was performed, but which does not appear to have afforded any material facilities, nor any more security against error than the common process of the pen. A similar attempt was made by Sir Samuel Moreland, who is described as having transferred to wheelwork the figures of *Napier's bones*, and as having made some additions to the machine of Pascal.*

Grillet, a French mechanician, made a like attempt with as little success. Another contrivance for mechanical calculation was made by Saunderson. Mechanical contrivances for performing particular arithmetical processes were also made about a century ago by Depleréne and Boitissendeau; but they were merely modifications of Pascal's, without varying or extending its objects. But one of the most remarkable attempts of this kind which has been made since that of Pascal, was a machine invented by Leibnitz, of which we are not aware that any detailed or intelligible description was ever published. Leibnitz described its mode of operation, and its results, in the "Berlin Miscellany,"† but he appears to have declined any description of its details. In a letter addressed by him to Bernoulli, in answer to a request of the latter that he would afford a description of the machinery, he says, "Descriptionem ejus dare accuratam res non facilis foret. De effectu ex eo judicaveris quod ad multiplicandum numerum sex figurarum, e.g., rotam quamdam tantum sexies gyrari necesse est, nulla alia opera mentis, nullis additionibus intervenientibus; quo facto, integrum absolutumque productum oculis objicietur."‡ He goes on to say that the process of division is performed independently of a succession of subtractions, such as that used by Pascal.

It appears that this machine was one of an extremely complicated nature, which would be attended with considerable expense of con-

---

* Equidem Morelandus in Anglia, tubæ stentoriæ author, Rhabdologiam ex baculis in cylindrulos transtulit, et additiones auxiliares peragit in adjuncta machina additionum Pascaliana.

† Tom. i., p. 317.

‡ *Com. Epist.*, tom. i., p. 289.

struction, and only fit to be used in cases where numerous and expensive calculations were necessary.* Leibnitz observes to his correspondent, who required whether it might not be brought into common use, "Non est facta pro his qui olera aut pisculos vendunt, sed pro observatoriis aut cameris computorum, aut aliis, qui sumptus facile ferunt et multo calculo egent." Nevertheless, it does not appear that this contrivance, of which the inventor states that he caused two models to be made, was ever applied to any useful purpose; nor indeed do the mechanical details of the invention appear ever to have been published.

Even had the mechanism of these machines performed all which their inventors expected from them, they would have been still altogether inapplicable for the purposes to which it is proposed that the calculating machinery of Mr. Babbage shall be applied. They were all constructed with a view to perform *particular arithmetical operations*, and in all of them the accuracy of the result depended more or less upon manipulation. The principle of the calculating machinery of Mr. Babbage is perfectly general in its nature, not depending on any particular arithmetical operation, and is equally applicable to numerical tables of every kind. This distinguishing characteristic was well expressed by Mr. Colebrooke in his address to the Astronomical Society on this invention.

The principle which essentially distinguishes Mr. Babbage's invention from all these is, that it proposes to calculate a series of numbers following any law, by the aid of differences, and that by setting a few figures at the outset, a long series of numbers is readily produced by a mechanical operation. The method of differences in a very wide sense is the mathematical principle of the contrivance. A machine to add a number of arbitrary figures together is no economy of time or trouble, since each individual figure must be placed in the machine; but it is otherwise when those figures follow some law. The insertion of a few at first determines the magnitude of the next, and those of the succeeding. It is this constant repetition of similar operations which renders the computation of tables a fit subject for the application of machinery. Mr. Babbage's invention puts an engine in the place of the computer; the question is set to the instrument, or the instrument is set to the question, and by simply giving it motion the solution is wrought, and a string of answers is exhibited.

But perhaps the greatest of its advantages is, that it prints what it calculates; and this completely precludes the possibility of error in

* Sed machinam esse sumptuosam et multarum rotarum instar horologii: Huygenius aliquoties admonuit ut absolvi curarem; quod non sine magno sumptu tædioque factum est, dum varie mihi cum opificibus fuit conflictandum.—*Com. Epist.*

these numerical results which pass into the hands of the public. "The usefulness of the instrument," says Mr. Colebrooke, "is thus more than doubled; for it not only saves time and trouble in transcribing results into a tabular form, and setting types for the printing of the table, but it likewise accomplishes the yet more important object of insuring accuracy, obviating numerous sources of error through the careless hands of transcribers and compositors."

Some solicitude will doubtless be felt respecting the present state of the calculating machinery, and the probable period of its completion. In the beginning of the year 1829, Government directed the Royal Society to institute such inquiries as would enable them to report upon the state to which it had then arrived; and also whether the progress made in its construction confirmed them in the opinion which they had formerly expressed—that it would ultimately prove adequate to the important object which it was intended to obtain. The Royal Society, in accordance with these directions, appointed a Committee to make the necessary inquiry, and report. This Committee consisted of Mr. Davies Gilbert, then President, the Secretaries, Sir John Herschel, Mr. Francis Baily, Mr. Brunel, engineer, Mr. Donkin, engineer, Mr. G. Rennie, engineer, Mr. Barton, Comptroller of the Mint, and Mr. Warburton, M.P. The voluminous drawings, the various tools, and the portion of the machinery then executed, underwent a close and elaborate examination by this Committee, who reported upon it to the Society.

They stated in their report, that they declined the consideration of the principle on which the practicability of the machinery depends, and of the public utility of the object which it proposes to attain; because they considered the former fully admitted, and the latter obvious to all who consider the immense advantage of accurate numerical tables in all matters of calculation, especially in those which relate to astronomy and navigation, and the great variety and extent of those which it is professedly the object of the machinery to calculate and print with perfect accuracy;—that absolute accuracy being one of the prominent pretensions of the undertaking, they had directed their attention especially to this point, by careful examination of the drawings and of the work already executed, and by repeated conferences with Mr. Babbage on the subject:—that the result of their inquiry was, that such precautions appeared to have been taken in every part of the contrivance, and so fully aware was the inventor of every circumstance which might by possibility produce error, that they had no hesitation in stating their belief that these precautions were effectual, and that whatever the machine would do, it would do truly.

They further stated, that the progress which Mr. Babbage had then made, considering the very great difficulties to be overcome in an undertaking of so novel a kind, fully equalled any expectations that could reasonably have been formed; and that although several years had elapsed since the commencement of the undertaking, yet when the necessity of constructing plans, sections, elevations, and working drawings of every part; of constructing, and in many cases inventing, tools and machinery of great expense and complexity, necessary to form with the requisite precision parts of the apparatus differing from any which had previously been introduced in ordinary mechanical works; of making many trials to ascertain the value of each proposed contrivance; of altering, improving, and simplifying the drawings;— that, considering all these matters, the Committee, instead of feeling surprise at the time which the work has occupied, felt more disposed to wonder at the possibility of accomplishing so much.

The Committee expressed their confident opinion of the adequacy of the machinery to work under all the friction and strain to which it can be exposed; of its durability, strength, solidity, and equilibrium; of the prevention of, or compensation for, wear by friction; of the accuracy of the various adjustments; and of the judgment and discretion displayed by the inventor, in his determination to admit into the mechanism nothing but the very best and most finished workmanship; as a contrary course would have been false economy, and might have led to the loss of the whole capital expended on it.

Finally, considering all that had come before them, and relying on the talent and skill displayed by Mr. Babbage as a mechanist in the progress of this arduous undertaking, not less for what remained, than on the matured and digested plan and admirable execution of what is completed, the Committee did not hesitate to express their opinion, that in the then state of the engine, they regarded it as likely to fulfil the expectations entertained of it by its inventor.

This report was printed in the commencement of the year 1829. From that time until the beginning of the year 1833, the progress of the work has been slow and interrupted. Meanwhile many unfounded rumours have obtained circulation as to the course adopted by Government in this undertaking; and as to the position in which Mr. Babbage stands with respect to it. We shall here state, upon authority on which the most perfect reliance may be placed, what have been the actual circumstances of the arrangement which has been made, and of the steps which have been already taken.

Being advised that the objects of the projected machinery were of paramount national importance to a maritime country, and that, from its nature, it could never be undertaken with advantage by any individual as a pecuniary speculation, Government determined to engage Mr. Babbage to construct the calculating engine for the nation. It was then thought that the work could be completed in two or three years; and it was accordingly undertaken on this understanding about the year 1821, and since then has been in progress. The execution of the workmanship was confided to an engineer by whom all the subordinate workmen were employed, and who supplied for the work the requisite tools and other machinery; the latter being his own property, and not that of Government. This engineer furnished, at intervals, his accounts, which were duly audited by proper persons appointed for that purpose. It was thought advisable—with a view, perhaps, to invest Mr. Babbage with a more strict authority over the subordinate agents—that the payments of these accounts of the engineer should pass through his hands. The amount was accordingly from time to time issued to him by the Treasury, and paid over to the engineer. This circumstance has given rise to reports, that he has received considerable sums of money as a remuneration for his skill and labour in inventing and constructing this machinery. Such reports are altogether destitute of truth. He has received, neither directly nor indirectly, any remuneration whatever;—on the contrary, owing to various official delays in the issues of money from the Treasury for the payment of the engineer, he has frequently been obliged to advance these payments himself, that the work might proceed without interruption. Had he not been enabled to do this from his private resources, it would have been impossible that the machinery could have arrived at its present advanced state.

It will be a matter of regret to every friend of science to learn, that, notwithstanding such assistance, the progress of the work has been suspended, and the workmen dismissed for more than a year and a half; nor does there at the present moment appear to be any immediate prospect of its being resumed. What the causes may be of a suspension so extraordinary, of a project of such great national and universal interest—in which the country has already invested a sum of such serious amount as £15,000—is a question which will at once suggest itself to every mind; and is one to which, notwithstanding frequent inquiries in quarters from which correct information might be expected, we have not been able to obtain any satisfactory answer. It is not true, we are assured, that the Government object to make the necessary payments, or even advances, to carry on the work. It is not true, we

also are assured, that any practical difficulty has arisen in the construction of the mechanism;—on the contrary, the drawings of all the
parts of it are completed, and may be inspected by any person appointed on the part of Government to examine them.*   Mr. Babbage
is known as a man of unwearied activity, and aspiring ambition.
Why, then, it may be asked, is it that he, seeing his present reputation
and future fame depending in so great a degree upon the successful
issue of this undertaking, has nevertheless allowed it to stand still for
so long a period, without distinctly pointing out to Government the
course which they should adopt to remove the causes of delay?   Had
he done this (which we consider to be equally due to the nation and
to himself), he would have thrown upon Government and its agents
the whole responsibility for the delay and consequent loss; but we
believe he has not done so.   On the contrary, it is said that he has of
late almost withdrawn from all interference on the subject, either with
the Government or the engineer.   Does not Mr. Babbage perceive
the inference which the world will draw from this course of conduct?
Does he not see that they will impute it to a distrust of his own power,
or even to a consciousness of his own inability to complete what he has
begun?   We feel assured that such is not the case; and we are anxious,
equally for the sake of science, and for Mr. Babbage's own reputation,
that the mystery—for such it must be regarded—should be cleared up;
and that all obstructions to the progress of the undertaking should
immediately be removed.   Does this supineness and apparent in-
difference, so incompatible with the known character of Mr. Babbage,
arise from any feeling of dissatisfaction at the existing arrangements
between himself and the Government?   If such be the actual cause
of the delay (and we believe that, in some degree, it is so), we cannot
refrain from expressing our surprise that he does not adopt the candid
and straightforward course of declaring the grounds of his discontent,
and explaining the arrangement which he desires to be adopted.   We
do not hesitate to say, that every reasonable accommodation and
assistance ought to be afforded him.   But if he will pertinaciously
abstain from this, to our minds, obvious and proper course, then it is
surely the duty of Government to appoint proper persons to inquire

* Government has erected a fire-proof building, in which it is intended that the
calculating machinery shall be placed when completed.   In this building are now
deposited the large collection of drawings, containing the designs, not only of the
part of the machinery which has been already constructed, but what is of much greater
importance, of those parts which have not yet been even modelled.   It is gratifying
to know that Government has shown a proper solicitude for the preservation of those
precious but perishable documents, the loss or destruction of which would, in the
event of the death of the inventor, render the completion of the machinery impracticable.

into and report upon the present state of the machinery; to ascertain the causes of its suspension; and to recommend such measures as may appear to be most effectual to insure its speedy completion.   If they do not by such means succeed in putting the project in a state of advancement, they will at least shift from themselves all responsibility for its suspension.

II

SKETCH OF THE ANALYTICAL ENGINE INVENTED BY CHARLES BABBAGE

By L. F. MENABREA

of Turin, Officer of the Military Engineers

from the *Bibliothèque Universelle de Genève*, October, 1842, No. 82

With notes upon the Memoir by the Translator

ADA AUGUSTA, COUNTESS OF LOVELACE

THOSE labours which belong to the various branches of the mathematical sciences, although on first consideration they seem to be the exclusive province of intellect, may, nevertheless, be divided into two distinct sections; one of which may be called the mechanical, because it is subjected to precise and invariable laws, that are capable of being expressed by means of the operations of matter; while the other, demanding the intervention of reasoning, belongs more specially to the domain of the understanding. This admitted, we may propose to execute, by means of machinery, the mechanical branch of these labours, reserving for pure intellect that which depends on the reasoning faculties. Thus the rigid exactness of those laws which regulate numerical calculations must frequently have suggested the employment of material instruments, either for executing the whole of such calculations or for abridging them; and thence have arisen several inventions having this object in view, but which have in general but partially attained it. For instance, the much-admired machine of Pascal is now simply an object of curiosity, which, whilst it displays the powerful intellect of its inventor, is yet of little utility in itself. Its powers extended no further than the execution of the first four*

---

* This remark seems to require further comment, since it is in some degree calculated to strike the mind as being at variance with the subsequent passage (page 231), where it is explained that *an engine which can effect these four operations* can in fact effect *every species of calculation*. The apparent discrepancy is stronger too in the translation than in the original, owing to its being impossible to render precisely into the English tongue all the niceties of distinction which the French idiom happens to admit of in the phrases used for the two passages we refer to. The explanation lies in this: that in the one case the execution of these four operations is the *fundamental starting-point*, and the object proposed for attainment by the machine is the *subsequent combination*

operations of arithmetic, and indeed were in reality confined to that of
the first two, since multiplication and division were the result of a series of
additions and subtractions.   The chief drawback hitherto on most of
such machines is, that they require the continual intervention of a human
agent to regulate their movements, and thence arises a source of errors;
so that, if their use has not become general for large numerical calculations, it is because they have not in fact resolved the double problem
which the question presents, that of *correctness* in the results, united
with *economy* of time.

Struck with similar reflections, Mr. Babbage has devoted some years
to the realization of a gigantic idea.   He proposed to himself nothing
less than the construction of a machine capable of executing not merely
arithmetical calculations, but even all those of analysis, if their laws
are known.   The imagination is at first astounded at the idea of such
an undertaking; but the more calm reflection we bestow on it, the less
impossible does success appear, and it is felt that it may depend on
the discovery of some principle so general, that, if applied to machinery,
the latter may be capable of mechanically translating the operations
which may be indicated to it by algebraical notation.   The illustrious
inventor having been kind enough to communicate to me some of his
views on this subject during a visit he made at Turin, I have, with his
approbation, thrown together the impressions they have left on my
mind.   But the reader must not expect to find a description of Mr.
Babbage's engine; the comprehension of this would entail studies of
much length; and I shall endeavour merely to give an insight into the end
proposed, and to develop the principles on which its attainment depends.

I must first premise that this engine is entirely different from that
of which there is a notice in the 'Treatise on the Economy of Machinery,' by the same author.   But as the latter gave rise† to the idea of

---

of these in every possible variety; whereas in the other case the execution of some *one*
of these four operations, selected at pleasure, is the *ultimatum*, the sole and utmost
result that can be proposed for attainment by the machine referred to, and which
result it cannot any further combine or work upon.   The one *begins* where the other
ends.   Should this distinction not now appear perfectly clear, it will become so on
perusing the rest of the Memoir, and the Notes that are appended to it.—NOTE BY
TRANSLATOR.

† The idea that the one engine is the offspring and has grown out of the other, is
an exceedingly natural and plausible supposition, until reflection reminds us that
no *necessary* sequence and connexion need exist between two such inventions, and that
they *may* be wholly independent.   M. Menabrea has shared this idea in common with
persons who have not his profound and accurate insight into the nature of either
engine.   In Note A. (see the Notes at the end of the Memoir) it will be found sufficiently explained, however, that this supposition is unfounded.   M. Menabrea's
opportunities were by no means such as could be adequate to afford him information
on a point like this, which would be naturally and almost unconsciously *assumed*, and
would scarcely suggest any inquiry with reference to it.—NOTE BY TRANSLATOR.

the engine in question, I consider it will be a useful preliminary briefly to recall what were Mr. Babbage's first essays, and also the circumstances in which they originated.

It is well known that the French government, wishing to promote the extension of the decimal system, had ordered the construction of logarithmical and trigonometrical tables of enormous extent. M. de Prony, who had been entrusted with the direction of this undertaking, divided it into three sections, to each of which was appointed a special class of persons. In the first section the formulæ were so combined as to render them subservient to the purposes of numerical calculation; in the second, these same formulæ were calculated for values of the variable, selected at certain successive distances; and under the third section, comprising about eighty individuals, who were most of them only acquainted with the first two rules of arithmetic, the values which were intermediate to those calculated by the second section were interpolated by means of simple additions and subtractions.

An undertaking similar to that just mentioned having been entered upon in England, Mr. Babbage conceived that the operations performed under the third section might be executed by a machine; and this idea he realized by means of mechanism, which has been in part put together, and to which the name Difference Engine is applicable, on account of the principle upon which its construction is founded. To give some notion of this, it will suffice to consider the series of whole square numbers, 1, 4, 9, 16, 25, 36, 49, 64, &c. By subtracting each of these from the succeeding one, we obtain a new series, which we will name the Series of First Differences, consisting of the numbers 3, 5, 7, 9, 11, 13, 15, &c. On subtracting from each of these the preceding one, we obtain the Second Differences, which are all constant and equal to 2. We may represent this succession of operations, and their results, in the table on the following page.

From the mode in which the last two columns B and C have been formed, it is easy to see, that if, for instance, we desire to pass from the number 5 to the succeeding one 7, we must add to the former the constant difference 2; similarly, if from the square number 9 we would pass to the following one 16, we must add to the former the difference 7, which difference is in other words the preceding difference 5, plus the constant difference 2; or again, which comes to the same thing, to obtain 16 we have only to add together the three numbers 2, 5, 9, placed obliquely in the direction ab. Similarly, we obtain the number 25 by summing up the three numbers placed in the oblique direction dc: commencing by the addition 2 + 7, we have the first difference 9 consecutively to 7; adding 16 to the 9 we have the square 25. We see then that the three numbers 2, 5, 9 being given, the whole series of successive square numbers, and that of their first differences likewise, may be obtained by means of simple additions.

A Column of Square Numbers	B First Differences	C Second Differences
1		
	3	
4		2 b
	5	
9		2 d
	7	
16		2
	9	
25		2
	11	
36		

Now, to conceive how these operations may be reproduced by a machine, suppose the latter to have three dials, designated as A, B, C, on each of which are traced, say a thousand divisions, by way of example, over which a needle shall pass. The two dials, C, B, shall have in addition a registering hammer, which is to give a number of strokes equal to that of the divisions indicated by the needle. For each stroke of the registering hammer of the dial C, the needle B shall advance one division; similarly, the needle A shall advance one division for every stroke of the registering hammer of the dial B. Such is the general disposition of the mechanism.

This being understood, let us, at the beginning of the series of operations we wish to execute, place the needle C on the division 2, the needle B on the division 5, and the needle A on the division 9. Let us allow the hammer of the dial C to strike; it will strike twice, and at the same time the needle B will pass over two divisions. The latter will then indicate the number 7, which succeeds the number 5 in the column of first differences. If we now permit the hammer of the dial B to strike in its turn, it will strike seven times, during which the needle A will advance seven divisions; these added to the nine already marked by it will give the number 16, which is the square number consecutive to 9. If we now recommence these operations, beginning with the needle C, which is always to be left on the division 2, we shall perceive that by repeating them indefinitely, we may successively reproduce the series of whole square numbers by means of a very simple mechanism.

The theorem on which is based the construction of the machine we have just been describing, is a particular case of the following more general theorem: that if in any polynomial whatever, the highest power of whose variable is m, this same variable be increased by equal degrees; the corresponding values of the polynomial then calculated, and the first, second, third, &c. differences of these be taken (as for the preceding series of squares); the mth differences will all be equal to each other. So that, in order to reproduce the series of values of the polynomial by means of a machine analogous to the one above described, it is sufficient that there be $(m+1)$ dials, having the mutual relations we have indicated. As the differences may be either positive or negative, the machine will have a contrivance for either advancing or retrograding each needle, according as the number to be algebraically added may have the sign *plus* or *minus*.

If from a polynomial we pass to a series having an infinite number of terms, arranged according to the ascending powers of the variable, it would at first appear, that in order to apply the machine to the calculation of the function represented by such a series, the mechanism must include an infinite number of dials, which would in fact render the thing impossible. But in many cases the difficulty will disappear, if we observe that for a great number of functions the series which represent them may be rendered convergent; so that, according to the degree of approximation desired, we may limit ourselves to the calculation of a certain number of terms of the series, neglecting the rest. By this method the question is reduced to the primitive case of a finite polynomial. It is thus that we can calculate the succession of the logarithms of numbers. But since, in this particular instance, the terms which had been originally neglected receive increments in a ratio so continually increasing for equal increments of the variable, that the degree of approximation required would ultimately be affected, it is necessary, at certain intervals, to calculate the value of the function by different methods, and then respectively to use the results thus obtained, as data whence to deduce, by means of the machine, the other intermediate values. We see that the machine here performs the office of the third section of calculators mentioned in describing the tables computed by order of the French government, and that the end originally proposed is thus fulfilled by it.

Such is the nature of the first machine which Mr. Babbage conceived. We see that its use is confined to cases where the numbers required are such as can be obtained by means of simple additions or subtractions; that the machine is, so to speak, merely the expression of one* particular theorem of analysis; and that, in short, its operations cannot be extended so as to embrace the solution of an infinity of other questions included within the domain of mathematical analysis. It was while contemplating the vast field which yet remained to be traversed, that Mr. Babbage, renouncing his original essays, conceived the plan of another system of mechanism whose operations should themselves possess all the generality of algebraical notation, and which, on this account, he denominates the *Analytical Engine*.

Having now explained the state of the question, it is time for me to develop the principle on which is based the construction of this latter machine. When analysis is employed for the solution of any problem, there are usually two classes of operations to execute: first, the numerical calculation of the various coefficients; and secondly, their distribution in relation to the quantities affected by them. If, for example, we have to obtain the product of two binomials $(a+bx)$ $(m+nx)$, the result will be represented by $am + (an+bm) x + bnx^2$, in which expression we must first calculate am, an, bm, bn; then take the sum of $an+bm$; and lastly, respectively distribute the coefficients thus obtained amongst the powers of the variable. In order to reproduce these operations by means of a machine, the latter must therefore possess two distinct sets of powers: first, that of executing numerical calculations; secondly, that of rightly distributing the values so obtained.

But if human intervention were necessary for directing each of these partial operations, nothing would be gained under the heads of correctness and economy of time; the machine must therefore have the additional requisite of executing by itself all the successive operations required for the solution of a problem proposed to it, when once the *primitive numerical data* for this same problem have been introduced. Therefore, since, from the moment that the nature of the calculation to be executed or of the problem to be resolved have been indicated to it, the machine is, by its own intrinsic power, of itself to go through all the intermediate operations which lead to the proposed result, it must exclude all methods of trial and guess-work, and can only admit the direct processes of calculation.†

It is necessarily thus; for the machine is not a thinking being, but simply an automaton which acts according to the laws imposed upon it. This being fundamental, one of the earliest researches its author had to undertake, was that of finding means for effecting the division

* See Note A, p. 245 ff.

† This must not be understood in too unqualified a manner. The engine is capable, under certain circumstances, of feeling about to discover which of two or more possible contingencies has occurred, and of then shaping its future course accordingly. —NOTE BY TRANSLATOR.

of one number by another without using the method of guessing indicated by the usual rules of arithmetic. The difficulties of effecting this combination were far from being among the least; but upon it depended the success of every other. Under the impossibility of my here explaining the process through which this end is attained, we must limit ourselves to admitting that the first four operations of arithmetic, that is addition, subtraction, multiplication and division, can be performed in a direct manner through the intervention of the machine. This granted, the machine is thence capable of performing every species of numerical calculation, for all such calculations ultimately resolve themselves into the four operations we have just named. To conceive how the machine can now go through its functions according to the laws laid down,. we will begin by giving an idea of the manner in which it materially represents numbers.

Let us conceive a pile or vertical column consisting of an indefinite number of circular discs, all pierced through their centres by a common axis, around which each of them can take an independent rotatory movement. If round the edge of each of these discs are written the ten figures which constitute our numerical alphabet, we may then, by arranging a series of these figures in the same vertical line, express in this manner any number whatever. It is sufficient for this purpose that the first disc represent units, the second tens, the third hundreds, and so on. When two numbers have been thus written on two distinct columns, we may propose to combine them arithmetically with each other, and to obtain the result on a third column. In general, if we have a series of columns* consisting of discs, which columns we will designate as V_0, V_1, V_2, V_3, V_4, &c., we may require, for instance, to divide the number written on the column V_1 by that on the column V_4, and to obtain the result on the column V_7. To effect this operation, we must impart to the machine two distinct arrangements; through the first it is prepared for executing *a division*, and through the second the columns it is to operate on are indicated to it, and also the column on which the result is to be represented. If this division is to be followed, for example, by the addition of two numbers taken on other columns, the two original arrangements of the machine must be simultaneously altered. If, on the contrary, a series of operations of the same nature is to be gone through, then the first of the original arrangements will remain, and the second alone must be altered. Therefore, the arrangements that may be communicated to

* See Note B, p. 258.

the various parts of the machine may be distinguished into two principal classes:

First, that relative to the *Operations*.
Secondly, that relative to the *Variables*.

By this latter we mean that which indicates the columns to be operated on. As for the operations themselves, they are executed by a special apparatus, which is designated by the name of *mill*, and which itself contains a certain number of columns, similar to those of the Variables. When two numbers are to be combined together, the machine commences by effacing them from the columns where they are written, that is, it places *zero*** on every disc of the two vertical lines on which the numbers were represented; and it transfers the numbers to the mill. There, the apparatus having been disposed suitably for the required operation, this latter is effected, and, when completed, the result itself is transferred to the column of Variables which shall have been indicated. Thus the mill is that portion of the machine which works, and the columns of Variables constitute that where the results are represented and arranged. After the preceding explanations, we may perceive that all fractional and irrational results will be represented in decimal fractions. Supposing each column to have forty discs, this extension will be sufficient for all degrees of approximation generally required.

It will now be inquired how the machine can of itself, and without having recourse to the hand of man, assume the successive dispositions suited to the operations. The solution of this problem has been taken from Jacquard's apparatus†, used for the manufacture of brocaded stuffs, in the following manner:—

Two species of threads are usually distinguished in woven stuffs; one is the *warp* or longitudinal thread, the other the *woof* or transverse thread, which is conveyed by the instrument called the shuttle, and which crosses the longitudinal thread or warp. When a brocaded stuff is required, it is necessary in turn to prevent certain threads from crossing the woof, and this according to a succession which is determined by the nature of the design that is to be reproduced. Formerly this process was lengthy and difficult, and it was requisite that the workman, by attending to the design which he was to copy, should himself regulate the movements the threads were to take. Thence

---

* Zero is not *always* substituted when a number is transferred to the mill. This is explained further on in the memoir, and still more fully in Note D, p. 265.—NOTE BY TRANSLATOR.

† See Note C, p. 264.

arose the high price of this description of stuffs, especially if threads of various colours entered into the fabric. To simplify this manufacture, Jacquard devised the plan of connecting each group of threads that were to act together, with a distinct lever belonging exclusively to that group. All these levers terminate in rods, which are united together in one bundle, having usually the form of a parallelopiped with a rectangular base. The rods are cylindrical, and are separated from each other by small intervals. The process of raising the threads is thus resolved into that of moving these various lever-arms in the requisite order. To effect this, a rectangular sheet of pasteboard is taken, somewhat larger in size than a section of the bundle of lever-arms. If this sheet be applied to the base of the bundle, and an advancing motion be then communicated to the pasteboard, this latter will move with it all the rods of the bundle, and consequently the threads that are connected with each of them. But if the pasteboard, instead of being plain, were pierced with holes corresponding to the extremities of the levers which meet it, then, since each of the levers would pass through the pasteboard during the motion of the latter, they would all remain in their places. We thus see that it is easy so to determine the position of the holes in the pasteboard, that, at any given moment, there shall be a certain number of levers, and consequently of parcels of threads, raised, while the rest remain where they were. Supposing this process is successively repeated according to a law indicated by the pattern to be executed, we perceive that this pattern may be reproduced on the stuff. For this purpose we need merely compose a series of cards according to the law required, and arrange them in suitable order one after the other; then, by causing them to pass over a polygonal beam which is so connected as to turn a new face for every stroke of the shuttle, which face shall then be impelled parallelly to itself against the bundle of lever-arms, the operation of raising the threads will be regularly performed. Thus we see that brocaded tissues may be manufactured with a precision and rapidity formerly difficult to obtain.

Arrangements analogous to those just described have been introduced into the Analytical Engine. It contains two principal species of cards: first, Operation cards, by means of which the parts of the machine are so disposed as to execute any determinate series of operations, such as additions, subtractions, multiplications, and divisions; secondly, cards of the Variables, which indicate to the machine the columns on which the results are to be represented. The cards, when put in motion, successively arrange the various portions of the machine according to the nature of the processes that are to be effected, and the machine at the same time executes these processes by means of the various pieces of mechanism of which it is constituted.

In order more perfectly to conceive the thing, let us select as an example the resolution of two equations of the first degree with two unknown quantities. Let the following be the two equations, in which x and y are the unknown quantities:—

$$\begin{cases} mx + ny = d \\ m'x + n'y = d'. \end{cases}$$

We deduce $x = \dfrac{dn' - d'n}{n'm - nm'}$, and for y an analogous expression. Let us continue to represent by V_0, V_1, V_2, &c. the different columns which contain the numbers, and let us suppose that the first eight columns have been chosen for expressing on them the numbers represented by m, n, d, m', n', d', n and n', which implies that $V_0 = m$, $V_1 = n$, $V_2 = d$, $V_3 = m'$, $V_4 = n'$, $V_5 = d'$, $V_6 = n$, $V_7 = n'$.

The series of operations commanded by the cards, and the results obtained, may be represented in the following table:—

Number of the operations	Operation-cards	Cards of the variables		Progress of the operations
	Symbols indicating the nature of the operations	Columns on which operations are to be performed	Columns which receive results of operations	
1	×	$V_2 \times V_4 =$	$V_8 \ldots\ldots$	$= dn'$
2	×	$V_5 \times V_1 =$	$V_9 \ldots\ldots$	$= d'n$
3	×	$V_4 \times V_0 =$	$V_{10} \ldots\ldots$	$= n'm$
4	×	$V_1 \times V_3 =$	$V_{11} \ldots\ldots$	$= nm'$
5	—	$V_8 - V_9 =$	$V_{12} \ldots\ldots$	$= dn' - d'n$
6	—	$V_{10} - V_{11} =$	$V_{13} \ldots\ldots$	$= n'm - nm'$
7	÷	$\dfrac{V_{12}}{V_{13}} =$	$V_{14} \ldots\ldots$	$= x = \dfrac{dn' - d'n}{n'm - nm'}$

Since the cards do nothing but indicate in what manner and on what columns the machine shall act, it is clear that we must still, in every particular case, introduce the numerical data for the calculation. Thus, in the example we have selected, we must previously inscribe the numerical values of m, n, d, m', n', d', in the order and on the columns indicated, after which the machine when put in action will give the value of the unknown quantity x for this particular case.   To obtain the value of y, another series of operations analogous to the preceding must be performed.   But we see that they will be only four in number, since the denominator of the expression for y, excepting the sign, is the same as that for x, and equal to $n'm - nm'$.   In the preceding table it will be remarked that the column for operations indicates four successive *multiplications*, two *subtractions*, and one *division*.   Therefore, if desired, we need only use three operation-cards; to manage which, it is sufficient to introduce into the machine an apparatus which shall, after the first multiplication, for instance, retain the card which relates to this operation, and not allow it to advance so as to be replaced by another one, until after this same operation shall have been four times repeated.   In the preceding example we have seen, that to find the value of x we must begin by writing the coefficients m, n, d, m', n', d', upon eight columns, thus repeating n and n' twice. According to the same method, if it were required to calculate y likewise, these coefficients must be written on twelve different columns. But it is possible to simplify this process, and thus to diminish the chances of errors, which chances are greater, the larger the number of the quantities that have to be inscribed previous to setting the machine in action.   To understand this simplification, we must remember that every number written on a column must, in order to be arithmetically combined with another number, be effaced from the column on which it is, and transferred to the *mill*.   Thus, in the example we have discussed, we will take the two coefficients m and n', which are each of them to enter into *two* different products, that is m into mn' and md', n' into mn' and $n'd$.   These coefficients will be inscribed on the columns V_0 and V_4.   If we commence the series of operations by the product of m into n', these numbers will be effaced from the columns V_0 and V_4, that they may be transferred to the mill, which will multiply them into each other, and will then command the machine to represent the result, say on the column V_6.   But as these numbers are each to be used again in another operation, they must again be inscribed somewhere; therefore, while the mill is working out their product, the machine will inscribe them anew on any two columns that may be indicated to it through the cards; and as, in the actual case, there is no reason why they should not resume their former places, we will suppose them again inscribed on V_0 and V_4, whence in short they would not finally disappear, to be reproduced no more, until they should have gone through all the combinations in which they might have to be used.

We see, then, that the whole assemblage of operations requisite for resolving the two* above equations of the first degree may be definitely represented in the following table:—

Columns on which are inscribed the primitive data	Number of the operations	Cards of the operations		Variable cards			Statement of results
		No. of the Oper'n-cards	Nature of each oper'n	Columns acted on by each operation	Columns that receive the result of each operation	Indication of change of value on any column	
$^1V_0 = m$	1	1	×	$^1V_0 \times {}^1V_4 = {}^1V_6 \ldots$		$\begin{cases}^1V_0 = {}^1V_0 \\ {}^1V_4 = {}^1V_4\end{cases}$	$^1V_6 = mn'$
$^1V_1 = n$	2	,,	×	$^1V_3 \times {}^1V_1 = {}^1V_7 \ldots$		$\begin{cases}^1V_3 = {}^1V_3 \\ {}^1V_1 = {}^1V_1\end{cases}$	$^1V_7 = m'n$
$^1V_2 = d$	3	,,	×	$^1V_2 \times {}^1V_4 = {}^1V_8 \ldots$		$\begin{cases}^1V_2 = {}^1V_2 \\ {}^1V_4 = {}^0V_4\end{cases}$	$^1V_3 = dn'$
$^1V_3 = m'$	4	,,	×	$^1V_5 \times {}^1V_1 = {}^1V_9 \ldots$		$\begin{cases}^1V_5 = {}^1V_5 \\ {}^1V_1 = {}^0V_1\end{cases}$	$^1V_9 = d'n$
$^1V_4 = n'$	5	,,	×	$^1V_0 \times {}^1V_5 = {}^1V_{10} \ldots$		$\begin{cases}^1V_0 = {}^0V_0 \\ {}^1V_5 = {}^0V_5\end{cases}$	$^1V_{10} = d'm$
$^1V_5 = d'$	6	,,	×	$^1V_2 \times {}^1V_3 = {}^1V_{11} \ldots$		$\begin{cases}^1V_2 = {}^0V_2 \\ {}^1V_3 = {}^0V_3\end{cases}$	$^1V_{11} = dm'$
	7	2	−	$^1V_6 - {}^1V_7 = {}^1V_{12} \ldots$		$\begin{cases}^1V_6 = {}^0V_6 \\ {}^1V_7 = {}^0V_7\end{cases}$	$^1V_{12} = mn' - m'n$
	8	,,	−	$^1V_8 - {}^1V_9 = {}^1V_{13} \ldots$		$\begin{cases}^1V_8 = {}^0V_8 \\ {}^1V_9 = {}^0V_9\end{cases}$	$^1V_{13} = dn' - d'n$
	9	,,	−	$^1V_{10} - {}^1V_{11} = {}^1V_{14} \ldots$		$\begin{cases}^1V_{10} = {}^0V_{10} \\ {}^1V_{11} = {}^0V_{11}\end{cases}$	$^1V_{14} = d'm - dm'$
	10	3	÷	$^1V_{13} \div {}^1V_{12} = {}^1V_{15} \ldots$		$\begin{cases}^1V_{13} = {}^0V_{13} \\ {}^1V_{12} = {}^1V_{12}\end{cases}$	$^1V_{15} = \dfrac{dn' - d'n}{mn' - m'n} = x$
	11	,,	÷	$^1V_{14} \div {}^1V_{12} = {}^1V_{16} \ldots$		$\begin{cases}^1V_{14} = {}^0V_{14} \\ {}^1V_{12} = {}^0V_{12}\end{cases}$	$^1V_{16} = \dfrac{d'm - dm'}{mn' - m'n} = y$
1	2	3	¿	5	6	7	8

In order to diminish to the utmost the chances of error in inscribing the numerical data of the problem, they are successively placed on one of the columns of the mill; then, by means of cards arranged for this purpose, these same numbers are caused to arrange themselves on the requisite columns, without the operator having to give his attention to it; so that his undivided mind may be applied to the simple inscription of these same numbers.

According to what has now been explained, we see that the collection of columns of Variables may be regarded as a *store* of numbers, accumulated there by the mill, and which, obeying the orders transmitted to the machine by means of the cards, pass alternately from the mill to the store and from the store to the mill, that they may undergo the transformations demanded by the nature of the calculation to be performed.

Hitherto no mention has been made of the *signs* in the results, and the machine would be far from perfect were it incapable of expressing and combining amongst each other positive and negative quantities.

* See Note D, p. 265.

To accomplish this end, there is, above every column, both of the mill and of the store, a disc, similar to the discs of which the columns themselves consist. According as the digit on this disc is even or uneven, the number inscribed on the corresponding column below it will be considered as positive or negative. This granted, we may, in the following manner, conceive how the signs can be algebraically combined in the machine. When a number is to be transferred from the store to the mill, and *vice versâ*, it will always be transferred with its sign, which will effected by means of the cards, as has been explained in what precedes. Let any two numbers then, on which we are to operate arithmetically, be placed in the mill with their respective signs. Suppose that we are first to add them together; the operation-cards will command the addition: if the two numbers be of the same sign, one of the two will be entirely effaced from where it was inscribed, and will go to add itself on the column which contains the other number; the machine will, during this operation, be able, by means of a certain apparatus, to prevent any movement in the disc of signs which belongs to the column on which the addition is made, and thus the result will remain with the sign which the two given numbers originally had. When two numbers have two different signs, the addition commanded by the card will be changed into a subtraction through the intervention of mechanisms which are brought into play by this very difference of sign. Since the subtraction can only be effected on the larger of the two numbers, it must be arranged that the disc of signs of the larger number shall not move while the smaller of the two numbers is being effaced from its column and subtracted from the other, whence the result will have the sign of this latter, just as in fact it ought to be. The combinations to which algebraical subtraction give rise, are analogous to the preceding. Let us pass on to multiplication. When two numbers to be multiplied are of the same sign, the result is positive; if the signs are different, the product must be negative. In order that the machine may act conformably to this law, we have but to conceive that on the column containing the product of the two given numbers, the digit which indicates the sign of that product has been formed by the mutual addition of the two digits that respectively indicated the signs of the two given numbers; it is then obvious that if the digits of the signs are both even, or both odd, their sum will be an even number, and consequently will express a positive number; but that if, on the contrary, the two digits of the signs are one even and the other odd, their sum will be an odd number, and will consequently express a negative number. In the case of division, instead of adding the digits of the discs, they must be subtracted one from the other, which will produce results analogous to the preceding; that is to say, that if these figures are both even or both uneven, the remainder of this subtraction will be even; and it will be uneven in the contrary case. When I speak of mutually adding or subtracting the numbers expressed by the digits of the signs, I merely mean that one of the sign-discs is made to advance or retrograde a number of divisions equal to that which is expressed by the digit on the other sign-disc. We see, then, from the preceding explanation, that it is possible mechanically to combine the signs of quantities so as to obtain results conformable to those indicated by algebra*.

The machine is not only capable of executing those numerical calculations which depend on a given algebraical formula, but it is also fitted for analytical calculations in which there are one or several variables to be considered. It must be assumed that the analytical expression to be operated on can be developed according to powers of the variable, or according to determinate functions of this same variable, such as circular functions, for instance; and similarly for the result that is to be attained. If we then suppose that above the columns of the store, we have inscribed the powers or the functions of the variable, arranged according to whatever is the prescribed law of development, the coefficients of these several terms may be respectively placed on the corresponding column below each. In this manner we shall have a representation of an analytical development; and, supposing the position of the several terms composing it to be invariable, the problem will be reduced to that of calculating their coefficients according to the laws demanded by the nature of the question. In order to make this more clear, we shall take the following† very simple example, in which we are to multiply $(a+bx^1)$ by $(A+B \cos^1 x)$. We shall begin by writing x^0, x^1, $\cos^0 x$, $\cos^1 x$, above the columns V_0, V_1, V_2, V_3; then since, from the form of the two functions to be combined, the terms which are to compose the products will be of the following nature, $x^0 . \cos^0 x$, $x^0 . \cos^1 x$, $x^1 . \cos^0 x$, $x^1 . \cos^1 x$, these will be inscribed above the columns V_4, V_5, V_6, V_7. The coefficients of x^0, x^1, $\cos^0 x$, $\cos^1 x$ being given, they will, by means of the mill, be passed to the

---

* Not having had leisure to discuss with Mr. Babbage the manner of introducing into his machine the combination of algebraical signs, I do not pretend here to expose the method he uses for this purpose; but I considered that I ought myself to supply the deficiency, conceiving that this paper would have been imperfect if I had omitted to point out one means that might be employed for resolving this essential part of the problem in question.

† See Note E, p. 271.

columns V_0, V_1, V_2 and V_3. Such are the primitive data of the problem. It is now the business of the machine to work out its solution, that is, to find the coefficients which are to be inscribed on V_4, V_5, V_6, V_7. To attain this object, the law of formation of these same coefficients being known, the machine will act through the intervention of the cards, in the manner indicated by the following table:—

* Columns above which are written the functions of the variable	Coefficients		Cards of the operations		Cards of the variables			
	Given	To be formed	No. of the Operations	Nature of the Operation	Columns on which operations are to be performed	Columns on which are to be inscribed the results of the operations	Indication of change of value on any column submitted to an operation	Results of the operations
$x^0 \ldots\ldots {}^1V_0$ a	,,	,,	,,	,,	,,	,,	,,	,,
$x^1 \ldots\ldots {}^1V_1$ b	,,	,,	,,	,,	,,	,,	,,	,,
$\cos^0 x \ldots {}^1V_2$ A	,,	,,	,,	,,	,,	,,	,,	,,
$\cos^1 x \ldots {}^1V_3$ B	,,	,,	,,	,,	,,	,,	,,	,,
$x^0 \cos^0 x \ldots {}^0V_4$ aA	1	×	${}^1V_0 × {}^1V_2 =$	${}^1V_4 \ldots\ldots$	$\left\{\begin{array}{l}{}^1V_0 = {}^1V_0\\ {}^1V_2 = {}^1V_2\end{array}\right\}$	${}^1V_4 = aA$ coefficients of $x^0 \cos^0 x$
$x^0 \cos^1 x \ldots {}^0V_5$ aB	2	×	${}^1V_0 × {}^1V_3 =$	${}^1V_5 \ldots\ldots$	$\left\{\begin{array}{l}{}^1V_0 = {}^0V_0\\ {}^1V_3 = {}^1V_3\end{array}\right\}$	${}^1V_5 = aB \quad \ldots \quad \ldots \quad x^0 \cos^1 x$
$x^1 \cos^0 x \ldots {}^0V_6$ bA	3	×	${}^1V_1 × {}^1V_2 =$	${}^1V_6 \ldots\ldots$	$\left\{\begin{array}{l}{}^1V_1 = {}^1V_1\\ {}^1V_2 = {}^0V_2\end{array}\right\}$	${}^1V_6 = bA \quad \ldots \quad \ldots \quad x^1 \cos^0 x$
$x^1 \cos^1 x \ldots {}^0V_7$ bB	4	×	${}^1V_1 × {}^1V_3 =$	${}^1V_7 \ldots\ldots$	$\left\{\begin{array}{l}{}^1V_1 = {}^0V_1\\ {}^1V_3 = {}^0V_3\end{array}\right\}$	${}^1V_7 = bB \quad \ldots \quad \ldots \quad x^1 \cos^1 x$

It will now be perceived that a general application may be made of the principle developed in the preceding example, to every species of process which it may be proposed to effect on series submitted to calculation. It is sufficient that the law of formation of the coefficients be known, and that this law be inscribed on the cards of the machine, which will then of itself execute all the calculations requisite for arriving at the proposed result. If, for instance, a recurring series were proposed, the law of formation of the coefficients being here uniform, the same operations which must be performed for one of them will be repeated for all the others; there will merely be a change in the locality of the operation, that is, it will be performed with different columns. Generally, since every analytical expression is susceptible of being expressed in a series ordered according to certain functions of the variable, we perceive that the machine will include all analytical calculations which can be definitively reduced to the

* For an explanation of the upper left-hand indices attached to the V's in this and in the preceding Table, we must refer the reader to Note D, amongst those appended to the memoir.—NOTE BY TRANSLATOR.

formation of coefficients according to certain laws, and to the distribution of these with respect to the variables.

We may deduce the following important consequence from these explanations, viz. that since the cards only indicate the nature of the operations to be performed, and the columns of Variables with which they are to be executed, these cards will themselves possess all the generality of analysis, of which they are in fact merely a translation. We shall now further examine some of the difficulties which the machine must surmount, if its assimilation to analysis is to be complete. There are certain functions which necessarily change in nature when they pass through zero or infinity, or whose values cannot be admitted when they pass these limits.   When such cases present themselves, the machine is able, by means of a bell, to give notice that the passage through zero or infinity is taking place, and it then stops until the attendant has again set it in action for whatever process it may next be desired that it shall perform.   If this process has been foreseen, then the machine, instead of ringing, will so dispose itself as to present the new cards which have relation to the operation that is to succeed the passage through zero and infinity.   These new cards may follow the first, but may only come into play contingently upon one or other of the two circumstances just mentioned taking place.

Let us consider a term of the form ab^n; since the cards are but a translation of the analytical formula, their number in this particular case must be the same, whatever be the value of n; that is to say, whatever be the number of multiplications required for elevating b to the nth power (we are supposing for the moment that n is a whole number).   Now, since the exponent n indicates that b is to be multiplied n times by itself, and all these operations are of the same nature, it will be sufficient to employ one single operation-card, viz. that which orders the multiplication.

But when n is given for the particular case to be calculated, it will be further requisite that the machine limit the number of its multiplications according to the given values.   The process may be thus arranged.   The three numbers a, b and n will be written on as many distinct columns of the store; we shall designate them V_0, V_1, V_2; the result ab^n will place itself on the column V_3.   When the number n has been introduced into the machine, a card will order a certain registering-apparatus to mark $(n-1)$, and will at the same time execute the multiplication of b by b.   When this is completed, it will be found that the registering-apparatus has effaced a unit, and that it only marks $(n-2)$; while the machine will now again order the number b written on the column V_1 to multiply itself with the product

b^2 written on the column V_3, which will give b^3. Another unit is then effaced from the registering-apparatus, and the same processes are continually repeated until it only marks zero. Thus the number b^n will be found inscribed on V_3, when the machine, pursuing its course of operations, will order the product of b^n by a; and the required calculation will have been completed without there being any necessity that the number of operation-cards used should vary with the value of n. If n were negative, the cards, instead of ordering the multiplication of a by b^n, would order its division; this we can easily conceive, since every number, being inscribed with its respective sign, is consequently capable of reacting on the nature of the operations to be executed. Finally, if n were fractional, of the form p/q, an additional column would be used for the inscription of q, and the machine would bring into action two sets of processes, one for raising b to the power p, the other for extracting the qth root of the number so obtained.

Again, it may be required, for example, to multiply an expression of the form $ax^m + bx^n$ by another $Ax^p + Bx^q$, and then to reduce the product to the least number of terms, if any of the indices are equal. The two factors being ordered with respect to x, the general result of the multiplication would be $Aax^{m+p} + Abx^{n+p} + Bax^{m+q} + Bbx^{n+q}$. Up to this point the process presents no difficulties; but suppose that we have $m=p$ and $n=q$, and that we wish to reduce the two middle terms to a single one $(Ab+Ba)x^{m+q}$. For this purpose, the cards may order $m+q$ and $n+p$ to be transferred into the mill, and there subtracted one from the other; if the remainder is nothing, as would be the case on the present hypothesis, the mill will order other cards to bring to it the coefficients Ab and Ba, that it may add them together and give them in this state as a coefficient for the single term $x^{n+p} = x^{m+q}$.

This example illustrates how the cards are able to reproduce all the operations which intellect performs in order to attain a determinate result, if these operations are themselves capable of being precisely defined.

Let us now examine the following expression:—

$$2 . \frac{2^2 . 4^2 . 6^2 . 8^2 . 10^2 \ldots (2n)^2}{1^2 . 3^2 . 5^2 . 7^2 . 9^2 \ldots (2n-1)^2 . (2n+1)^2},$$

which we know becomes equal to the ratio of the circumference to the diameter, when n is infinite. We may require the machine not only to perform the calculation of this fractional expression, but further to give indication as soon as the value becomes identical with that of the ratio of the circumference to the diameter when n is infinite, a case in which the computation would be impossible. Observe that we should thus require of the machine to interpret a result not of itself evident, and that this is not amongst its attributes, since it is no thinking being. Nevertheless, when the cos of $n = 1/0$ has been foreseen, a card may immediately order the substitution of the value of π (π being the ratio of the circumference to the diameter), without going through the series of calculations indicated. This would merely require that the machine contain a special card, whose office it should be to place the number π in a direct and independent manner on the column indicated to it. And here we should introduce the mention of a third species of cards, which may be called *cards of numbers*. There are certain numbers, such as those expressing the ratio of the circumference to the diameter, the Numbers of Bernoulli, &c., which frequently present themselves in calculations. To avoid the necessity for computing them every time they have to be used, certain cards may be combined specially in order to give these numbers ready made into the mill, whence they afterwards go and place themselves on those columns of the store that are destined for them. Through this means the machine will be susceptible of those simplifications afforded by the use of numerical tables. It would be equally possible to introduce, by means of these cards, the logarithms of numbers; but perhaps it might not be in this case either the shortest or the most appropriate method; for the machine might be able to perform the same calculations by other more expeditious combinations, founded on the rapidity with which it executes the first four operations of arithmetic. To give an idea of this rapidity, we need only mention that Mr. Babbage believes he can, by his engine, form the product of two numbers, each containing twenty figures, in *three minutes*.

Perhaps the immense number of cards required for the solution of any rather complicated problem may appear to be an obstacle; but this does not seem to be the case. There is no limit to the number of cards that can be used. Certain stuffs require for their fabrication not less than *twenty thousand* cards, and we may unquestionably far exceed even this quantity*.

Resuming what we have explained concerning the Analytical Engine, we may conclude that it is based on two principles: the first, consisting in the fact that every arithmetical calculation ultimately depends on four principal operations—addition, subtraction, multiplication, and division; the second, in the possibility of reducing every analytical calculation to that of the coefficients for the several terms of a series. If this last principle be true, all the operations of analysis come within the domain of the engine. To take another point of view: the

* See Note F, p. 281.

use of the cards offers a generality equal to that of algebraical formulæ, since such a formula simply indicates the nature and order of the operations requisite for arriving at a certain definite result, and similarly the cards merely command the engine to perform these same operations; but in order that the mechanisms may be able to act to any purpose, the numerical data of the problem must in every particular case be introduced. Thus the same series of cards will serve for all questions whose sameness of nature is such as to require nothing altered excepting the numerical data. In this light the cards are merely a translation of algebraical formulæ, or, to express it better, another form of analytical notation.

Since the engine has a mode of acting peculiar to itself, it will in every particular case be necessary to arrange the series of calculations conformably to the means which the machine possesses; for such or such a process which might be very easy for a calculator may be long and complicated for the engine, and *vice versâ*.

Considered under the most general point of view, the essential object of the machine being to calculate, according to the laws dictated to it, the values of numerical coefficients which it is then to distribute appropriately on the columns which represent the variables, it follows that the interpretation of formulæ and of results is beyond its province, unless indeed this very interpretation be itself susceptible of expression by means of the symbols which the machine employs. Thus, although it is not itself the being that reflects, it may yet be considered as the being which executes the conceptions of intelligence*. The cards receive the impress of these conceptions, and transmit to the various trains of mechanism composing the engine the orders necessary for their action. When once the engine shall have been constructed, the difficulty will be reduced to the making out of the cards; but as these are merely the translation of algebraical formulæ, it will, by means of some simple notations, be easy to consign the execution of them to a workman. Thus the whole intellectual labour will be limited to the preparation of the formulæ, which must be adapted for calculation by the engine.

Now, admitting that such an engine can be constructed, it may be inquired: what will be its utility? To recapitulate; it will afford the following advantages:—First, rigid accuracy. We know that numerical calculations are generally the stumbling-block to the solution of problems, since errors easily creep into them, and it is by no means always easy to detect these errors. Now the engine, by the very nature of its mode of acting, which requires no human intervention during the course of its operations, presents every species of security under the head of

* See Note G, p. 284.

correctness: besides, it carries with it its own check; for at the end of every operation it prints off, not only the results, but likewise the numerical data of the question; so that it is easy to verify whether the question has been correctly proposed.   Secondly, economy of time: to convince ourselves of this, we need only recollect that the multiplication of two numbers, consisting each of twenty figures, requires at the very utmost three minutes.   Likewise, when a long series of identical computations is to be performed, such as those required for the formation of numerical tables, the machine can be brought into play so as to give several results at the same time, which will greatly abridge the whole amount of the processes.   Thirdly, economy of intelligence: a simple arithmetical computation requires to be performed by a person possessing some capacity; and when we pass to more complicated calculations, and wish to use algebraical formulæ in particular cases, knowledge must be possessed which presupposes preliminary mathematical studies of some extent.   Now the engine, from its capability of performing by itself all these purely material operations, spares intellectual labour, which may be more profitably employed.   Thus the engine may be considered as a real manufactory of figures, which will lend its aid to those many useful sciences and arts that depend on numbers.   Again, who can foresee the consequences of such an invention?   In truth, how many precious observations remain practically barren for the progress of the sciences, because there are not powers sufficient for computing the results !   And what discouragement does the perspective of a long and arid computation cast into the mind of a man of genius, who demands time exclusively for meditation, and who beholds it snatched from him by the material routine of operations!   Yet it is by the laborious route of analysis that he must reach truth; but he cannot pursue this unless guided by numbers; for without numbers it is not given us to raise the veil which envelopes the mysteries of nature.   Thus the idea of constructing an apparatus capable of aiding human weakness in such researches, is a conception which, being realized, would mark a glorious epoch in the history of the sciences.   The plans have been arranged for all the various parts, and for all the wheel-work, which compose this immense apparatus, and their action studied; but these have not yet been fully combined together in the drawings* and mechanical notation†.   The

* This sentence has been slightly altered in the translation in order to express more exactly the present state of the engine.—NOTE BY TRANSLATOR.

† The notation here alluded to is a most interesting and important subject, and would have well deserved a separate and detailed Note upon it amongst those appended to the Memoir.   It has, however, been impossible, within the space allotted, even to touch upon so wide a field.—NOTE BY TRANSLATOR.

confidence which the genius of Mr. Babbage must inspire, affords legitimate ground for hope that this enterprise will be crowned with success; and while we render homage to the intelligence which directs it, let us breathe aspirations for the accomplishment of such an undertaking.

NOTES BY THE TRANSLATOR

Note A

THE PARTICULAR function whose integral the Difference Engine was constructed to tabulate, is

$$\Delta^7 u_z = 0.$$

The purpose which that engine has been specially intended and adapted to fulfil, is the computation of nautical and astronomical tables. The integral of

$$\Delta^7 u_z = 0$$

being $u_z = a + bx + cx^2 + dx^3 + ex^4 + fx^5 + gx^6,$

the constants a, b, c, &c. are represented on the seven columns of discs, of which the engine consists.   It can therefore tabulate *accurately* and to an *unlimited extent*, all series whose general term is comprised in the above formula; and it can also tabulate *approximatively* between *intervals of greater or less extent*, all other series which are capable of tabulation by the Method of Differences.

The Analytical Engine, on the contrary, is not merely adapted for *tabulating* the results of one particular function and of no other, but for *developing and tabulating* any function whatever. In fact the engine may be described as being the material expression of any indefinite function of any degree of generality and complexity, such as for instance,

$$F(x, y, z \log x, \sin y, x^p, \&c.),$$

which is, it will be observed, a function of all other possible functions of any number of quantities.

In this, which we may call the *neutral* or *zero* state of the engine, it is ready to receive at any moment, by means of cards constituting a portion of its mechanism (and applied on the principle of those used in the Jacquard-loom), the impress of whatever *special* function we may desire to develope or to tabulate. These cards contain within themselves (in a manner explained in the Memoir itself, pages 232-234) the law of development of the particular function that may be under consideration, and they compel the mechanism to act accordingly in a certain corresponding order. One of the simplest cases would be, for example, to suppose that

$$F(x,y,z, \text{ &c. &c.})$$

is the particular function

$$\Delta^n u_z = 0$$

which the Difference Engine tabulates for values of n only up to 7. In this case the cards would order the mechanism to go through that succession of operations which would tabulate

$$u_z = a + bx + cx^2 + \ldots + mx^{n-1},$$

where n might be any number whatever.

These cards, however, have nothing to do with the regulation of the particular *numerical* data. They merely determine the *operations** to be effected, which operations may of course be performed on an infinite variety of particular numerical values, and do not bring out any definite numerical results unless the numerical data of the problem have been impressed on the requisite portions of the train of mechanism. In the above example, the first essential step towards an arithmetical result would be the substitution of specific numbers for n, and for the other primitive quantities which enter into the function.

Again, let us suppose that for F we put two complete equations of the fourth degree between x and y. We must then express on the cards the law of elimination for such equations. The engine would follow out those laws, and would ultimately give the equation of one variable which results from such elimination. Various *modes* of elimination might be selected; and of course the cards must be made out accordingly. The following is one mode that might be adopted. The engine is able to multiply together any two functions of the form

$$a + bx + cx^2 + \ldots + px^n.$$

This granted, the two equations may be arranged according to the powers of y, and the coefficients of the powers of y may be arranged

---

* We do not mean to imply that the *only* use made of the Jacquard cards is that of regulating the algebraical *operations*; but we mean to explain that *those* cards and portions of mechanism which regulate these *operations* are wholly independent of those which are used for other purposes. M. Menabrea explains that there are *three* classes of cards used in the engine for three distinct sets of objects, viz. *Cards of the Operations*, *Cards of the Variables*, and certain *Cards of Numbers*. (See pages 233 and 242.)

according to powers of x. The elimination of y will result from the successive multiplications and subtractions of several such functions. In this, and in all other instances, as was explained above, the particular *numerical* data and the *numerical* results are determined by means and by portions of the mechanism which act quite independently of those that regulate the *operations*.

In studying the action of the Analytical Engine, we find that the peculiar and independent nature of the considerations which in all mathematical analysis belong to *operations*, as distinguished from *the objects operated upon* and from the *results* of the operations performed upon those objects, is very strikingly defined and separated.

It is well to draw attention to this point, not only because its full appreciation is essential to the attainment of any very just and adequate general comprehension of the powers and mode of action of the Analytical Engine, but also because it is one which is perhaps too little kept in view in the study of mathematical science in general. It is, however, impossible to confound it with other considerations, either when we trace the manner in which that engine attains its results, or when we prepare the data for its attainment of those results. It were much to be desired, that when mathematical processes pass through the human brain instead of through the medium of inanimate mechanism, it were equally a necessity of things that the reasonings connected with *operations* should hold the same just place as a clear and well-defined branch of the subject of analysis, a fundamental but yet independent ingredient in the science, which they must do in studying the engine. The confusion, the difficulties, the contradictions which, in consequence of a want of accurate distinctions in this particular, have up to even a recent period encumbered mathematics in all those branches involving the consideration of negative and impossible quantities, will at once occur to the reader who is at all versed in this science, and would alone suffice to justify dwelling somewhat on the point, in connexion with any subject so peculiarly fitted to give forcible illustration of it as the Analytical Engine. It may be desirable to explain, that by the word *operation*, we mean *any process which alters the mutual relation of two or more things*, be this relation of what kind it may. This is the most general definition, and would include all subjects in the universe. In abstract mathematics, of course operations alter those particular relations which are involved in the considerations of number and space, and the *results* of operations are those peculiar results which correspond to the nature of the subjects of operation. But the science of operations, as derived from mathematics more especially, is a science of itself, and has its own abstract truth and value; just as logic has its own peculiar truth and value, independently of the subjects to which we may apply its reasonings and processes. Those who are accustomed to some of the more modern views of the above subject, will know that a few fundamental relations being true, certain other combinations of relations must of necessity follow; combinations unlimited in variety and extent if the deductions from the primary relations be carried on far enough. They will also be aware that one main reason why the separate nature of the science of operations has been little felt, and in general little dwelt on, is the *shifting* meaning of many of the symbols used in mathematical notation. First, the symbols of *operation* are frequently *also* the symbols of the *results* of operations. We may say that these symbols are apt to have both a *retrospective* and a *prospective* signification. They may signify either relations that are the consequences of a series of processes already performed, or relations that are yet to be effected through certain processes. Secondly, figures, the symbols of *numerical magnitude*, are frequently *also* the symbols of *operations*, as when they are the indices of powers. Wherever terms have a shifting meaning, independent sets of considerations are liable to become complicated together, and reasonings and results are frequently falsified. Now in the Analytical Engine, the operations which come under the first of the above heads are ordered and combined by means of a notation and of a train of mechanism which belong exclusively to themselves; and with respect to the second head, whenever numbers meaning *operations* and not *quantities* (such as the indices of powers) are inscribed on any column or set of columns, those columns immediately act in a wholly separate and independent manner, becoming connected with the *operating mechanism* exclusively, and re-acting upon this. They never come into combination with numbers upon any other columns meaning *quantities*; though, of course, if there are numbers meaning *operations* upon *n* columns, these may *combine amongst each other*, and will often be required to do so, just as numbers meaning *quantities* combine with each other in any variety. It might have been arranged that all numbers meaning *operations* should have appeared on some separate portion of the engine from that which presents numerical *quantities*; but the present mode is in some cases more simple, and offers in reality quite as much distinctness when understood.

The operating mechanism can even be thrown into action independently of any object to operate upon (although of course no *result* could then be developed). Again, it might act upon other things besides *number*, were objects found whose mutual fundamental relations could be expressed by those of the abstract science of operations, and which should be also susceptible of adaptations to the action of the operating notation and mechanism of the engine. Supposing, for instance, that the fundamental relations of pitched sounds in the science of harmony and of musical composition were susceptible of such expression and adaptations, the engine might compose elaborate and scientific pieces of music of any degree of complexity or extent.

The Analytical Engine is an *embodying of the science of operations*, constructed with peculiar reference to abstract number as the subject of those operations. The Difference Engine is the embodying of *one particular and very limited set of operations*, which (see the notation used in Note B) may be expressed thus $(+, +, +, +, +, +)$, or thus, $6(+)$. Six repetitions of the one operation, $+$, is, in fact, the whole sum and object of that engine. It has seven columns, and a number on any column can add itself to a number on the next column to its *right-hand*. So that, beginning with the column furthest to the left, six additions can be effected, and the result appears on the seventh column, which is the last on the right-hand. The *operating* mechanism of this engine acts in as separate and independent a manner as that of the Analytical Engine; but being susceptible of only one unvarying and restricted combination, it has little force or interest in illustration of the distinct nature of the *science of operations*. The importance of regarding the Analytical Engine under this point of view will, we think, become more and more obvious as the reader proceeds with M. Menabrea's clear and masterly article. The calculus of operations is likewise in itself a topic of so much interest, and has of late years been so much more written on and thought on than formerly, that any bearing which that engine, from its mode of constitution, may possess upon the illustration of this branch of mathematical science should not be overlooked. Whether the inventor of this engine had any such views in his mind while working out the invention, or whether he may subsequently ever have regarded it under this phase, we do not know; but it is one that forcibly occurred to ourselves on becoming acquainted with the means through which analytical combinations are actually attained by the mechanism. We cannot forbear suggesting one practical result which it appears to us must be greatly facilitated by the independent manner in which the engine orders and combines its *operations*: we allude to the attainment of those combinations into which *imaginary quantities* enter. This is a branch of its processes into which we have not had the opportunity of inquiring, and our conjecture therefore as to the principle on which we conceive the accomplishment of such results may have been made to depend, is very probably not in accordance with the fact, and less subservient for the purpose than some other principles, or at least requiring the cooperation of others. It seems to us obvious, however, that where operations are so independent in their mode of acting, it must be easy, by means of a few simple provisions, and additions in arranging the mechanism, to bring out a *double* set of *results*, viz.—1st, the *numerical magnitudes* which are the results of operations performed on *numerical data*. (These results are the *primary* object of the engine.) 2ndly, the *symbolical results* to be attached to those numerical results, which symbolical results are not less the necessary and logical consequences of operations performed upon *symbolical data*, than are numerical results when the data are numerical*.

If we compare together the powers and the principles of construction of the Difference and of the Analytical Engines, we shall perceive that the capabilities of the latter are immeasurably more extensive than those of the former, and that they in fact hold to each other the same relationship as that of analysis to arithmetic. The Difference Engine can effect but one particular series of operations, viz. that required for tabulating the integral of the special function

$$\Delta^n u_z = 0;$$

and as it can only do this for values of n up to 7†, it cannot be considered as being the most *general* expression even of *one particular* function, much less as being the expression of any and all possible functions of all degrees of generality. The Difference Engine can in reality (as has been already partly explained) do nothing but *add*; and any other processes, not excepting those of simple subtraction, multiplication and division, can be performed by it only just to that extent in which it is possible, by judicious mathematical arrangement and artifices, to reduce them to a *series of additions*. The method of

---

* In fact, such an extension as we allude to would merely constitute a further and more perfected development of any system introduced for making the proper combinations of the signs *plus* and *minus*. How ably M. Menabrea has touched on this restricted case is pointed out in Note B, p. 258.

† The machine might have been constructed so as to tabulate for a higher value of n than seven. Since, however, every unit added to the value of n increases the extent of the mechanism requisite, there would on this account be a limit beyond which it could not be practically carried. Seven is sufficiently high for the calculation of all ordinary tables.

The fact that, in the Analytical Engine, the same extent of mechanism suffices for the solution of $\Delta^n u_z = 0$, whether $n=7$, $n=100,000$, or $n=$ any number whatever, at once suggests how entirely distinct must be the *nature of the principles* through whose application matter has been enabled to become the working agent of abstract mental operations in each of these engines respectively; and it affords an equally obvious presumption, that in the case of the Analytical Engine, not only are those principles in themselves of a higher and more comprehensive description, but also such as must vastly extend the *practical* value of the engine whose basis they constitute.

differences is, in fact, a method of additions; and as it includes within its means a larger number of results attainable by *addition* simply, than any other mathematical principle, it was very appropriately selected as the basis on which to construct *an Adding Machine*, so as to give to the powers of such a machine the widest possible range.   The Analytical Engine, on the contrary, can either add, subtract, multiply or divide with equal facility; and performs each of these four operations in a direct manner, without the aid of any of the other three.   This one fact implies everything; and it is scarcely necessary to point out, for instance, that while the Difference Engine can merely *tabulate*, and is incapable of *developing*, the Analytical Engine can *either tabulate or develope*.

The former engine is in its nature strictly *arithmetical*, and the results it can arrive at lie within a very clearly defined and restricted range, while there is no finite line of demarcation which limits the powers of the Analytical Engine.   These powers are co-extensive with our knowledge of the laws of analysis itself, and need be bounded only by our acquaintance with the latter.   Indeed we may consider the engine as the *material and mechanical representative* of analysis, and that our actual working powers in this department of human study will be enabled more effectually than heretofore to keep pace with our theoretical knowledge of its principles and laws, through the complete control which the engine gives us over the *executive manipulation* of algebraical and numerical symbols.

Those who view mathematical science, not merely as a vast body of abstract and immutable truths, whose intrinsic beauty, symmetry and logical completeness, when regarded in their connexion together as a whole, entitle them to a prominent place in the interest of all profound and logical minds, but as possessing a yet deeper interest for the human race, when it is remembered that this science constitutes the language through which alone we can adequately express the great facts of the natural world, and those unceasing changes of mutual relationship which, visibly or invisibly, consciously or unconsciously to our immediate physical perceptions, are interminably going on in the agencies of the creation we live amidst: those who thus think on mathematical truth as the instrument through which the weak mind of man can most effectually read his Creator's works, will regard with especial interest all that can tend to facilitate the translation of its principles into explicit practical forms.

The distinctive characteristic of the Analytical Engine, and that which has rendered it possible to endow mechanism with such extensive faculties as bid fair to make this engine the executive right-hand of abstract algebra, is the introduction into it of the principle which Jacquard devised for regulating, by means of punched cards, the most complicated patterns in the fabrication of brocaded stuffs. It is in this that the distinction between the two engines lies. Nothing of the sort exists in the Difference Engine. We may say most aptly, that the Analytical Engine *weaves algebraical patterns* just as the Jacquard-loom weaves flowers and leaves. Here, it seems to us, resides much more of originality than the Difference Engine can be fairly entitled to claim. We do not wish to deny to this latter all such claims. We believe that it is the only proposal or attempt ever made to construct a calculating machine *founded on the principle of successive orders of differences*, and capable of *printing off its own results*; and that this engine surpasses its predecessors, both in the extent of the calculations which it can perform, in the facility, certainty and accuracy with which it can effect them, and in the absence of all necessity for the intervention of human intelligence *during the performance of its calculations*. Its nature is, however, limited to the strictly arithmetical, and it is far from being the first or only scheme for constructing *arithmetical* calculating machines with more or less of success.

The bounds of *arithmetic* were however outstepped the moment the idea of applying the cards had occurred; and the Analytical Engine does not occupy common ground with mere "calculating machines." It holds a position wholly its own; and the considerations it suggests are most interesting in their nature. In enabling mechanism to combine together *general* symbols in successions of unlimited variety and extent, a uniting link is established between the operations of matter and the abstract mental processes of the *most abstract* branch of mathematical science. A new, a vast, and a powerful language is developed for the future use of analysis, in which to wield its truths so that these may become of more speedy and accurate practical application for the purposes of mankind than the means hitherto in our possession have rendered possible. Thus not only the mental and the material, but the theoretical and the practical in the mathematical world, are brought into more intimate and effective connexion with each other. We are not aware of its being on record that anything partaking in the nature of what is so well designated the *Analytical* Engine has been hitherto proposed, or even thought of, as a practical possibility, any more than the idea of a thinking or of a reasoning machine.

We will touch on another point which constitutes an important distinction in the modes of operating of the Difference and Analytical Engines. In order to enable the former to do its business, it is necessary to put into its columns the series of numbers constituting the first terms of the several orders of differences for whatever is the particular table under consideration. The machine then works *upon* these as its data. But these data must themselves have been already computed through a series of calculations by a human head. Therefore that engine can only produce results depending on data which have been arrived at by the explicit and actual working out of processes that are in their nature different from any that come within the sphere of its own powers. In other words, an *analysing* process must have been gone through by a human mind in order to obtain the data upon which the engine then *synthetically* builds its results. The Difference Engine is in its character exclusively *synthetical*, while the Analytical Engine is equally capable of analysis or of synthesis.

It is true that the Difference Engine can calculate to a much greater extent with these few preliminary data, than the data themselves required for their own determination. The table of squares, for instance, can be calculated to any extent whatever, when the numbers *one* and *two* are furnished; and a very few differences computed at any part of a table of logarithms would enable the engine to calculate many hundreds or even thousands of logarithms. Still the circumstance of its requiring, as a previous condition, that any function whatever shall have been numerically worked out, makes it very inferior in its nature and advantages to an engine which, like the Analytical Engine, requires merely that we should know the *succession and distribution of the operations* to be performed; without there being any occasion*, in order to obtain data on which it can work, for our ever having gone through either the same particular operations which it is itself to effect, or any others. Numerical data must of course be given it, but they are mere arbitrary ones; not data that could only be arrived at through a systematic and necessary series of previous numerical calculations, which is quite a different thing.

To this it may be replied, that an analysing process must equally have been performed in order to furnish the Analytical Engine with the necessary *operative* data; and that herein may also lie a possible source of error. Granted that the actual mechanism is unerring in its processes, the *cards* may give it wrong orders. This is unquestionably the case; but there is much less chance of error, and likewise far less expenditure of time and labour, where operations only, and the distribution of these operations, have to be made out, than where explicit numerical results are to be attained. In the case of the Analytical Engine we have undoubtedly to lay out a certain capital

* This subject is further noticed in Note F, p. 281.

of analytical labour in one particular line; but this is in order that the engine may bring us in a much larger return in another line.   It should be remembered also that the cards, when once made out for any formula, have all the generality of algebra, and include an infinite number of particular cases.

We have dwelt considerably on the distinctive peculiarities of each of these engines, because we think it essential to place their respective attributes in strong relief before the apprehension of the public; and to define with clearness and accuracy the wholly different nature of the principles on which each is based, so as to make it self-evident to the reader (the mathematical reader at least) in what manner and degree the powers of the Analytical Engine transcend those of an engine, which, like the Difference Engine, can only work out such results as may be derived from *one restricted and particular series of processes*, such as those included in $\Delta^n u_z = 0$.   We think this of import- ance, because we know that there exists considerable vagueness and inaccuracy in the mind of persons in general on the subject.   There is a misty notion amongst most of those who have attended at all to it, that *two* "calculating machines" have been successively invented by the same person within the last few years; while others again have never heard but of the one original "calculating machine," and are not aware of there being any extension upon this.   For either of these two classes of persons the above considerations are appropriate. While the latter require a knowledge of the fact that there *are two* such inventions, the former are not less in want of accurate and well- defined information on the subject.   No very clear or correct ideas prevail as to the characteristics of each engine, or their respective advantages or disadvantages; and in meeting with those incidental allusions, of a more or less direct kind, which occur in so many publica- tions of the day, to these machines, it must frequently be matter of doubt *which* "calculating machine" is referred to, or whether *both* are included in the general allusion.

We are desirous likewise of removing two misapprehensions which we know obtain, to some extent, respecting these engines.   In the first place it is very generally supposed that the Difference Engine, after it had been completed up to a certain point, *suggested* the idea of the Analytical Engine; and that the second is in fact the improved offspring of the first, and *grew out* of the existence of its predecessor, through some natural or else accidental combination of ideas suggested by this one.   Such a supposition is in this instance contrary to the facts; although it seems to be almost an obvious inference, wherever two inventions, similar in their nature and objects, succeed each other closely in order of *time*, and strikingly in order of *value*; more especially when the same individual is the author of both. Nevertheless the ideas which led to the Analytical Engine occurred in a manner wholly independent of any that were connected with the Difference Engine. These ideas are indeed in their own intrinsic nature independent of the latter engine, and might equally have occurred had it never existed nor been even thought of at all.

The second of the misapprehensions above alluded to relates to the well-known suspension, during some years past, of all progress in the construction of the Difference Engine. Respecting the circumstances which have interfered with the actual completion of either invention, we offer no opinion; and in fact are not possessed of the data for doing so, had we the inclination. But we know that some persons suppose these obstacles (be they what they may) to have arisen *in consequence* of the subsequent invention of the Analytical Engine while the former was in progress. We have ourselves heard it even *lamented* that an idea should ever have occurred at all, which had turned out to be merely the means of arresting what was already in a course of successful execution, without substituting the superior invention in its stead. This notion we can contradict in the most unqualified manner. The progress of the Difference Engine had long been suspended, before there were even the least crude glimmerings of any invention superior to it. Such glimmerings, therefore, and their subsequent development, were in no way the original *cause* of that suspension; although, where difficulties of some kind or other evidently already existed, it was not perhaps calculated to remove or lessen them that an invention should have been meanwhile thought of, which, while including all that the first was capable of, possesses powers so extended as to eclipse it altogether.

We leave it for the decision of each individual (*after he has possessed himself* of competent information as to the characteristics of each engine) to determine how far it ought to be matter of regret that such an accession has been made to the powers of human science, even if it *has* (which we greatly doubt) increased to a certain limited extent some already existing difficulties that had arisen in the way of completing a valuable but lesser work. We leave it for each to satisfy himself as to the wisdom of desiring the obliteration (were that now possible) of all records of the more perfect invention, in order that the comparatively limited one might be finished. The Difference Engine would doubtless fulfil all those practical objects which it was originally destined for. It would certainly calculate all the tables that are more directly necessary for the physical purposes of life, such as nautical and other computations. Those who incline to very strictly utilitarian views may perhaps feel that the peculiar powers of the Analytical Engine bear upon questions of abstract and speculative science, rather than upon those involving every-day and ordinary human interests. These persons being likely to possess but little sympathy, or possibly acquaintance, with any branches of science which they do not find to be *useful* (according to *their* definition of that word), may conceive that the undertaking of that engine, now that the other one is already in progress, would be a barren and unproductive laying out of yet more money and labour; in fact, a work of supererogation. Even in the utilitarian aspect, however, we do not doubt that very valuable practical results would be developed by the extended faculties of the Analytical Engine; some of which results we think we could now hint at, had we the space; and others, which it may not yet be possible to foresee, but which would be brought forth by the daily increasing requirements of science, and by a more intimate practical acquaintance with the powers of the engine, were it in actual existence.

On general grounds, both of an *a priori* description as well as those founded on the scientific history and experience of mankind, we see strong presumptions that such would be the case. Nevertheless all will probably concur in feeling that the completion of the Difference Engine would be far preferable to the non-completion of any calculating engine at all. With whomsoever or wheresoever may rest the present causes of difficulty that apparently exist towards either the completion of the old engine, or the commencement of the new one, we trust they will not ultimately result in this generation's being acquainted with these inventions through the medium of pen, ink and paper merely; and still more do we hope, that for the honour of our country's reputation in the future pages of history, these causes will not lead to the completion of the undertaking by some *other* nation or government. This could not but be matter of just regret; and equally so, whether the obstacles may have originated in private interests and feelings, in considerations of a more public description, or in causes combining the nature of both such solutions.

We refer the reader to the 'Edinburgh Review' of July 1834, for a very able account of the Difference Engine.* The writer of the article we allude to has selected as his prominent matter for exposition, a wholly different view of the subject from that which M. Menabrea has chosen. The former chiefly treats it under its mechanical aspect, entering but slightly into the mathematical principles of which that

* [See page 163 of this volume.]

engine is the representative, but giving, in considerable length, many details of the mechanism and contrivances by means of which it tabulates the various orders of differences. M. Menabrea, on the contrary, exclusively developes the analytical view; taking it for granted that mechanism is able to perform certain processes, but without attempting to explain *how*; and devoting his whole attention to explanations and illustrations of the manner in which analytical laws can be so arranged and combined as to bring every branch of that vast subject within the grasp of the assumed powers of mechanism. It is obvious that, in the invention of a calculating engine, these two branches of the subject are equally essential fields of investigation, and that on their mutual adjustment, one to the other, must depend all success. They must be made to meet each other, so that the weak points in the powers of either department may be compensated by the strong points in those of the other. They are indissolubly connected, though so different in their intrinsic nature, that perhaps the same mind might not be likely to prove equally profound or successful in both. We know those who doubt whether the powers of mechanism will in practice prove adequate in all respects to the demands made upon them in the working of such complicated trains of machinery as those of the above engines, and who apprehend that unforeseen practical difficulties and disturbances will arise in the way of accuracy and of facility of operation. The Difference Engine, however, appears to us to be in a great measure an answer to these doubts. It is complete as far as it goes, and it does work with all the anticipated success. The Analytical Engine, far from being more complicated, will in many respects be of simpler construction; and it is a remarkable circumstance attending it, that with very *simplified* means it is so much more powerful.

The article in the 'Edinburgh Review' was written some time previous to the occurrence of any ideas such as afterwards led to the invention of the Analytical Engine; and in the nature of the Difference Engine there is much less that would invite a writer to take exclusively, or even prominently, the mathematical view of it, than in that of the Analytical Engine; although mechanism has undoubtedly gone much further to meet mathematics, in the case of this engine, than of the former one. Some publication embracing the *mechanical* view of the Analytical Engine is a desideratum which we trust will be supplied before long.

Those who may have the patience to study a moderate quantity of rather dry details will find ample compensation, after perusing the article of 1834, in the clearness with which a succinct view will have been attained of the various practical steps through which mechanism can accomplish certain processes; and they will also find themselves still further capable of appreciating M. Menabrea's more comprehensive and generalized memoir. The very difference in the style and object of these two articles makes them peculiarly valuable to each other; at least for the purposes of those who really desire something more than a merely superficial and popular comprehension of the subject of calculating engines. A. A. L.

Note B

That portion of the Analytical Engine here alluded to is called the storehouse. It contains an indefinite number of the columns of discs described by M. Menabrea. The reader may picture to himself a pile of rather large draughtsmen heaped perpendicularly one above another to a considerable height, each counter having the digits from 0 to 9 inscribed on its *edge* at equal intervals; and if he then conceives that the counters do not actually lie one upon another so as to be in contact, but are fixed at small intervals of vertical distance on a common axis which passes perpendicularly through their centres, and around which each disc can *revolve horizontally* so that any required digit amongst those inscribed on its margin can be brought into view, he will have a good idea of one of these columns. The *lowest* of the discs on any column belongs to the units, the next above to the tens, the next above this to the hundreds, and so on. Thus, if we wished to inscribe 1345 on a column of the engine, it would stand thus:—

1
3
4
5

In the Difference Engine there are seven of these columns placed side by side in a row, and the working mechanism extends behind them: the general form of the whole mass of machinery is that of a quadrangular prism (more or less approaching to the cube); the results always appearing on that perpendicular face of the engine which contains the columns of discs, opposite to which face a spectator may place himself. In the Analytical Engine there would be many more of these columns, probably at least two hundred. The precise form and arrangement which the whole mass of its mechanism will assume is not yet finally determined.

We may conveniently represent the columns of discs on paper in a diagram like the following :—

$$V_1 \quad V_2 \quad V_3 \quad V_4 \quad \&c.$$

V_1	V_2	V_3	V_4	
○	○	○	○	&c.
0	0	0	0	
0	0	0	0	
0	0	0	0	&c.
0	0	0	0	
□	□	□	□	&c.

The V's are for the purpose of convenient reference to any column, either in writing or speaking, and are consequently numbered. The reason why the letter V is chosen for the purpose in preference to any other letter, is because these columns are designated (as the reader will find in proceeding with the Memoir) the *Variables*, and sometimes the *Variable columns*, or the *columns of Variables*. The origin of this appellation is, that the values on the columns are destined to change, that is to *vary*, in every conceivable manner. But it is necessary to guard against the natural misapprehension that the columns are only intended to receive the values of the *variables* in an analytical formula, and not of the *constants*. The columns are called Variables on a ground wholly unconnected with the *analytical* distinction between constants and variables. In order to prevent the possibility of confusion, we have, both in the translation and in the notes, written Variable with a capital letter when we use the word to signify a *column of the engine*, and variable with a small letter when we mean the *variable of a formula*. Similarly, *Variable-cards* signify any cards that belong to a column of the engine.

To return to the explanation of the diagram: each circle at the top is intended to contain the algebraic sign + or −, either of which can be substituted* for the other, according as the number represented on the column below is positive or negative. In a similar manner any other purely *symbolical* results of algebraical processes might be made

---

* A fuller account of the manner in which the *signs* are regulated is given in M. Menabrea's Memoir, pages 237, 238. He himself expresses doubts (in a note of his own at the bottom of the latter page) as to his having been likely to hit on the precise methods really adopted; his explanation being merely a conjectural one. That it *does* accord precisely with the fact is a remarkable circumstance, and affords a convincing proof how completely M. Menabrea has been imbued with the true spirit of the invention. Indeed the whole of the above Memoir is a striking production, when we consider that M. Menabrea had had but very slight means for obtaining any adequate ideas respecting the Analytical Engine. It requires however a considerable acquaintance with the abstruse and complicated nature of such a subject, in order fully to appreciate the penetration of the writer who could take so just and comprehensive a view of it upon such limited opportunity.

to appear in these circles.   In Note A. the practicability of developing *symbolical* with no less ease than *numerical* results has been touched on. The zeros beneath the *symbolic* circles represent each of them a disc, supposed to have the digit 0 presented in front.   Only four tiers of zeros have been figured in the diagram, but these may be considered as representing thirty or forty, or any number of tiers of discs that may be required.   Since each disc can present any digit, and each circle any sign, the discs of every column may be so adjusted* as to express any positive or negative number whatever within the limits of the machine; which limits depend on the *perpendicular* extent of the mechanism, that is, on the number of discs to a column.

Each of the squares below the zeros is intended for the inscription of any *general* symbol or combination of symbols we please; it being understood that the number represented on the column immediately above is the numerical value of that symbol, or combination of symbols.   Let us, for instance, represent the three quantities a, n, x, and let us further suppose that $a=5$, $n=7$, $x=98$.   We should have—

V_1	V_2	V_3	V_4 &c.
$+\dagger$	$+$	$+$	$+$
0	0	0	0
0	0	0	0 &c.
0	0	9	0
5	7	8	0 &c.
$\boxed{a}$	$\boxed{n}$	$\boxed{x}$	$\boxed{}$

We may now combine these symbols in a variety of ways, so as to form any required function or functions of them, and we may then inscribe each such function below brackets, every bracket uniting together those quantities (and those only) which enter into the function inscribed below it.   We must also, when we have decided on the particular function whose numerical value we desire to calculate, assign another column to the right-hand for receiving the *results*, and must inscribe the function in the square below this column.   In the above instance we might have any one of the following functions:—

$$ax^n, \quad x^{an}, \quad a.n.x, \quad \frac{a}{n}x, \quad a+n+x, \quad \text{&c. &c.}$$

---

* This adjustment is done by hand merely.

† It is convenient to omit the circles whenever the signs + or − can be actually represented.

Let us select the first.   It would stand as follows, previous to calculation:—

V_1	V_2	V_3	V_4 &c.
$+$	$+$	$+$	$+$
0	0	0	0 &c.
0	0	0	0
0	0	9	0
5	7	8	0 &c.
$\boxed{a}$	$\boxed{n}$	$\boxed{x}$	$\boxed{ax^n}$ &c.

$$\underbrace{}_{ax_n}$$

The data being given, we must now put into the engine the cards proper for directing the operations in the case of the particular function chosen.   These operations would in this instance be,—

First, six multiplications in order to get x^n ($=98^7$ for the above particular data).

Secondly, one multiplication in order then to get $a.x^n$ ($=5\cdot98^7$).

In all, seven multiplications to complete the whole process.   We may thus represent them:—

$$(\times, \times, \times, \times, \times, \times, \times), \quad \text{or} \quad 7\,(\times).$$

The multiplications would, however, at successive stages in the solution of the problem, operate on pairs of numbers, derived from *different* columns.   In other words, the *same operation* would be performed on different *subjects of operation*.   And here again is an illustration of the remarks made in the preceding Note on the independent manner in which the engine directs its *operations*.   In determining the value of ax^n, the *operations* are *homogeneous*, but are distributed amongst different *subjects of operation*, at successive stages of the computation. It is by means of certain punched cards, belonging to the Variables themselves, that the action of the operations is so *distributed* as to suit each particular function.   The *Operation-cards* merely determine the succession of operations in a general manner.   They in fact throw all that portion of the mechanism included in the *mill* into a series of different *states*, which we may call the *adding state*, or the *multiplying state*, &c. respectively.   In each of these states the mechanism is ready to act in the way peculiar to that state, on any pair of numbers which may be permitted to come within its sphere of action.   Only *one* of these operating states of the mill can exist at a time; and the nature of the mechanism is also such that only *one pair of numbers* can be received and acted on at a time.   Now, in order to secure that the mill shall receive a constant supply of the proper pairs of numbers in succession, and that it shall also rightly locate the result of an operation performed upon any pair, each Variable has cards of its own belonging to it. It has, first, a class of cards whose business it is to *allow* the number on the Variable to pass into the mill, there to be operated upon.   These cards may be called the *Supplying-cards*.   They furnish the mill with its proper food.   Each Variable has, secondly, another class of cards, whose office it is to allow the Variable to *receive* a number *from* the mill. These cards may be called the *Receiving-cards*.   They regulate the location of results, whether temporary or ultimate results.   The Variable-cards in general (including both the preceding classes) might, it appears to us, be even more appropriately designated the Distributive-cards, since it is through their means that the action of the operations, and the results of this action, are rightly *distributed*.

There are *two varieties* of the *Supplying* Variable-cards, respectively adapted for fulfilling two distinct subsidiary purposes: but as these modifications do not bear upon the present subject, we shall notice them in another place.

In the above case of *ax^n*, the Operation-cards merely order seven multiplications, that is, they order the mill to be in the *multiplying state* seven successive times (without any reference to the particular columns whose numbers are to be acted upon).   The proper Distributive Variable-cards step in at each successive multiplication, and cause the distributions requisite for the particular case.

For x^{an}　　　the operations would be　　　　34 ($\times$)

... $a.n.x$　　...　　...　　...　　($\times$, $\times$), or 2 ($\times$)

... $\dfrac{a}{n}.x$　　...　　...　　...　　($\div$, $\times$)

... $a+n+x$　　...　　...　　...　　(+, +), or 2 (+)

The engine might be made to calculate all these in succession.   Having completed *ax^n*, the function *x^{an}* might be written under the brackets instead of *ax^n*, and a new calculation commenced (the appropriate Operation and Variable-cards for the new function of course coming into play).   The results would then appear on V_5.   So on for any number of different functions of the quantities *a*, *n*, *x*.   Each *result* might either permanently remain on its column during the succeeding calculations, so that when all the functions had been computed, their values would simultaneously exist on V_4, V_5, V_6, &c.; or each result might (after being printed off, or used in any specified manner) be effaced, to make way for its successor.   The square under V_4 ought, for the latter arrangement, to have the functions ax^n, x^{an}, anx, &c. successively inscribed in it.

Let us now suppose that we have *two* expressions whose values have been computed by the engine independently of each other (each having its own group of columns for data and results). Let them be ax^n, and bpy. They would then stand as follows on the columns:—

V_1	V_2	V_3	V_4	V_5	V_6	V_7	V_8	V_9
+	+	+	+	+	+	+	+	+
0	0	0	0	0	0	0	0	0
0	0	0	0	0	0	0	0	0
0	0	0	0	0	0	0	0	0
0	0	0	0	0	0	0	0	0
a	n	x	ax^n	b	p	y	bpy	ax^n
								$\overline{bpy}$

We may now desire to combine together these two *results*, in any manner we please; in which case it would only be necessary to have an additional card or cards, which should order the requisite operations to be performed with the numbers on the two result-columns V_4 and V_8, and the *result of these further operations* to appear on a new column, V_9. Say that we wish to divide ax^n by bpy. The numerical value of this division would then appear on the column V_9, beneath which we have inscribed $\frac{ax^n}{bpy}$. The whole series of operations from the beginning would be as follows (n being $=7$):

$$\{7(\times),\ 2\ (\times),\ \div\},\quad \text{or}\quad \{9(\times),\ \div\}.$$

This example is introduced merely to show that we may, if we please, retain separately and permanently any *intermediate* results (like ax^n, bpy) which occur in the course of processes having an ulterior and more complicated result as their chief and final object $\left(\text{like } \frac{ax^n}{bpy}\right)$.

Any group of columns may be considered as representing a *general* function, until a *special* one has been implicitly impressed upon them through the introduction into the engine of the Operation and Variable-cards made out for a *particular* function. Thus, in the preceding example, V_1, V_2, V_3, V_5, V_6, V_7 represent the *general* function $\phi(a, n, b, p, x, y)$ until the function $\frac{ax^n}{bpy}$ has been determined on, and *implicitly* expressed by the placing of the right cards in the engine. The actual working of the mechanism, as regulated by these cards, then *explicitly* developes the value of the function. The inscription of a function under the brackets, and in the square under the result-column, in no way influences the processes or the results, and is merely a memorandum for the observer, to remind him of what is going on. It is the Operation and the Variable-cards only which in reality determine the function. Indeed it should be distinctly kept in mind, that the inscriptions within *any* of the squares are quite independent of the mechanism or workings of the engine, and are nothing but arbitrary memorandums placed there at pleasure to assist the spectator.

The further we analyse the manner in which such an engine performs its processes and attains its results, the more we perceive how distinctly it places in a true and just light the mutual relations and connexion of the various steps of mathematical analysis; how clearly it separates those things which are in reality distinct and independent, and unites those which are mutually dependent.   A. A. L.

Note C

Those who may desire to study the principles of the Jacquard-loom in the most effectual manner, viz. that of practical observation, have only to step into the Adelaide Gallery or the Polytechnic Institution. In each of these valuable repositories of scientific *illustration*, a weaver is constantly working at a Jacquard-loom, and is ready to give any information that may be desired as to the construction and modes of acting of his apparatus. The volume on the manufacture of silk, in Lardner's Cyclopædia, contains a chapter on the Jacquard-loom, which may also be consulted with advantage.

The mode of application of the cards, as hitherto used in the art of weaving, was not found, however, to be sufficiently powerful for all the simplifications which it was desirable to attain in such varied and complicated processes as those required in order to fulfil the purposes of an Analytical Engine. A method was devised of what was technically designated *backing* the cards in certain groups according to certain laws. The object of this extension is to secure the possibility of bringing any particular card or set of cards into use *any number of times successively* in the solution of one problem. Whether this power shall be taken advantage of or not, in each particular instance, will depend on the nature of the operations which the problem under consideration may require. The process is alluded to by M. Menabrea in page 239, and it is a very important simplification. It has been proposed to use it for the reciprocal benefit of that art, which, while it has itself no apparent connexion with the domains of abstract science, has yet proved so valuable to the latter, in suggesting the principles which, in their new and singular field of application, seem likely to place *algebraical* combinations not less completely within the province of mechanism, than are all those varied intricacies of which *intersecting threads* are susceptible.   By the introduction of the system of *backing* into the Jacquard-loom itself, patterns which should possess symmetry, and follow regular laws of any extent, might be woven by means of comparatively few cards.

Those who understand the mechanism of this loom will perceive that the above improvement is easily effected in practice, by causing the prism over which the train of pattern-cards is suspended to revolve *backwards* instead of *forwards*, at pleasure, under the requisite circumstances; until, by so doing, any particular card, or set of cards, that has done duty once, and passed on in the ordinary regular succession, is brought back to the position it occupied just before it was used the preceding time.   The prism then resumes its *forward* rotation, and thus brings the card or set of cards in question into play a second time. This process may obviously be repeated any number of times.

A. A. L.

Note D

We have represented the solution of these two equations, with every detail, in a diagram* similar to those used in Note B; but additional explanations are requisite, partly in order to make this more complicated case perfectly clear, and partly for the comprehension of certain indications and notations not used in the preceding diagrams.   Those who may wish to understand Note G completely, are recommended to pay particular attention to the contents of the present Note, or they will not otherwise comprehend the similar notation and indications when applied to a much more complicated case.

In all calculations, the columns of Variables used may be divided into three classes:—

1st. Those on which the data are inscribed:

2ndly. Those intended to receive the final results:

3rdly. Those intended to receive such intermediate and temporary combinations of the primitive data as are not to be permanently retained, but are merely needed for *working with*, in order to attain the ultimate results.   Combinations of this kind might properly be called

* See the diagram on page 266.

DIAGRAM BELONGING TO NOTE D

		Variables for Data						Working Variables									Variables for Results	
Number of Operations	Nature of Operations	$^1V_0$	$^1V_1$	$^1V_2$	$^1V_3$	$^1V_4$	$^1V_5$	$^0V_6$	$^0V_7$	$^0V_8$	$^0V_9$	$^0V_{10}$	$^0V_{11}$	$^0V_{12}$	$^0V_{13}$	$^0V_{14}$	$^0V_{15}$	$^0V_{16}$
		m	n	d	m'	n'	d'											
1	$\times$	m	n	d	m'	n'	d'	mn'										
2	$\times$		0	d	m'	n'	d'		$m'n$									
3	$\times$			0	m'	n'	d'			dn'								
4	$\times$				0	n'	d'				$d'n$							
5	$\times$						d' 0					$d'm$						
6	$\times$							0	0	0	0	0	dm'					
7	$-$							0	0				0	$(mn'-m'n)$				
8	$-$									0	0			0	$(dn'-d'n)$			
9	$-$											0	0		0	$(d'm-dm')$		
10	$\div$													$(mn'-m'n)$	0		$\dfrac{dn'-d'n}{mn'-m'n}=x$	
11	$\div$													$(mn'-m'n)$		0	$\cdots$	$\dfrac{d'm-dm'}{mn'-m'n}=y$

Results:

$$\frac{dn'-d'n}{mn'-m'n}=x \qquad \frac{d'm-dm'}{mn'-m'n}=y$$

secondary data. They are in fact so many *successive stages* towards the final result. The columns which receive them are rightly named *Working-Variables*, for their office is in its nature purely *subsidiary* to other purposes. They develope an intermediate and transient class of results, which unite the original data with the final results.

The Result-Variables sometimes partake of the nature of Working-Variables. It frequently happens that a Variable destined to receive a final result is the recipient of one or more intermediate values successively, in the course of the processes. Similarly, the Variables for data often become Working-Variables, or Result-Variables, or even both in succession. It so happens, however, that in the case of the present equations the three sets of offices remain throughout perfectly separate and independent.

It will be observed, that in the squares below the *Working*-Variables nothing is inscribed. Any one of these Variables is in many cases destined to pass through various values successively during the performance of a calculation (although in these particular equations no instance of this occurs). Consequently no *one fixed* symbol, or combination of symbols, should be considered as properly belonging to a merely *Working*-Variable; and as a general rule their squares are left blank. Of course in this, as in all other cases where we mention a *general* rule, it is understood that many particular exceptions may be expedient.

In order that all the indications contained in the diagram may be completely understood, we shall now explain two or three points, not hitherto touched on. When the value on any Variable is called into use, one of two consequences may be made to result. Either the value may *return* to the Variable after it has been used, in which case it is ready for a second use if needed; or the Variable may be made zero. (We are of course not considering a third case, of not unfrequent occurrence, in which the same Variable is destined to receive the *result* of the very operation which it has just supplied with a number.) Now the ordinary rule is, that the value *returns* to the Variable; unless it has been foreseen that no use for that value can recur, in which case zero is substituted. At the *end* of a calculation, therefore, every column ought as a general rule to be zero, excepting those for results. Thus it will be seen by the diagram, that when m, the value on V_0, is used for the second time by Operation 5, V_0 becomes 0, since m is not again needed; that similarly, when $(mn' - m'n)$, on V_{12}, is used for the third time by Operation 11, V_{12} becomes zero, since $(mn' - m'n)$ is not again needed. In order to provide for the one or the other of the courses above indicated, there are *two* varieties of the *Supplying* Variable-cards. One of these varieties has provisions which cause the number given off from any Variable to *return* to that Variable after doing its duty in the mill. The other variety has provisions which cause *zero* to be substituted on the Variable, for the number given off. These two varieties are distinguished, when needful, by the respective appellations of the *Retaining* Supply-cards and the *Zero* Supply-cards. We see that the *primary* office (see Note B.) of both these varieties of cards is the same; they only differ in their *secondary* office.

Every Variable thus has belonging to it *one* class of *Receiving* Variable-cards and *two* classes of *Supplying* Variable-cards. It is plain however that only the *one* or the *other* of these two latter classes can be used by any one Variable for *one* operation; never *both* simultaneously; their respective functions being mutually incompatible.

It should be understood that the Variable-cards are not placed in *immediate contiguity* with the columns. Each card is connected by means of wires with the column it is intended to act upon.

Our diagram ought in reality to be placed side by side with M. Menabrea's corresponding table, so as to be compared with it, line for line belonging to each operation. But it was unfortunately inconvenient to print them in this desirable form. The diagram is, in the main, merely another manner of indicating the various relations denoted in M. Menabrea's table. Each mode has some advantages and some disadvantages. Combined, they form a complete and accurate method of registering every step and sequence in all calculations performed by the engine.

No notice has yet been taken of the *upper* indices which are added to the left of each V in the diagram; an addition which we have also taken the liberty of making to the V's in M. Menabrea's tables of pages 236, 239, since it does not *alter* anything therein represented by him, but merely *adds* something to the previous indications of those tables. The *lower* indices are obviously indices of *locality* only, and are wholly independent of the operations performed or of the results obtained, their value continuing unchanged during the performance of calculations. The *upper* indices, however, are of a different nature. Their office is to indicate any *alteration* in the value which a Variable represents; and they are of course liable to changes during the processes of a calculation. Whenever a Variable has only zeros upon it, it is called ^0V; the moment a value appears on it (whether that value be placed there arbitrarily, or appears in the natural course of a calculation), it becomes ^1V. If this value gives place to another value, the Variable becomes ^2V, and so forth. Whenever a *value* again gives place to *zero*, the Variable again becomes ^0V, even if it have been ^nV the moment before. If a *value* then again be substituted, the Variable becomes ^{n+1}V (as it would have done if it had not passed through the intermediate ^0V); &c. &c.   Just before any calculation is commenced, and after the data have been given, and everything adjusted and prepared for setting the mechanism in action, the upper indices of the Variables for data are all unity, and those for the Working and Result-variables are all zero.   In this state the diagram represents them*.

There are several advantages in having a set of indices of this nature; but these advantages are perhaps hardly of a kind to be immediately perceived, unless by a mind somewhat accustomed to trace the successive steps by means of which the engine accomplishes its purposes.   We have only space to mention in a general way, that the whole notation of the tables is made more consistent by these indices, for they are able to mark a *difference* in certain cases, where there would otherwise be an apparent *identity* confusing in its tendency. In such a case as $V_n = V_p + V_n$ there is more clearness and more consistency with the usual laws of algebraical notation, in being able to write $^{m+1}V_n = {}^qV_p + {}^mV_n$.   It is also obvious that the indices furnish a powerful means of tracing back the derivation of any result; and of registering various circumstances concerning that *series of successive substitutions*, of which every *result* is in fact merely the final consequence; circumstances that may in certain cases involve relations which it is important to observe, either for purely analytical reasons, or for practically adapting the workings of the engine to their occurrence. The series of substitutions which lead to the equations of the diagram are as follow:—

$$\text{(1.)}\quad\quad\text{(2.)}\quad\quad\quad\text{(3.)}\quad\quad\quad\quad\quad\text{(4.)}$$

$$^{1}V^{*}_{16} = \frac{^{1}V_{14}}{^{1}V_{12}} = \frac{^{1}V_{10} - {}^{1}V_{11}}{^{1}V_{6} - {}^{1}V_{7}} = \frac{^{1}V_{0}\cdot{}^{1}V_{5} - {}^{1}V_{2}\cdot{}^{1}V_{3}}{^{1}V_{0}\cdot{}^{1}V_{4} - {}^{1}V_{3}\cdot{}^{1}V_{1}} = \frac{d'm - dm'}{mn' - m'n}$$

$$\text{(1.)}\quad\quad\text{(2.)}\quad\quad\quad\text{(3.)}\quad\quad\quad\quad\quad\text{(4.)}$$

$$^{1}V_{15} = \frac{^{1}V_{13}}{^{1}V_{12}} = \frac{^{1}V_{8} - {}^{1}V_{9}}{^{1}V_{6} - {}^{1}V_{7}} = \frac{^{1}V_{2}\cdot{}^{1}V_{4} - {}^{1}V_{5}\cdot{}^{1}V_{1}}{^{1}V_{0}\cdot{}^{1}V_{4} - {}^{1}V_{3}\cdot{}^{1}V_{1}} = \frac{dn' - d'n}{mn' - m'n}$$

There are *three* successive substitutions for each of these equations. The formulæ (2.), (3.) and (4.) are *implicitly* contained in (1.), which latter we may consider as being in fact the *condensed* expression of any of the former.   It will be observed that every succeeding substitution must contain *twice* as many V's as its predecessor.   So that if a problem

---

* We recommend the reader to trace the successive substitutions backwards from (1) to (4), in M. Menabrea's Table.   This he will easily do by means of the upper and lower indices, and it is interesting to observe how each V successively ramifies (so to speak) into two other V's in some other column of the Table, until at length the V's of the original data are arrived at.

require n substitutions, the successive series of numbers for the V's in the whole of them will be 2, 4, 8, 16...2^n.

The substitutions in the preceding equations happen to be of little value towards illustrating the power and uses of the upper indices, for, owing to the nature of these particular equations, the indices are all unity throughout. We wish we had space to enter more fully into the relations which these indices would in many cases enable us to trace.

M. Menabrea incloses the three centre columns of his table under the general title *Variable-cards*. The V's however in reality all represent the actual *Variable-columns* of the engine, and not the cards that belong to them. Still the title is a very just one, since it is through the special action of certain Variable-cards (when *combined* with the more generalized agency of the Operation-cards) that every one of the particular relations he has indicated under that title is brought about.

Suppose we wish to ascertain how often any *one* quantity, or combination of quantities, is brought into use during a calculation. We easily ascertain *this*, from the inspection of any vertical column or columns of the diagram in which that quantity may appear. Thus, in the present case, we see that all the data, and all the intermediate results likewise, are used twice, excepting $(mn' - m'n)$, which is used three times.

The *order* in which it is possible to perform the operations for the present example, enables us to effect all the eleven operations of which it consists with only *three Operation cards*; because the problem is of such a nature that it admits of each *class* of operations being performed in a group together; all the multiplications one after another, all the subtractions one after another, &c. The operations are $\{6(\times),\ 3(-),\ 2(\div)\}$.

Since the very definition of an operation implies that there must be *two* numbers to act upon, there are of course *two Supplying* Variable-cards necessarily brought into action for every operation, in order to furnish the two proper numbers. (See Note B.) Also, since every operation must produce a *result*, which must be placed *somewhere*, each operation entails the action of a *Receiving* Variable-card, to indicate the proper locality for the result. Therefore, at least three times as many Variable-cards as there are *operations* (not *Operation-cards*, for these, as we have just seen, are by no means always as numerous as the *operations*) are brought into use in every calculation. Indeed, under certain contingencies, a still larger proportion is requisite; such, for example, would probably be the case when the same result has to appear on more than one Variable simultaneously (which is not unfrequently a provision necessary for subsequent purposes in a calculation), and in some other cases which we shall not here specify.

We see therefore that a great disproportion exists between the amount of *Variable* and of *Operation*-cards requisite for the working of even the simplest calculation.

All calculations do not admit, like this one, of the operations of the same nature being performed in groups together. Probably very few do so without exceptions occurring in one or other stage of the progress; and some would not admit it at all. The *order* in which the operations shall be performed in every particular case is a very interesting and curious question, on which our space does not permit us fully to enter. In almost every computation a great *variety* of arrangements for the succession of the processes is possible, and various considerations must influence the selection amongst them for the purposes of a Calculating Engine. One essential object is to choose that arrangement which shall tend to reduce to a minimum the *time* necessary for completing the calculation.

It must be evident how multifarious and how mutually complicated are the considerations which the working of such an engine involve. There are frequently several distinct *sets of effects* going on simultaneously; all in a manner independent of each other, and yet to a greater or less degree exercising a mutual influence. To adjust each to every other, and indeed even to perceive and trace them out with perfect correctness and success, entails difficulties whose nature partakes to a certain extent of those involved in every question where *conditions* are very numerous and inter-complicated; such as for instance the estimation of the mutual relations amongst *statistical* phænomena, and of those involved in many other classes of facts.   A. A. L.

Note E

This example has evidently been chosen on account of its brevity and simplicity, with a view merely to explain the *manner* in which the engine would proceed in the case of an *analytical calculation containing variables*, rather than to illustrate the *extent of its powers* to solve cases of a difficult and complex nature. The equations of page 234 are in fact a more complicated problem than the present one.

We have not subjoined any diagram of its development for this new example, as we did for the former one, because this is unnecessary after the full application already made of those diagrams to the illustration of M. Menabrea's excellent tables.

It may be remarked that a slight discrepancy exists between the formulæ

$$(a + bx^1)$$
$$(A + B \cos^1 x)$$

given in the Memoir as the *data* for calculation, and the *results* of the calculation as developed in the last division of the table which accompanies it. To agree perfectly with this latter, the data should have been given as

$$(ax^0 + bx^1)$$
$$(A \cos^0 x + B \cos^1 x).$$

The following is a more complicated example of the manner in which the engine would compute a trigonometrical function containing variables. To multiply

$$A + A_1 \cos \theta + A_2 \cos 2\theta + A_3 \cos 3\theta + \ldots$$

by $B + B_1 \cos \theta.$

Let the resulting products be represented under the general form

$$C_0 + C_1 \cos \theta + C_2 \cos 2\theta + C_3 \cos 3\theta + \ldots \quad . \quad . \quad (1.)$$

This trigonometrical series is not only in itself very appropriate for illustrating the processes of the engine, but is likewise of much practical interest from its frequent use in astronomical computations. Before proceeding further with it, we shall point out that there are three very distinct classes of ways in which it may be desired to deduce numerical values from any analytical formula.

First. We may wish to find the collective numerical value of the *whole formula*, without any reference to the quantities of which that formula is a function, or to the particular mode of their combination and distribution, of which the formula is the result and representative. Values of this kind are of a strictly arithmetical nature in the most limited sense of the term, and retain no trace whatever of the processes through which they have been deduced. In fact, any one such numerical value may have been attained from an *infinite variety* of data, or of problems. The values for x and y in the two equations (see Note D.) come under this class of numerical results.

Secondly. We may propose to compute the collective numerical value of *each term* of a formula, or of a series, and to keep these results separate. The engine must in such a case appropriate as many columns to *results* as there are terms to compute.

Thirdly. It may be desired to compute the numerical value of various *subdivisions of each term*, and to keep all these results separate. It may be required, for instance, to compute each coefficient separately from its variable, in which particular case the engine must appropriate *two* result-columns to *every term that contains both a variable and coefficient*.

There are many ways in which it may be desired in special cases to distribute and keep separate the numerical values of different parts of an algebraical formula; and the power of effecting such distributions to any extent is essential to the *algebraical* character of the Analytical Engine. Many persons who are not conversant with mathematical studies, imagine that because the business of the engine is to give its results in *numerical notation*, the *nature of its processes* must consequently be *arithmetical* and *numerical*, rather than *algebraical* and *analytical*. This is an error. The engine can arrange and combine its numerical quantities exactly as if they were *letters* or any other *general* symbols; and in fact it might bring out its results in algebraical *notation*, were provisions made accordingly. It might develope three sets of results simultaneously, viz. *symbolic* results (as already alluded to in Notes A. and B.); *numerical* results (its chief and primary object); and *algebraical* results in *literal* notation. This latter however has not been deemed a necessary or desirable addition to its powers, partly because the necessary arrangements for effecting it would increase the complexity and extent of the mechanism to a degree that would not be commensurate with the advantages, where the main object of the invention is to translate into *numerical* language general formulæ of analysis already known to us, or whose laws of formation are known to us. But it would be a mistake to suppose that because its *results* are given in the *notation* of a more restricted science, its *processes* are therefore restricted to those of that science. The object of the engine is in fact to give the *utmost practical efficiency* to the resources of *numerical interpretations* of the higher science of analysis, while it uses the processes and combinations of this latter.

To return to the trigonometrical series. We shall only consider the first four terms of the factor $(A + A_1 \cos \theta + \&c.)$, since this will be sufficient to show the method. We propose to obtain separately the numerical value of *each coefficient* C_0, C_1, &c. of (1.). The direct multiplication of the two factors gives

$$\left. \begin{array}{l} BA + BA_1 \cos \theta + BA_2 \quad \cos 2\theta + BA_3 \quad \cos 3\theta + \dots \dots \dots \dots \dots \\ + B_1A \cos \theta + B_1A_1 \cos \theta . \cos \theta + B_1A_2 \cos 2\theta . \cos \theta + B_1A_3 \cos 3\theta . \cos \theta \end{array} \right\} \quad . \quad (2.)$$

a result which would stand thus on the engine:—

Variables for Data

V_0	V_1	V_2	V_3	$\dots\dots$ V_{10}	V_{11}
A	A_1	A_2	A_3	B	B_1
	$\cos \theta$	$\cos 2\theta$	$\cos 3\theta$		$\cos \theta$

Variables for Results

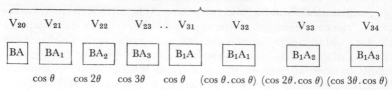

The variable belonging to each coefficient is written below it, as we have done in the diagram, by way of memorandum. The only further reduction which is at first apparently possible in the preceding result, would be the addition of V_{21} to V_{31} (in which case B_1A should be effaced from V_{31}). The whole operations from the beginning would then be—

First Series of Operations	*Second Series of Operations*	*Third Series, which contains only one (final) operation*
$^1V_{10} \times {}^1V_0 = {}^1V_{20}$	$^1V_{11} \times {}^1V_0 = {}^1V_{31}$	$^1V_{21} + {}^1V_{31} = {}^2V_{21}$, and
$^1V_{10} \times {}^1V_1 = {}^1V_{21}$	$^1V_{11} \times {}^1V_1 = {}^1V_{32}$	V_{31} becomes $= 0$.
$^1V_{10} \times {}^1V_2 = {}^1V_{22}$	$^1V_{11} \times {}^1V_2 = {}^1V_{33}$	
$^1V_{10} \times {}^1V_3 = {}^1V_{23}$	$^1V_{11} \times {}^1V_3 = {}^1V_{34}$	

We do not enter into the same detail of *every* step of the processes as in the examples of Notes D. and G., thinking it unnecessary and tedious to do so. The reader will remember the meaning and use of the upper and lower indices, &c., as before explained.

To proceed: we know that

$$\cos n\theta . \cos \theta = \tfrac{1}{2} \cos \overline{n+1}\theta + \tfrac{1}{2}\overline{n-1}.\theta \qquad . \quad . \quad (3.)$$

Consequently, a slight examination of the second line of (2.) will show that by making the proper substitutions, (2.) will become

BA	$+BA_1 \quad .\cos \theta$	$+BA_2 \quad .\cos 2\theta$	$+BA_3 \quad .\cos 3\theta$	
	$+B_1A \quad .\cos \theta$			
$+\tfrac{1}{2}B_1A_1$		$+\tfrac{1}{2}B_1A_1.\cos 2\theta$		
	$+\tfrac{1}{2}B_1A_2.\cos \theta$		$+\tfrac{1}{2}B_1A_2.\cos 3\theta$	
		$+\tfrac{1}{2}B_1A_3.\cos 2\theta$		$+\tfrac{1}{2}B_1A_3.\cos 4\theta$
C_0	C_1	C_2	C_3	C_4

These coefficients should respectively appear on

| V_{20} | V_{21} | V_{22} | V_{23} | V_{24}. |

We shall perceive, if we inspect the particular arrangement of the results in (2.) on the Result-columns as represented in the diagram, that, in order to effect this transformation, each successive coefficient upon V_{32}, V_{33}, &c. (beginning with V_{32}), must through means of proper cards be divided by *two**; and that one of the halves thus obtained must be added to the coefficient on the Variable which precedes it by ten columns, and the other half to the coefficient on the Variable which precedes it by twelve columns; V_{32}, V_{33}, &c. themselves becoming zeros during the process.

This series of operations may be thus expressed:—

Fourth Series

$$\begin{cases} ^1V_{32} \div 2 + {}^1V_{22} = {}^2V_{22} = BA_2 + \tfrac{1}{2}B_1A_1 \\ ^1V_{32} \div 2 \mid {}^1V_{20} - {}^2V_{20} = BA + \tfrac{1}{2}B_1A_1 \dots\dots = C_0 \end{cases}$$

$$\begin{cases} ^1V_{33} \div 2 + {}^1V_{23} = {}^2V_{23} = BA_3 + \tfrac{1}{2}B_1A_2 \dots\dots = C_3\dagger \\ ^1V_{33} \div 2 + {}^2V_{21} = {}^3V_{21} = BA_1 + \quad B_1A + \tfrac{1}{2}B_1A_2 = C_1 \end{cases}$$

$$\begin{cases} ^1V_{34} \div 2 + {}^0V_{24} = {}^1V_{24} = \tfrac{1}{2}B_1A_3 \dots\dots\dots = C_4 \\ ^1V_{34} \div 2 + {}^2V_{22} = {}^3V_{22} = BA_2 + \tfrac{1}{2}B_1A_1 + \tfrac{1}{2}B_1A_3 = C_2. \end{cases}$$

The calculation of the coefficients C_0, C_1, &c. of (1.) would now be completed, and they would stand ranged in order on V_{20}, V_{21}, &c. It will be remarked, that from the moment the fourth series of operations is ordered, the Variables V_{31}, V_{32}, &c. cease to be *Result*-Variables, and become mere *Working*-Variables.

The substitution made by the engine of the processes in the second side of (3.) for those in the first side is an excellent illustration of the manner in which we may arbitrarily order it to substitute any function, number, or process, at pleasure, for any other function, number or process, on the occurrence of a specified contingency.

We will now suppose that we desire to go a step further, and to obtain the numerical value of each *complete* term of the product (1.); that is, of each *coefficient and variable united*, which for the $(n+1)$th term would be $C_n \cdot \cos n\theta$.

We must for this purpose place the variables themselves on another set of columns, V_{41}, V_{42}, &c., and then order their successive multiplication by V_{21}, V_{22}, &c., each for each. There would thus be a final series of operations as follows:—

---

* This division would be managed by ordering the number 2 to appear on any separate new column which should be conveniently situated for the purpose, and then directing this column (which is in the strictest sense a *Working*-Variable) to divide itself successively with V_{32}, V_{33}, &c.

† It should be observed, that were the rest of the factor $(A + A \cos \theta + \&c.)$ taken into account, instead of *four* terms only, C_3 would have the additional term $\tfrac{1}{2}B_1A_4$; and C_4 the two additional terms, BA_4, $\tfrac{1}{2}B_1A_5$. This would indeed have been the case had even *six* terms been multiplied.

Fifth and Final Series of Operations

$$^2V_{20} \times {}^0V_{40} = {}^1V_{40}$$
$$^3V_{21} \times {}^0V_{41} = {}^1V_{41}$$
$$^3V_{22} \times {}^0V_{42} = {}^1V_{42}$$
$$^2V_{23} \times {}^0V_{43} = {}^1V_{43}$$
$$^1V_{24} \times {}^0V_{44} = {}^1V_{44}$$

(N.B. that V_{40} being intended to receive the coefficient on V_{20} which has *no* variable, will only have cos 0θ $(=1)$ inscribed on it, preparatory to commencing the fifth series of operations.)

From the moment that the fifth and final series of operations is ordered, the Variables V_{20}, V_{21}, &c. then in their turn cease to be *Result*-Variables and become mere *Working*-Variables; V_{40}, V_{41}, &c. being now the recipients of the ultimate results.

We should observe, that if the variables cos θ, cos 2θ, cos 3θ, &c. are furnished, they would be placed directly upon V_{41}, V_{42}, &c., like any other data. If not, a separate computation might be entered upon in a separate part of the engine, in order to calculate them, and place them on V_{41}, &c.

We have now explained how the engine might compute (1.) in the most direct manner, supposing we knew nothing about the *general* term of the resulting series. But the engine would in reality set to work very differently, whenever (as in this case) we *do* know the law for the general term.

The first two terms of (1.) are

$$(BA + \tfrac{1}{2}B_1A_1) + (\overline{BA_1 + B_1A + \tfrac{1}{2}B_1A_2} . \cos \theta) \qquad (4.)$$

and the general term for all after these is

$$(BA_n + \tfrac{1}{2}B_1 . \overline{A_{n-1} + A_{n+2}}) \cos n\theta \qquad . \qquad . \qquad (5.)$$

which is the coefficient of the $(n+1)$th term. The engine would calculate the first two terms by means of a separate set of suitable Operation-cards, and would then need another set for the third term; which last set of Operation-cards would calculate all the succeeding terms *ad infinitum*, merely requiring certain new Variable-cards for each term to direct the operations to act on the proper columns. The following would be the successive sets of operations for computing the coefficients of $n+2$ terms:—

$$(\times, \times, \div, +), \quad (\times, \times, \times, \div, +, +), \quad n(\times, +, \times, \div, +).$$

Or we might represent them as follows, according to the numerical order of the operations:—

$$(1, 2...4), \quad (5, 6...10), \quad n(11, 12...15).$$

The brackets, it should be understood, point out the relation in which the operations may be *grouped*, while the comma marks *succession*. The symbol + might be used for this latter purpose, but this would be liable to produce confusion, as + is also necessarily used to represent one class of the actual operations which are the subject of that succession. In accordance with this meaning attached to the comma, care must be taken when any one group of operations recurs more than once, as is represented above by $n(11...15)$, not to insert a comma after the number or letter prefixed to that group. n, $(11...15)$ would stand for *an operation n, followed by the group of operations* $(11...15)$; instead of denoting *the number of groups which are to follow each other.*

Wherever a *general term* exists, there will be a *recurring group* of operations, as in the above example. Both for brevity and for distinctness, a *recurring group* is called a *cycle*. A *cycle* of operations, then, must be understood to signify any *set of operations* which is repeated *more than once*. It is equally a *cycle*, whether it be repeated *twice* only, or an indefinite number of times; for it is the fact of a *repetition occurring at all* that constitutes it such. In many cases of analysis there is a *recurring group* of one or more *cycles*; that is, a *cycle of a cycle*, or a *cycle of cycles*. For instance: suppose we wish to divide a series by a series,

$$(1.) \qquad \frac{a+bx+cx^2+...}{a'+b'x+c'x^2+...},$$

it being required that the result shall be developed, like the dividend and the divisor, in successive powers of x. A little consideration of (1.), and of the steps through which algebraical division is effected, will show that (if the denominator be supposed to consist of p terms) the first partial quotient will be completed by the following operations:—

$$(2.) \qquad \{(\div), p(\times, -)\} \quad \text{or} \quad \{(1), p(2, 3)\},$$

that the second partial quotient will be completed by an exactly similar set of operations, which acts on the remainder obtained by the first set, instead of on the original dividend. The whole of the processes therefore that have been gone through, by the time the *second* partial quotient has been obtained, will be,—

$$(3.) \qquad 2\{(\div), p(\times, -)\} \quad \text{or} \quad 2\{(1), p(2, 3)\},$$

which is a cycle that includes a cycle, or a cycle of the second order. The operations for the *complete* division, supposing we propose to obtain n terms of the series constituting the quotient, will be,—

$$(4.) \qquad n\{(\div), p(\times, -)\} \quad \text{or} \quad n\{(1), p(2, 3)\}.$$

It is of course to be remembered that the process of algebraical division in reality continues *ad infinitum*, except in the few exceptional cases which admit of an exact quotient being obtained.   The number n in the formula (4.) is always that of the number of terms we propose to ourselves to obtain; and the nth partial quotient is the coefficient of the $(n-1)$th power of x.

There are some cases which entail *cycles of cycles of cycles*, to an indefinite extent.   Such cases are usually very complicated, and they are of extreme interest when considered with reference to the engine. The algebraical development in a series of the nth function of any given function is of this nature.   Let it be proposed to obtain the nth function of

(5.) $\phi(a, b, c, ..., x)$, x being the variable.

We should premise, that we suppose the reader to understand what is meant by an nth function.   We suppose him likewise to comprehend distinctly the difference between developing *an* nth *function algebraically*, and merely *calculating an* nth *function arithmetically*.   If he does not, the following will be by no means very intelligible; but we have not space to give any preliminary explanations.   To proceed: the law, according to which the successive functions of (5.) are to be developed, must of course first be fixed on.   This law may be of very various kinds.   We may propose to obtain our results in successive *powers* of x, in which case the general form would be

$$C + C_1 x + C_2 x^2 + \text{ \&c.} ;$$

or in successive powers of n itself, the index of the function we are ultimately to obtain, in which case the general form would be

$$C + C_1 n + C_2 n^2 + \text{ \&c.,}$$

and x would only enter in the coefficients.   Again, other functions of x or of n instead of *powers* might be selected.   It might be in addition proposed, that the coefficients themselves should be arranged according to given functions of a certain quantity.   Another mode would be to make equations arbitarily amongst the coefficients only, in which case the several functions, according to either of which it might be possible to develope the nth function of (5.), would have to be determined from the combined consideration of these equations and of (5.) itself.

The *algebraical* nature of the engine (so strongly insisted on in a previous part of this Note) would enable it to follow out any of these various modes indifferently; just as we recently showed that it can distribute and separate the numerical results of any one prescribed series of processes, in a perfectly arbitrary manner.   Were it otherwise, the engine could merely *compute the arithmetical* nth *function*, a result which, like any other purely arithmetical results, would be simply a collective number, bearing no traces of the data or the processes which had led to it.

Secondly, the *law* of development for the *n*th function being selected, the next step would obviously be to develope (5.) itself, according to this law. This result would be the first function, and would be obtained by a determinate series of processes. These in most cases would include amongst them one or more *cycles* of operations.

The third step (which would consist of the various processes necessary for effecting the actual substitution of the series constituting the *first function*, for the *variable* itself) might proceed in either of two ways. It might make the substitution either wherever *x* occurs in the original (5.), or it might similarly make it wherever *x* occurs in the first function itself which is the equivalent of (5.). In some cases the former mode might be best, and in others the latter.

Whichever is adopted, it must be understood that the result is to appear arranged in a series following the law originally prescribed for the development of the *n*th function. This result constitutes the second function; with which we are to proceed exactly as we did with the first function, in order to obtain the third function, and so on, $n-1$ times, to obtain the *n*th function. We easily perceive that since every successive function is arranged in a series *following the same law*, there would (after the *first* function is obtained) be a *cycle of a cycle of a cycle*, &c. of operations*, one, two, three, up to $n-1$ times, in order to get the *n*th function. We say, *after the first function is obtained*, because (for reasons on which we cannot here enter) the *first* function might in many cases be developed through a set of processes peculiar to itself, and not recurring for the remaining functions.

We have given but a very slight sketch of the principal *general* steps which would be requisite for obtaining an *n*th function of such a formula as (5.). The question is so exceedingly complicated, that perhaps few persons can be expected to follow, to their own satisfaction, so brief and general a statement as we are here restricted to on this subject. Still it is a very important case as regards the engine, and suggests ideas peculiar to itself, which we should regret to pass wholly without allusion. Nothing could be more interesting than to follow out, in every detail, the solution by the engine of such a case as the

---

* A cycle that includes *n* other cycles, successively *contained one within another*, is called a cycle of the $n+1$th order. A cycle may simply *include* many other cycles, and yet only be of the second order. If a series follows a certain law for a certain number of terms, and then another law for another number of terms, there will be a cycle of operations for every new law; but these cycles will not be *contained one within another*,—they merely *follow each other*. Therefore their number may be infinite without influencing the *order* of a cycle that includes a repetition of such a series.

above; but the time, space and labour this would necessitate, could only suit a very extensive work.

To return to the subject of *cycles* of operations: some of the notation of the integral calculus lends itself very aptly to express them: (2.) might be thus written:—

(6.) $\qquad (\div), \sum(+1)^p (\times, -)$, or (1), $\sum(+1)^p (2, 3)$,

where p stands for the variable; $(+1)^p$ for the function of the variable, that is, for ϕp; and the limits are from 1 to p, or from 0 to $p-1$, each increment being equal to unity. Similarly, (4.) would be,—

(7.) $\qquad\qquad \sum(+1)^n \{(\div), \sum(+1)^p (\times, -)\}$

the limits of n being from 1 to n, or from 0 to $n-1$,

(8.) $\qquad$ or $\sum(+1)^n \{(1), \sum(+1)^p (2, 3)\}$.

Perhaps it may be thought that this notation is merely a circuitous way of expressing what was more simply and as effectually expressed before; and, in the above example, there may be some truth in this. But there is another description of cycles which *can* only effectually be expressed, in a condensed form, by the preceding notation. We shall call them *varying cycles*. They are of frequent occurrence, and include successive cycles of operations of the following nature:—

(9.) $\quad p(1, 2...m), \overline{p-1}(1, 2...m), \overline{p-2}(1, 2...m)...\overline{p-n}(1, 2...m)$,

where each cycle contains the same group of operations, but in which the number of repetitions of the group varies according to a fixed rate, with every cycle. (9.) can be well expressed as follows:—

(10.) $\qquad \sum p(1, 2...m)$, the limits of p being from $p-n$ to p.

Independent of the intrinsic advantages which we thus perceive to result in certain cases from this use of the notation of the integral calculus, there are likewise considerations which make it interesting, from the connections and relations involved in this new application. It has been observed in some of the former Notes, that the processes used in analysis form a logical system of much higher generality than the applications to number merely. Thus, when we read over any algebraical formula, considering it exclusively with reference to the processes of the engine, and putting aside for the moment its abstract signification as to the relations of quantity, the symbols $+$, $\times$, &c. in reality represent (as their immediate and proximate effect, when the formula is applied to the engine) that a certain prism which is a part of the mechanism (see Note C.) turns a new face, and thus presents a new card to act on the bundles of levers of the engine; the new card being perforated with holes, which are arranged according to the peculiarities of the operation of addition, or of multiplication, &c. Again, the *numbers* in the preceding formula (8.), each of them really represents one of these very pieces of card that are hung over the prism.

Now in the use made in the formulæ (7.), (8.) and (10.), of the notation of the integral calculus, we have glimpses of a similar new application of the language of the *higher* mathematics. $\sum$, in reality, here indicates that when a certain number of cards have acted in succession, the prism over which they revolve must *rotate backwards*, so as to bring those cards into their former position; and the limits 1 to n, 1 to p, &c., regulate how often this backward rotation is to be repeated.   A. A. L.

Note F

There is in existence a beautiful woven portrait of Jacquard, in the fabrication of which 24,000 cards were required.

The power of *repeating* the cards, alluded to by M. Menabrea in page 239, and more fully explained in Note C., reduces to an immense extent the number of cards required. It is obvious that this mechanical improvement is especially applicable wherever *cycles* occur in the mathematical operations, and that, in preparing data for calculations by the engine, it is desirable to arrange the order and combination of the processes with a view to obtain them as much as possible *symmetrically* and in cycles, in order that the mechanical advantages of the *backing* system may be applied to the utmost. It is here interesting to observe the manner in which the value of an *analytical* resource is *met* and *enhanced* by an ingenious *mechanical* contrivance. We see in it an instance of one of those mutual *adjustments* between the purely mathematical and the mechanical departments, mentioned in Note A. as being a main and essential condition of success in the invention of a calculating engine. The nature of the resources afforded by such adjustments would be of two principal kinds. In some cases, a difficulty (perhaps in itself insurmountable) in the one department would be overcome by facilities in the other; and sometimes (as in the present case) a strong point in the one would be rendered still stronger and more available by combination with a corresponding strong point in the other.

As a mere example of the degree to which the combined systems of cycles and of backing can diminish the *number* of cards requisite, we shall choose a case which places it in strong evidence, and which has likewise the advantage of being a perfectly different *kind* of problem from those that are mentioned in any of the other Notes. Suppose it be required to eliminate nine variables from ten simple equations of the form—

$$ax_0 + bx_1 + cx_2 + dx_3 + \ldots = p \qquad (1.)$$

$$a^1x_0 + b^1x_1 + c^1x_2 + d^1x_3 + \ldots = p' \qquad (2.)$$

$$\text{\&c.} \qquad \text{\&c.} \qquad \text{\&c.} \qquad \text{\&c.}$$

We should explain, before proceeding, that it is not our object to consider this problem with reference to the actual arrangement of the data on the Variables of the engine, but simply as an abstract question of the *nature* and *number* of the *operations* required to be performed during its complete solution.

The first step would be the elimination of the first unknown quantity x_0 between the first two equations. This would be obtained by the form—

$$(a^1a - aa^1)x_0 + (a^1b - ab^1)x_1 + (a^1c - ac^1)x_2 +$$
$$+ (a^1d - ad')x_3 + \ldots\ldots\ldots\ldots\ldots\ldots\ldots\ldots\ldots = a^1p - ap^1,$$

for which the operations 10 ($\times$, $\times$, $-$) would be needed. The second step would be the elimination of x_0 between the second and third equations, for which the operations would be precisely the same. We should then have had altogether the following operations:—

$$10(\times, \times, -), 10(\times, \times, -) = 20(\times, \times, -).$$

Continuing in the same manner, the total number of operations for the complete elimination of x_0 between all the successive pairs of equations would be—

$$9.10(\times, \times, -) = 90(\times, \times, -).$$

We should then be left with nine simple equations of nine variables from which to eliminate the next variable x_1, for which the total of the processes would be—

$$8.9(\times, \times, -) = 72(\times, \times, -).$$

We should then be left with eight simple equations of eight variables from which to eliminate x_2, for which the processes would be—

$$7.8(\times, \times, -) = 56(\times, \times, -),$$

and so on. The total operations for the elimination of all the variables would thus be—

$$9.10 + 8.9 + 7.8 + 6.7. + 5.6 + 4.5 + 3.4 + 2.3 + 1.2 = 330.$$

So that *three* Operation-cards would perform the office of 330 such cards.

If we take n simple equations containing $n-1$ variables, n being a number unlimited in magnitude, the case becomes still more obvious, as the same three cards might then take the place of thousands or millions of cards.

We shall now draw further attention to the fact, already noticed, of its being by no means necessary that a formula proposed for solution should ever have been actually worked out, as a condition for enabling the engine to solve it. Provided we know the *series of operations* to be gone through, that is sufficient. In the foregoing instance this will be obvious enough on a slight consideration. And it is a circumstance which deserves particular notice, since herein may reside a latent value of such an engine almost incalculable in its possible ultimate results. We already know that there are functions whose numerical value it is of importance for the purposes both of abstract and of practical science to ascertain, but whose determination requires processes so lengthy and so complicated, that, although it is possible to arrive at them through great expenditure of time, labour and money, it is yet on these accounts practically almost unattainable; and we can conceive there being some results which it may be *absolutely impossible* in practice to attain with any accuracy, and whose precise determination it may prove highly important for some of the future wants of science, in its manifold, complicated and rapidly-developing fields of inquiry, to arrive at.

Without, however, stepping into the region of conjecture, we will mention a particular problem which occurs to us at this moment as being an apt illustration of the use to which such an engine may be turned for determining that which human brains find it difficult or impossible to work out unerringly. In the solution of the famous problem of the Three Bodies, there are, out of about 295 coefficients of lunar perturbations given by M. Clausen (Astro^e. Nachrichten, No. 406) as the result of the calculations by Burg, of two by Damoiseau, and of one by Burckhardt, fourteen coefficients that differ in the nature of their algebraic sign; and out of the remainder there are only 101 (or about one-third) that agree precisely both in signs and in amount. These discordances, which are generally small in individual magnitude, may arise either from an erroneous determination of the abstract coefficients in the development of the problem, or from discrepancies in the data deduced from observation, or from both causes combined. The former is the most ordinary source of error in astronomical computations, and this the engine would entirely obviate.

We might even invent laws for series or formulæ in an arbitrary manner, and set the engine to work upon them, and thus deduce numerical results which we might not otherwise have thought of obtaining; but this would hardly perhaps in any instance be productive of any great practical utility, or calculated to rank higher than as a kind of philosophical amusement. A. A. L.

Note G

It is desirable to guard against the possibility of exaggerated ideas that might arise as to the powers of the Analytical Engine. In considering any new subject, there is frequently a tendency, first, to *overrate* what we find to be already interesting or remarkable; and, secondly, by a sort of natural reaction, to *undervalue* the true state of the case, when we do discover that our notions have surpassed those that were really tenable.

The Analytical Engine has no pretensions whatever to *originate* anything. It can do whatever we *know how to order it* to perform. It can *follow* analysis; but it has no power of *anticipating* any analytical relations or truths. Its province is to assist us in making *available* what we are already acquainted with. This it is calculated to effect primarily and chiefly of course, through its executive faculties; but it is likely to exert an *indirect* and reciprocal influence on science itself in another manner. For, in so distributing and combining the truths and the formulæ of analysis, that they may become most easily and rapidly amenable to the mechanical combinations of the engine, the relations and the nature of many subjects in that science are necessarily thrown into new lights, and more profoundly investigated. This is a decidedly indirect, and a somewhat *speculative*, consequence of such an invention. It is however pretty evident, on general principles, that in devising for mathematical truths a new form in which to record and throw themselves out for actual use, views are likely to be induced, which should again react on the more theoretical phase of the subject. There are in all extensions of human power, or additions to human knowledge, various *collateral* influences, besides the main and primary object attained.

To return to the executive faculties of this engine: the question must arise in every mind, are they *really* even able to *follow* analysis in its whole extent? No reply, entirely satisfactory to all minds, can be given to this query, excepting the actual existence of the engine, and actual experience of its practical results. We will however sum up for each reader's consideration the chief elements with which the engine works:—

1. It performs the four operations of simple arithmetic upon any numbers whatever.

2. By means of certain artifices and arrangements (upon which we cannot enter within the restricted space which such a publication as the present may admit of), there is no limit either to the *magnitude* of the *numbers* used, or to the *number of quantities* (either variables or constants) that may be employed.

3. It can combine these numbers and these quantities either algebraically or arithmetically, in relations unlimited as to variety, extent, or complexity.

4. It uses algebraic *signs* according to their proper laws, and developes the logical consequences of these laws.

5. It can arbitrarily substitute any formula for any other; effacing the first from the columns on which it is represented, and making the second appear in its stead.

6. It can provide for singular values. Its power of doing this is referred to in M. Menabrea's memoir, page 240, where he mentions the passage of values through zero and infinity. The practicability of causing it arbitrarily to change its processes at any moment, on the occurrence of any specified contingency (of which its substitution of $(\frac{1}{2}\cos.\overline{n+1}\theta + \frac{1}{2}\cos.\overline{n-1}\theta)$ for $(\cos.n\theta.\cos.\theta)$, explained in Note E., is in some degree an illustration), at once secures this point.

The subject of integration and of differentiation demands some notice. The engine can effect these processes in either of two ways:—

First. We may order it, by means of the Operation and of the Variable-cards, to go through the various steps by which the required *limit* can be worked out for whatever function is under consideration.

Secondly. It may (if we know the form of the limit for the function in question) effect the integration or differentiation by direct* substitution. We remarked in Note B., that any *set* of columns on which numbers are inscribed, represents merely a *general* function of the several quantities, until the special function have been impressed by means of the Operation and Variable-cards. Consequently, if instead of requiring the value of the function, we require that of its integral, or of its differential coefficient, we have merely to order whatever particular combination of the ingredient quantities may constitute

---

* The engine cannot of course compute limits for perfectly *simple* and *uncompounded* functions, except in this manner. It is obvious that it has no power of representing or of manipulating with any but *finite* increments or decrements; and consequently that wherever the computation of limits (or of any other functions) depends upon the *direct* introduction of quantities which either increase or decrease *indefinitely*, we are absolutely beyond the sphere of its powers. Its nature and arrangements are remarkably adapted for taking into account all *finite* increments or decrements (however small or large), and for developing the true and logical modifications of form or value dependent upon differences of this nature. The engine may indeed be considered as including the whole Calculus of Finite Differences; many of whose theorems would be especially and beautifully fitted for development by its processes, and would offer peculiarly interesting considerations. We may mention, as an example, the calculation of the Numbers of Bernoulli by means of the *Differences of Zero*.

that integral or that coefficient.   In ax^n, for instance, instead of the quantities

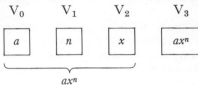

being ordered to appear on V_3 in the combination ax^n, they would be ordered to appear in that of

$$anx^{n-1}.$$

They would then stand thus:—

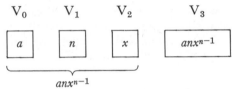

Similarly, we might have $\dfrac{a}{n+1}\,x^{n+1}$, the integral of ax^n.

An interesting example for following out the processes of the engine would be such a form as

$$\int \frac{x^n\,dx}{\sqrt{a^2 - x^2}},$$

or any other cases of integration by successive reductions, where an integral which contains an operation repeated n times can be made to depend upon another which contains the same $n-1$ or $n-2$ times, and so on until by continued reduction we arrive at a certain *ultimate* form, whose value has then to be determined.

The methods in Arbogast's *Calcul des Dérivations* are peculiarly fitted for the notation and the processes of the engine.   Likewise the whole of the Combinatorial Analysis, which consists first in a purely numerical calculation of indices, and secondly in the distribution and combination of the quantities according to laws prescribed by these indices.

We will terminate these Notes by following up in detail the steps through which the engine could compute the Numbers of Bernoulli, this being (in the form in which we shall deduce it) a rather complicated example of its powers.   The simplest manner of computing these numbers would be from the direct expansion of

$$\frac{x}{\epsilon^x - 1} = \frac{1}{1 + \frac{x}{2} + \frac{x^2}{2.3} + \frac{x^3}{2.3.4} + \&c.} \qquad . \quad . \quad . \quad (1.)$$

which is in fact a particular case of the development of

$$\frac{a + bx + cx^2 + \&c.}{a' + b'x + c'x^2 + \&c.}$$

mentioned in Note E.   Or again, we might compute them from the well-known form

$$B_{2n-1} = 2 \cdot \frac{1.2.3...2n}{(2\pi)^{2n}} \left\{ 1 + \frac{1}{2^{2n}} + \frac{1}{3^{2n}} + ... \right\} \qquad . \quad . \quad (2.)$$

or from the form

$$B_{2n-1} = \frac{\pm 2^n}{(2^{2n} - 1) 2^{n-1}} \left\{ \begin{array}{l} \frac{1}{2} n^{2n-1} \\[2mm] -(n-1)^{2n-1} \left\{ 1 + \frac{1}{2} \cdot \frac{2n}{1} \right\} \\[2mm] +(n-2)^{2n-1} \left\{ 1 + \frac{2n}{1} + \frac{1}{2} \cdot \frac{2n.(2n-1)}{1.2} \right\} \\[2mm] -(n-3)^{2n-1} \left\{ \begin{array}{l} 1 + \frac{2n}{1} + \frac{2n.2n-1}{1.2} + \\[2mm] + \frac{1}{2} \cdot \frac{2n.(2n-1).(2n-2)}{1.2.3} \end{array} \right\} \\[2mm] + \quad ... \qquad ... \qquad ... \qquad ... \end{array} \right\} \quad (3.)$$

or from many others.   As however our object is not simplicity or facility of computation, but the illustration of the powers of the engine, we prefer selecting the formula below, marked (8.)   This is derived in the following manner:—

If in the equation

$$\frac{x}{\epsilon^x - 1} = 1 - \frac{x}{2} + B_1 \frac{x^2}{2} + B_3 \frac{x^4}{2.3.4} + B_5 \frac{x^6}{2.3.4.5.6} + ... \qquad . \quad (4.)$$

(in which B_1, B_3..., &c. are the Numbers of Bernoulli), we expand the denominator of the first side in powers of x, and then divide both numerator and denominator by x, we shall derive

$$1 = \left(1 - \frac{x}{2} + B_1 \frac{x^2}{2} + B_3 \frac{x^4}{2.3.4} + ... \right) \left(1 + \frac{x}{2} + \frac{x^2}{2.3} + \frac{x^3}{2.3.4} ... \right) \quad (5.)$$

If this latter multiplication be actually performed, we shall have a series of the general form

$$1 + D_1 x + D_2 x^2 + D_3 x^3 + \dots \qquad \dots \qquad (6.)$$

in which we see, first, that all the coefficients of the powers of x are severally equal to zero; and secondly, that the general form for D_{2n}, the coefficient of the $2n+1$th *term* (that is of x^{2n} any *even* power of x), is the following:—

$$\left. \begin{array}{l} \dfrac{1}{2.3\dots2n+1} - \dfrac{1}{2}\dfrac{1}{2.3\dots2n} + \dfrac{B_1}{2}\cdot\dfrac{1}{2.3\dots2n-1} + \dfrac{B_3}{2.3.4}\dfrac{1}{2.3\dots2n-3} + \\[3mm] + \dfrac{B_5}{2.3.4.5.6}\cdot\dfrac{1}{2.3\dots2n-5} + \dots + \dfrac{B_{2n-1}}{2.3\dots2n}\cdot1 = 0 \end{array} \right\} \quad (7.)$$

Multiplying every term by $(2.3\dots2n)$, we have

$$\left. \begin{array}{l} 0 = -\dfrac{1}{2}\dfrac{2n-1}{2n+1} + B_1\left(\dfrac{2n}{2}\right) + B_3\left(\dfrac{2n.2n-1.2n-2}{2.3.4}\right) + \\[3mm] + B_5\left(\dfrac{2n.2n-1\dots2n-4}{2.3.4.5.6}\right) + \dots + B_{2n-1} \end{array} \right\} \quad (8.)$$

which it may be convenient to write under the general form:—

$$0 = A_0 + A_1 B_1 + A_3 B_3 + A_5 B_5 + \dots + B_{2n-1} \qquad \dots \qquad (9.)$$

A_1, A_3, &c. being those functions of n which respectively belong to B_1, B_3, &c.

We might have derived a form nearly similar to (8.), from D_{2n-1} the coefficient of any *odd* power of x in (6.); but the general form is a little different for the coefficients of the *odd* powers, and not quite so convenient.

On examining (7.) and (8.), we perceive that, when these formulæ are isolated from (6.), whence they are derived, and considered in themselves separately and independently, n may be any whole number whatever; although when (7.) occurs *as one of the* D's in (6.), it is obvious that n is then not arbitrary, but is always a certain function of the *distance of that* D *from the beginning.* If that distance be $= d$, then

$$2n+1 = d, \quad \text{and} \quad n = \frac{d-1}{2} \text{ (for any *even* power of } x)$$

$$2n = d, \quad \text{and} \quad n = \frac{d}{2} \text{ (for any *odd* power of } x).$$

It is with the *independent* formula (8.) that we have to do. Therefore it must be remembered that the conditions for the value of n are now modified, and that n is a perfectly *arbitrary* whole number. This circumstance, combined with the fact (which we may easily perceive) that whatever n is, every term of (8.) after the $(n+1)$th is $=0$, and that the $(n+1)$th term itself is always $=B_{2n-1}\cdot\dfrac{1}{1}=B_{2n-1}$, enables us to find the value (either numerical or algebraical) of any nth Number of Bernoulli B_{2n-1}, *in terms of all the preceding ones*, if we but know the values of B_1, B_3...B_{2n-3}. We append to this Note a Diagram and Table, containing the details of the computation for B_7 (B_1, B_3, B_5 being supposed given).

On attentively considering (8.), we shall likewise perceive that we may derive from it the numerical value of *every* Number of Bernoulli in succession, from the very beginning, *ad infinitum*, by the following series of computations:—

1st Series.—Let $n=1$, and calculate (8.) for this value of n. The result is B_1.

2nd Series.—Let $n=2$. Calculate (8.) for this value of n, substituting the value of B_1 just obtained. The result is B_3.

3rd Series.—Let $n=3$. Calculate (8.) for this value of n, substituting the values of B_1, B_3 before obtained. The result is B_5. And so on, to any extent.

The diagram* represents the columns of the engine when just prepared for computing B_{2n-1} (in the case of $n=4$); while the table beneath them presents a complete simultaneous view of all the successive changes which these columns then severally pass through in order to perform the computation. (The reader is referred to Note D. for explanations respecting the nature and notation of such tables.)

Six numerical *data* are in this case necessary for making the requisite combinations. These data are 1, 2, $n(=4)$, B_1, B_3, B_5. Were $n=5$, the additional datum B_7 would be needed. Were $n=6$, the datum B_9 would be needed; and so on. Thus the actual *number of data* needed will always be $n+2$, for $n=n$; and out of these $n+2$ data, $\overline{(n+2-3)}$ of them are successive Numbers of Bernoulli. The reason why the Bernoulli Numbers used as data are nevertheless placed on *Result*-columns in the diagram, is because they may properly be supposed to have been previously computed in succession by the *engine* itself; under which circumstances each B will appear as a *result*, previous to being used as a *datum* for computing the succeeding B. Here then is an instance (of the kind alluded to in Note D.) of the same Variables

---

* See the diagram at the end of these Notes, p. 296-7.

filling more than one office in turn. It is true that if we consider our computation of B_7 as a perfectly *isolated* calculation, we may conclude B_1, B_3, B_5 to have been arbitrarily placed on the columns; and it would then perhaps be more consistent to put them on V_4, V_5, V_6 as data and not results. But we are not taking this view. On the contrary, we suppose the engine to be *in the course of* computing the Numbers to an indefinite extent, from the very beginning; and that we merely single out, by way of example, *one amongst* the successive but distinct series of computations it is thus performing. Where the B's are fractional, it must be understood that they are computed and appear in the notation of *decimal* fractions. Indeed this is a circumstance that should be noticed with reference to all calculations. In any of the examples already given in the translation and in the Notes, some of the *data*, or of the temporary or permanent results, might be fractional, quite as probably as whole numbers. But the arrangements are so made, that the nature of the processes would be the same as for whole numbers.

In the above table and diagram we are not considering the *signs* of any of the B's, merely their numerical magnitude. The engine would bring out the sign for each of them correctly of course, but we cannot enter on *every* additional detail of this kind as we might wish to do. The circles for the signs are therefore intentionally left blank in the diagram.

Operation-cards 1, 2, 3, 4, 5, 6 prepare $-\dfrac{1}{2} \cdot \dfrac{2n-1}{2n+1}$. Thus, Card 1 multiplies *two* into *n*, and the three *Receiving* Variable-cards belonging respectively to V_4, V_5, V_6, allow the result $2n$ to be placed on each of these latter columns (this being a case in which a triple receipt of the result is needed for subsequent purposes); we see that the upper indices of the two Variables used, during Operation 1, remain unaltered.

We shall not go through the details of every operation singly, since the table and diagram sufficiently indicate them; we shall merely notice some few peculiar cases.

By Operation 6, a *positive* quantity is turned into a *negative* quantity, by simply subtracting the quantity from a column which has only zero upon it. (The sign at the top of V_8 would become − during this process.)

Operation 7 will be unintelligible, unless it be remembered that if we were calculating for $n=1$ instead of $n=4$, Operation 6 would have completed the computation of B_1 itself; in which case the engine, instead of continuing its processes, would have to put B_1 on V_{21};

and then either to stop altogether, or to begin Operations 1, 2...7 all over again for value of $n(=2)$, in order to enter on the computation of B_3; (having however taken care, previous to this recommencement, to make the number on V_3 equal to *two*, by the addition of unity to the former $n=1$ on that column). Now Operation 7 must either bring out a result equal to zero (if $n=1$); or a result *greater* than *zero*, as in the present case; and the engine follows the one or the other of the two courses just explained, contingently on the one or the other result of Operation 7. In order fully to perceive the necessity of this *experimental* operation, it is important to keep in mind what was pointed out, that we are not treating a perfectly isolated and independent computation, but one out of a series of antecedent and prospective computations.

Cards 8, 9, 10 produce $-\dfrac{1}{2}\cdot\dfrac{2n-1}{2n+1}+B_1\dfrac{2n}{2}$. In Operation 9 we see an example of an upper index which again becomes a value after having passed from preceding values to zero. V_{11} has successively been ${}^0V_{11}$, ${}^1V_{11}$, ${}^2V_{11}$, ${}^0V_{11}$, ${}^3V_{11}$; and, from the nature of the office which V_{11} performs in the calculation, its index will continue to go through further changes of the same description, which, if examined, will be found to be regular and periodic.

Card 12 has to perform the same office as Card 7 did in the preceding section; since, if n had been $=2$, the 11th operation would have completed the computation of B_3.

Cards 13 to 20 make A_3. Since A_{2n-1} always consists of $2n-1$ factors, A_3 has three factors; and it will be seen that Cards 13, 14, 15, 16 make the second of these factors, and then multiply it with the first; and that 17, 18, 19, 20 make the third factor, and then multiply this with the product of the two former factors.

Card 23 has the office of Cards 11 and 7 to perform, since if n were $=3$, the 21st and 22nd operations would complete the computation of B_5. As our case is B_7, the computation will continue one more stage; and we must now direct attention to the fact, that in order to compute A_7 it is merely necessary precisely to repeat the group of Operations 13 to 20; and then, in order to complete the computation of B_7, to repeat Operations 21, 22.

It will be perceived that every unit added to n in B_{2n-1}, entails an additional repetition of operations (13...23) for the computation of B_{2n-1}. Not only are all the *operations* precisely the same however for every such repetition, but they require to be respectively supplied with numbers from the very *same pairs of columns*; with only the one exception of Operation 21, which will of course need B_5 (from V_{23}) instead of B_3 (from V_{22}). This identity in the *columns* which supply the requisite numbers must not be confounded with identity in the *values* those columns have upon them and give out to the mill. Most of those values undergo alterations during a performance of the operations (13...23), and consequently the columns present a new set of values for the *next* performance of (13...23) to work on.

At the termination of the *repetition* of operations (13...23) in computing B_7, the alterations in the values on the Variables are, that

$$V_6 = 2n-4 \text{ instead of } 2n-2.$$
$$V_7 = 6 \dots\dots\dots\dots 4.$$
$$V_{10} = 0 \dots\dots\dots\dots 1.$$
$$V_{13} = A_0 + A_1B_1 + A_3B_3 + A_5B_5 \text{ instead of } A_0 + A_1B_1 + A_3B_3.$$

In this state the only remaining processes are, first, to transfer the value which is on V_{13} to V_{24}; and secondly, to reduce V_6, V_7, V_{13} to zero, and to add* *one* to V_3, in order that the engine may be ready to commence computing B_9. Operations 24 and 25 accomplish these purposes. It may be thought anomalous that Operation 25 is represented as leaving the upper index of V_3 still = unity; but it must be remembered that these indices always begin anew for a separate calculation, and that Operation 25 places upon V_3 the *first* value *for the new calculation.*

It should be remarked, that when the group (13...23) is *repeated*, changes occur in some of the *upper* indices during the course of the repetition: for example, $^3V_6$ would become $^4V_6$ and $^5V_6$.

We thus see that when $n=1$, nine Operation-cards are used; that when $n=2$, fourteen Operation-cards are used; and that when $n>2$, twenty-five Operation-cards are used; but that no *more* are needed, however great n may be; and not only this, but that these same twenty-five cards suffice for the successive computation of all the Numbers from B_1 to B_{2n-1} inclusive. With respect to the number of *Variable*-cards, it will be remembered, from the explanations in previous Notes, that an average of three such cards to each *operation* (not however to each Operation-*card*) is the estimate. According to this, the computation of B_1 will require twenty-seven Variable-cards; B_3 forty-two such cards; B_5 seventy-five; and for every succeeding B after B_5, there would be thirty-three additional Variable-cards (since each

---

* It is interesting to observe, that so complicated a case as this calculation of the Bernoullian Numbers nevertheless presents a remarkable simplicity in one respect; viz. that during the processes for the computation of *millions* of these Numbers, no other arbitrary modification would be requisite in the arrangements, excepting the above simple and uniform provision for causing one of the data periodically to receive the finite increment unity.

repetition of the group (13...23) adds eleven to the number of operations required for computing the previous B). But we must now explain, that whenever there is a *cycle of operations*, and if these merely require to be supplied with numbers from the *same pairs of columns*, and likewise each operation to place its *result* on the *same* column for every repetition of the whole group, the process then admits of a *cycle of Variable-cards* for effecting its purposes. There is obviously much more symmetry and simplicity in the arrangements, when cases do admit of repeating the Variable as well as the Operation-cards. Our present example is of this nature. The only exception to a *perfect identity* in *all* the processes and columns used, for every repetition of Operations (13...23), is, that Operation 21 always requires one of its factors from a new column, and Operation 24 always puts its result on a new column. But as these variations follow the same law at each repetition (Operation 21 always requiring its factor from a column *one* in advance of that which it used the previous time, and Operation 24 always putting its result on the column *one* in advance of that which received the previous result), they are easily provided for in arranging the recurring group (or cycle) of Variable-cards.

We may here remark, that the average estimate of three Variable-cards coming into use to each operation, is not to be taken as an absolutely and literally correct amount for all cases and circumstances. Many special circumstances, either in the nature of a problem, or in the arrangements of the engine under certain contingencics, influence and modify this average to a greater or less extent; but it is a very safe and correct *general* rule to go upon. In the preceding case it will give us seventy-five Variable-cards as the total number which will be necessary for computing any B after B_3. This is very nearly the precise amount really used, but we cannot here enter into the minutiæ of the few particular circumstances which occur in this example (as indeed at some one stage or other of probably most computations) to modify slightly this number.

It will be obvious that the very *same* seventy-five Variable-cards may be repeated for the computation of every succeeding Number, just on the same principle as admits of the repetition of the thirty-three Variable-cards of Operations (13...23) in the computation of any *one* Number. Thus there will be a *cycle of a cycle* of Variable-cards.

If we now apply the notation for cycles, as explained in Note E., we may express the operations for computing the Numbers of Bernoulli in the following manner:—

$(1...7)$, $(24, 25)$ .... gives B_1 $\quad = 1$st number; $(n$ being $= 1)$.

$(1...7)$, $(8...12)$, $(24, 25)$ ......B_3 $\quad = 2$nd ......; $(n ..... = 2)$.

$(1...7)$, $(8...12)$, $(13...23)$,

$\quad (24, 25)$ ....B_5 $\quad = 3$rd ......; $(n ..... = 3)$.

$(1...7)$, $(8...12)$, $2(13...23)$,

$\quad (24, 25)$ ....B_7 $\quad = 4$th ......; $(n ..... = 4)$.

....

....

$(1...7)$, $(8...12)$, $\Sigma(+1)^{n-2}$

$\quad (13...23)$, $(24, 25)$ ....B_{2n-1} $\quad = n$th ......; $(n ..... = n)$.

Again,

$$(1...7),\ (24, 25),\ \underset{\text{limits 1 to } n}{\Sigma(+1)^n} \left\{ (1...7),\ (8...12),\ \underset{\text{limits 0 to } (n+2)}{\Sigma(n+2)(13...23)},\ (24, 25) \right\}$$

represents the total operations for computing every number in succession, from B_1 to B_{2n-1} inclusive.

In this formula we see a *varying cycle* of the *first* order, and an ordinary cycle of the *second* order. The latter cycle in this case includes in it the varying cycle.

On inspecting the ten Working-Variables of the diagram, it will be perceived, that although the *value* on any one of them (excepting V_4 and V_5) goes through a series of changes, the *office* which each performs is in this calculation *fixed* and *invariable*. Thus V_6 always prepares the *numerators* of the factors of any A; V_7 the *denominators*. V_8 always receives the $(2n-3)$th factor of A_{2n-1}, and V_9 the $(2n-1)$th. V_{10} always decides which of two courses the succeeding processes are to follow, by feeling for the value of n through means of a subtraction; and so on; but we shall not enumerate further. It is desirable in all calculations so to arrange the processes, that the *offices* performed by the Variables may be as uniform and fixed as possible.

Supposing that it was desired not only to tabulate B_1, B_3, &c., but A_0, A_1, &c.; we have only then to appoint another series of Variables, V_{41}, V_{42}, &c., for receiving these latter results as they are successively produced upon V_{11}. Or again, we may, instead of this, or in addition to this second series of results, wish to tabulate the value of each successive *total* term of the series (8.), viz. A_0, A_1B_1, A_3B_3, &c. We have then merely to multiply each B with each corresponding A, as produced, and to place these successive products on Result-columns appointed for the purpose.

The formula (8.) is interesting in another point of view. It is one particular case of the general Integral of the following Equation of Mixed Differences:—

$$\frac{d^2}{dx^2}\left(z_{n+1}\, x^{2n+2}\right) = (2n+1)(2n+2)\, z^n x^{2n}$$

for certain special suppositions respecting z, x and n.

The *general* integral itself is of the form,

$$z_n = f(n).x + f_1(n) + f_2(n).x^{-1} + f_3(n).x^{-3} + \ldots$$

and it is worthy of remark, that the engine might (in a manner more or less similar to the preceding) calculate the value of this formula upon most *other* hypotheses for the functions in the integral with as much, or (in many cases) with more ease than it can formula (8.).

A. A. L.

Number of Operation	Nature of Operation	Variables acted upon	Variables receiving results	Indication of change in the value on any Variable	Statement of Results	$^1V_1$ 0 0 0 1	$^1V_2$ 0 0 0 2	$^1V_3$ 0 0 0 4	$^0V_4$ 0 0 0 0	$^0V_5$ 0 0 0 0
					Data	1	2	n	▢	▢
1	$\times$	$^1V_2 \times {}^1V_3$	$^1V_4,\ {}^1V_5,\ {}^1V_6$	$\left\{\begin{matrix}{}^1V_2={}^1V_2\\{}^1V_3={}^1V_3\end{matrix}\right\}$	$= 2n$	...	2	n	$2n$	$2n$
2	$-$	$^1V_4 - {}^1V_1$	$^2V_4$	$\left\{\begin{matrix}{}^1V_4={}^2V_4\\{}^1V_1={}^1V_1\end{matrix}\right\}$	$= 2n-1$	1	...	...	$2n-1$	
3	$+$	$^1V_5 + {}^1V_1$	$^2V_5$	$\left\{\begin{matrix}{}^1V_5={}^2V_5\\{}^1V_1={}^1V_1\end{matrix}\right\}$	$= 2n+1$	1	...	...	...	$2n+1$
4	$\div$	$^2V_5 \div {}^2V_4$	$^1V_{11}$	$\left\{\begin{matrix}{}^2V_5={}^0V_5\\{}^2V_4={}^0V_4\end{matrix}\right\}$	$= \dfrac{2n-1}{2n+1}$	...	...	...	0	0
5	$\div$	$^1V_{11} \div {}^1V_2$	$^2V_{11}$	$\left\{\begin{matrix}{}^1V_{11}={}^2V_{11}\\{}^1V_2={}^1V_2\end{matrix}\right\}$	$= \dfrac{1}{2}\dfrac{2n-1}{2n+1}$	...	2	...	...	...
6	$-$	$^0V_{13} - {}^2V_{11}$	$^1V_{13}$	$\left\{\begin{matrix}{}^2V_{11}={}^0V_{11}\\{}^0V_{13}={}^1V_{13}\end{matrix}\right\}$	$= -\dfrac{1}{2}\dfrac{2n-1}{2n+1} = A_0$	...				
7	$-$	$^1V_3 - {}^1V_1$	$^1V_{10}$	$\left\{\begin{matrix}{}^1V_3={}^1V_3\\{}^1V_1={}^1V_1\end{matrix}\right\}$	$= n-1(=3)$	1	...	n	...	...
8	$+$	$^1V_2 + {}^0V_7$	$^1V_7$	$\left\{\begin{matrix}{}^1V_2={}^1V_2\\{}^0V_7={}^1V_7\end{matrix}\right\}$	$= 2+0 = 2$	...	2	...	...	...
9	$\div$	$^1V_6 \div {}^1V_7$	$^3V_{11}$	$\left\{\begin{matrix}{}^1V_6={}^1V_6\\{}^0V_{11}={}^3V_{11}\end{matrix}\right\}$	$= \dfrac{2n}{2} = A_1$	...				
10	$\times$	$^1V_{21} \times {}^3V_{11}$	$^1V_{12}$	$\left\{\begin{matrix}{}^1V_{21}={}^1V_{21}\\{}^3V_{11}={}^3V_{11}\end{matrix}\right\}$	$= B_1 \cdot \dfrac{2n}{2} = B_1A_1$	...				
11	$+$	$^1V_{12} + {}^1V_{13}$	$^2V_{13}$	$\left\{\begin{matrix}{}^1V_{12}={}^0V_{12}\\{}^1V_{13}={}^2V_{13}\end{matrix}\right\}$	$= -\dfrac{1}{2}\dfrac{2n-1}{2n+1}+B_1 \cdot \dfrac{2n}{2}$	...				
12	$-$	$^1V_{10} - {}^1V_1$	$^2V_{10}$	$\left\{\begin{matrix}{}^1V_{10}={}^2V_{10}\\{}^1V_1={}^1V_1\end{matrix}\right\}$	$= n-2(=2)$	1	...	...	...	
13	$-$	$^1V_6 - {}^1V_1$	$^2V_6$	$\left\{\begin{matrix}{}^1V_6={}^2V_6\\{}^1V_1={}^1V_1\end{matrix}\right\}$	$= 2n-1$	1	...	...	...	
14	$+$	$^1V_1 + {}^1V_7$	$^2V_7$	$\left\{\begin{matrix}{}^1V_1={}^1V_1\\{}^1V_7={}^2V_7\end{matrix}\right\}$	$= 2+1 = 3$	1	...			
15	$\div$	$^2V_6 \div {}^2V_7$	$^1V_8$	$\left\{\begin{matrix}{}^2V_6={}^2V_6\\{}^2V_7={}^2V_7\end{matrix}\right\}$	$= \dfrac{2n-1}{3}$	...				
16	$\times$	$^1V_8 \times {}^3V_{11}$	$^4V_{11}$	$\left\{\begin{matrix}{}^1V_8={}^0V_8\\{}^3V_{11}={}^4V_{11}\end{matrix}\right\}$	$= \dfrac{2n}{2} \cdot \dfrac{2n-1}{3}$	...	...	...		
17	$-$	$^2V_6 - {}^1V_1$	$^3V_6$	$\left\{\begin{matrix}{}^2V_6={}^3V_6\\{}^1V_1={}^1V_1\end{matrix}\right\}$	$= 2n-2$	1	...			
18	$+$	$^1V_1 + {}^2V_7$	$^3V_7$	$\left\{\begin{matrix}{}^2V_7={}^3V_7\\{}^1V_1={}^1V_1\end{matrix}\right\}$	$= 3+1 = 4$	1	...			
19	$+$	$^3V_6 \div {}^3V_7$	$^1V_9$	$\left\{\begin{matrix}{}^3V_6={}^3V_6\\{}^3V_7={}^3V_7\end{matrix}\right\}$	$= \dfrac{2n-2}{4}$	...				
20	$\times$	$^1V_9 \times {}^4V_{11}$	$^5V_{11}$	$\left\{\begin{matrix}{}^1V_9={}^0V_9\\{}^4V_{11}={}^5V_{11}\end{matrix}\right\}$	$= \dfrac{2n}{2} \cdot \dfrac{2n-1}{3} \cdot \dfrac{2n-2}{4} = A_3$	...	...	...		
21	$\times$	$^1V_{22} \times {}^5V_{11}$	$^0V_{12}$	$\left\{\begin{matrix}{}^1V_{22}={}^1V_{22}\\{}^0V_{12}={}^2V_{12}\end{matrix}\right\}$	$= B_3 \cdot \dfrac{2n}{2} \cdot \dfrac{2n-1}{3} \cdot \dfrac{2n-2}{3} = B_3A_3$	...	...	...		
22	$+$	$^2V_{12} + {}^2V_{13}$	$^3V_{13}$	$\left\{\begin{matrix}{}^2V_{12}={}^0V_{12}\\{}^2V_{13}={}^3V_{13}\end{matrix}\right\}$	$= A_0+B_1A_1+B_3A_3$	...	...	...		
23	$-$	$^2V_{10} - {}^1V_1$	$^3V_{10}$	$\left\{\begin{matrix}{}^2V_{10}={}^3V_{10}\\{}^1V_1={}^1V_1\end{matrix}\right\}$	$= n-3(=1)$	1	...			
					Here follows a repetition					
24	$+$	$^4V_{13} + {}^0V_{24}$	$^1V_{24}$	$\left\{\begin{matrix}{}^4V_{13}={}^0V_{13}\\{}^0V_{24}={}^1V_{24}\end{matrix}\right\}$	$= B_7$	...	...	...	...	...
25	$+$	$^1V_1 + {}^1V_3$	$^1V_3$	$\left\{\begin{matrix}{}^1V_1={}^1V_1\\{}^1V_3={}^1V_3\\{}^5V_6={}^0V_6\\{}^5V_7={}^0V_7\end{matrix}\right.$	$= n+1 = 4+1 = 5$ .... by a Variable-card. by a Variable-card.	1	...	$n+1$	...	...

Engine of the Numbers of Bernoulli

Working Variables								Result Variables			
$^0V_6$ 0 0 0 0	$^0V_7$ 0 0 0 0	$^0V_8$ 0 0 0 0	$^0V_9$ 0 0 0 0	$^0V_{10}$ 0 0 0 0	$^0V_{11}$ 0 0 0 0	$^0V_{12}$ 0 0 0 0	$^0V_{13}\ldots$ 0 0 0 0	$^1V_{21}$ B_1 in a dec. fract. 0 0 0 0	$^1V_{22}$ B_3 in a dec. fract. 0 0 0 0	$^1V_{23}$ B_5 in a dec. fract. 0 0 0 0	$^0V_{24}\ldots$ 0 0 0 0
								B_1	B_3	B_3	B_7
$2n$											
$\ldots$	$\ldots$	$\ldots$	$\ldots$	$\ldots$	$\dfrac{2n-1}{2n+1}$						
$\ldots$	$\ldots$	$\ldots$	$\ldots$	$\ldots$	$\dfrac{1}{2}\dfrac{2n-1}{2n+1}$						
$\ldots$	$\ldots$	$\ldots$	$\ldots$	$\ldots$	0	$\ldots$	$-\dfrac{1}{2}\cdot\dfrac{2n-1}{2n+1}=A_0$				
$\ldots$	$\ldots$	$\ldots$	$\ldots$	$n-1$							
$\ldots$	2										
$2n$	2	$\ldots$	$\ldots$	$\ldots$	$\dfrac{2n}{2}=A_1$						
$\ldots$	$\ldots$	$\ldots$	$\ldots$	$\ldots$	$\dfrac{2n}{2}=A_1$	$B_1\cdot\dfrac{2n}{2}=B_1A_1$	$\ldots\ldots\ldots$	B_1			
$\ldots$	$\ldots$	$\ldots$	$\ldots$	$\ldots$	$\ldots\ldots\ldots$	0	$\left\{-\dfrac{1}{2}\cdot\dfrac{2n-1}{2n+1}+B_1\cdot\dfrac{2n}{2}\right\}$				
$\ldots$	$\ldots$	$\ldots$	$\ldots$	$n-2$							
$2n-1$											
$\ldots$	3										
$2n-1$	3	$\dfrac{2n-1}{3}$									
$\ldots$	$\ldots$	0	$\ldots$	$\ldots$	$\dfrac{2n}{2}\cdot\dfrac{2n-1}{3}$						
$2n-2$											
$\ldots$	4										
$2n-2$	4	$\ldots$	$\dfrac{2n-2}{4}$	$\ldots$	$\left\{\dfrac{2n}{2}\dfrac{2n-1}{3}\cdot\dfrac{2n-2}{3}=A_3\right\}$						
$\ldots$	$\ldots$	$\ldots$	0								
$\ldots$	$\ldots$	$\ldots$	$\ldots$	$\ldots$	0	B_3A_3	$\ldots\ldots\ldots$	$\ldots$	B_3		
$\ldots$	$\ldots$	$\ldots$	$\ldots$	$\ldots$	$\ldots\ldots\ldots$	0	$\left\{A_3+B_1A_1+B_3A_3\right\}$				
$\ldots$	$\ldots$	$\ldots$	$\ldots$	$n-3$							

of Operations thirteen to twenty-three

Working Variables								Result Variables			
$\ldots$	$\ldots$	$\ldots$	$\ldots$	$\ldots$	$\ldots\ldots\ldots$	$\ldots\ldots\ldots$	$\ldots\ldots\ldots$	$\ldots$	$\ldots$	$\ldots$	B
0	0										

III

ON THE APPLICATION OF MACHINERY TO THE PURPOSE OF CALCULATING AND PRINTING MATHEMATICAL TABLES

By CHARLES BABBAGE

A Letter to Sir Humphry Davy, Bart., President
of the Royal Society, July 3, 1822

My DEAR SIR,

THE great interest you have expressed in the success of that system of contrivances which has lately occupied a considerable portion of my attention, induces me to adopt this channel for stating more generally the principles on which they proceed, and for pointing out the probable extent and important consequences to which they appear to lead. Acquainted as you were with this inquiry almost from its commencement, much of what I have now to say cannot fail to have occurred to your own mind: you will, however, permit me to restate it for the consideration of those with whom the principles and the machinery are less familiar.

The intolerable labour and fatiguing monotony of a continued repetition of similar arithmetical calculations, first excited the desire, and afterwards suggested the idea, of a machine, which, by the aid of gravity or any other moving power, should become a substitute for one of the lowest operations of human intellect. It is not my intention in the present Letter to trace the progress of this idea, or the means which I have adopted for its execution; but I propose stating some of their general applications, and shall commence with describing the powers of several engines which I have contrived: of that part which is already executed I shall speak more in the sequel.

The first engine of which drawings were made was one which is capable of computing any table by the aid of differences, whether they are positive or negative, or of both kinds. With respect to the number of the order of differences, the nature of the machinery did

298

not in my own opinion, nor in that of a skilful mechanic whom I consulted, appear to be restricted to any very limited number; and I should venture to construct one with ten or a dozen orders with perfect confidence.   One remarkable property of this machine is, that the greater the number of differences the more the engine will outstrip the most rapid calculator.

By the application of certain parts of no great degree of complexity, this may be converted into a machine for extracting the roots of equations, and consequently the roots of numbers: and the extent of the approximation depends on the magnitude of the machine.

Of a machine for multiplying any number of figures (m) by any other number (n), I have several sketches; but it is not yet brought to that degree of perfection which I should wish to give it before it is to be executed.

I have also certain principles by which, if it should be desirable, a table of prime numbers might be made, extending from 0 to ten millions.

Another machine, whose plans are much more advanced than several of those just named, is one for constructing tables which have no order of differences constant.

A vast variety of equations of finite differences may by its means be solved, and a variety of tables, which could be produced in successive parts by the first machine I have mentioned, could be calculated by the latter one with a still less exertion of human thought.   Another and very remarkable point in the structure of this machine is, that it will calculate tables governed by laws which have not been hitherto shown to be explicitly determinable, or that it will solve equations for which analytical methods of solution have not yet been contrived.

Supposing these engines executed, there would yet be wanting other means to ensure the accuracy of the printed tables to be produced by them.

The errors of the persons employed to copy the figures presented by the engines would first interfere with their correctness.   To remedy this evil, I have contrived means by which the machines themselves shall take from several boxes containing type, the numbers which they calculate, and place them side by side, thus becoming at the same time a substitute for the compositor and the computer: by which means all error in copying as well as in printing is removed.

There are, however, two sources of error which have not yet been guarded against.   The ten boxes with which the engine is provided contain each about three thousand types; any box having of course, only those of one number in it.   It may happen that the person employed in filling these boxes shall accidentally place a wrong type in some of them; as for instance, the number 2 in the box which ought only to contain 7's.   When these boxes are delivered to the superintendent of the engine, I have provided a simple and effectual means by which he shall, in less than half an hour, ascertain whether, amongst these 30,000 types, there be any individual misplaced, or even inverted. The other cause of error to which I have alluded, arises from the type falling out when the page has been set up: this I have rendered impossible by means of a similar kind.

The quantity of errors from carelessness in correcting the press, even in tables of the greatest credit, will scarcely be believed, except by those who have had constant occasion for their use.   A friend of mine, whose skill in practical as well as theoretical astronomy is well known, produced to me a copy of the tables published by order of the French Board of Longitude, containing those of the Sun by Delambre, and of the Moon by Burg, in which he had corrected above *five hundred errors*: most of these appear to be errors of the press; and it is somewhat remarkable, that in turning over the leaves in the fourth page I opened we observed a new error before unnoticed.   These errors are so much more dangerous, because independent computers using the same tables will agree in the same errors.

To bring to perfection the various machinery which I have contrived, would require an expense both of time and money, which can be known only to those who have themselves attempted to execute mechanical inventions.   Of the greater part of that which has been mentioned, I have at present contented myself with sketches on paper, accompanied by short memorandums, by which I might at any time more fully develop the contrivances; and where any new principles are introduced, I have had models executed, in order to examine their actions.   For the purpose of demonstrating the practicability of these views, I have chosen the engine for differences, and have constructed one of them, which will produce any tables whose second differences are constant. Its size is the same as that which I should propose for any more extensive one of the same kind: the chief difference would be, that in one intended for use there would be a greater repetition of the same parts in order to adapt it to the calculation of a larger number of figures.   Of the action of this engine, you have yourself had opportunities of judging, and I will only at present mention a few trials which have since been made by some scientific gentlemen to whom it has been shown, in order to determine the rapidity with which it calculates. The computed table is presented to the eye at two opposite sides of the machine; and a friend having undertaken to write down the numbers as they appeared, it proceeded to make a table from the formula $x^2 + x + 41$. In the earlier numbers my friend, in writing quickly, rather more than kept pace with the engine; but as soon as four figures were required, the machine was at least equal in speed to the writer.

In another trial it was found that thirty numbers of the same table were calculated in two minutes and thirty seconds: as these contained eighty-two figures, the engine produced thirty-three every minute.

In another trial it produced figures at the rate of forty-four in a minute. As the machine may be made to move uniformly by a weight, this rate might be maintained for any length of time, and I believe few writers would be found to copy with equal speed for many hours together. Imperfect as a first machine generally is, and suffering as this particular one does from great defect in the workmanship, I have every reason to be satisfied with the accuracy of its computations; and by the few skilful mechanics to whom I have in confidence shown it, I am assured that its principles are such that it may be carried to any extent. In fact, the parts of which it consists are few but frequently repeated, resembling in this respect the arithmetic to which it is applied, which, by the aid of a few digits often repeated, produces all the wide variety of number. The wheels of which it consists are numerous, but few move at the same time; and I have employed a principle by which any small error that may arise from accident or bad workmanship is corrected as soon as it is produced, in such a manner as effectually to prevent any accumulation of small errors from producing a wrong figure in the calculation.

Of those contrivances by which the composition is to be effected, I have made many experiments and several models; the results of these leave me no reason to doubt of success, which is still further confirmed by a working model that is just finished.

As the engine for calculating tables by the method of differences is the only one yet completed, I shall, in my remarks on the utility of such instruments, confine myself to a statement of the powers which that method possesses.

I would however premise, that if any one shall be of opinion, notwithstanding all the precaution I have taken and means I have employed to guard against the occurrence of error, that it may still be possible for it to arise, the method of differences enables me to determine its existence. Thus, if proper numbers are placed at the outset in the engine, and if it has composed a page of any kind of table, then by comparing the last number it has set up with that number previously calculated, if they are found to agree, the whole page must be correct:

should any disagreement occur, it would scarcely be worth the trouble of looking for its origin, as the shortest plan would be to make the engine recalculate the whole page, and nothing would be lost but a few hours' labour of the moving power.

Of the variety of tables which such an engine could calculate, I shall mention but a few. The tables of powers and products published at the expense of the Board of Longitude, and calculated by Dr. Hutton, were solely executed by the method of differences; and other tables of the roots of numbers have been calculated by the same gentleman on similar principles.

As it is not my intention in the present instance to enter into the theory of differences, a field far too wide for the limits of this letter, and which will probably be yet further extended in consequence of the machinery I have contrived, I shall content myself with describing the course pursued in one of the most stupendous monuments of arithmetical calculation which the world has yet produced, and shall point out the mode in which it was conducted, and what share of mental labour would have been saved by the employment of such an engine as I have contrived.

The tables to which I allude are those calculated under the direction of M. Prony, by order of the French Government,—a work which will ever reflect the highest credit on the nation which patronized and on the scientific men who executed it. The tables computed were the following.

1. The natural sines of each 10,000 of the quadrant calculated to twenty-five figures with seven or eight orders of differences.

2. The logarithmic sines of each 100,000 of the quadrant calculated to fourteen decimals with five orders of differences.

3. The logarithm of the ratios of the sines to their arcs of the first 5,000 of the 100,000ths of the quadrant calculated to fourteen decimals with three orders of differences.

4. The logarithmic tangents corresponding to the logarithmic sines calculated to the same extent.

5. The logarithms of the ratios of the tangents to their arcs calculated in the same manner as the logarithms of the ratios of the sines to their arcs.

6. The logarithms of numbers from 1 to 10,000 calculated to nineteen decimals.

7. The logarithms of all numbers from 10,000 to 200,000 calculated to fourteen figures with five orders of differences.

Such are the tables which have been calculated, occupying in their present state seventeen large folio volumes. It will be observed that the trigonometrical tables are adapted to the decimal system, which has not been generally adopted even by the French, and which has not been at all employed in this country. But, notwithstanding this objection, such was the opinion entertained of their value, that a distinguished member of the English Board of Longitude was not long since commissioned by our Government to make a proposal to the Board of Longitude of France to print an abridgment of these tables at the joint expense of the two countries; and five thousand pounds were named as the sum our Government was willing to advance for this purpose. It is gratifying to record this disinterested offer, so far above those little jealousies which frequently interfere between nations long rivals, and manifesting so sincere a desire to render useful to mankind the best materials of science in whatever country they might be produced. Of the reasons why this proposal was declined by our neighbours, I am at present uninformed: but, from a personal acquaintance with many of the distinguished foreigners to whom it was referred, I am convinced that it was received with the same good feelings as those which dictated it.

I will now endeavour shortly to state the manner in which this enormous mass of computation was executed; one table of which (that of the logarithms of numbers) must contain about eight millions of figures.

The calculators were divided into three sections. The first section comprised five or six mathematicians of the highest merit, amongst whom were M. Prony and M. Legendre. These were occupied entirely with the analytical part of the work; they investigated and determined on the formulæ to be employed.

The second section consisted of seven or eight skilful calculators habituated both to analytical and arithmetical computations. These received the formulæ from the first section, converted them into numbers, and furnished to the third section the proper differences at the stated intervals.

They also received from that section the calculated results, and compared the two sets, which were computed independently for the purpose of verification.

The third section, on whom the most laborious part of the operations devolved, consisted of from sixty to eighty persons, few of them possessing a knowledge of more than the first rules of arithmetic: these received from the second class certain numbers and differences, with which, by additions and subtractions in a prescribed order, they completed the whole of the tables above mentioned.

I will now examine what portion of this labour might be dispensed with, in case it should be deemed advisable to compute these or any similar tables of equal extent by the aid of the engine I have referred to.

In the first place, the labour of the first section would be considerably reduced, because the formulæ used in the great work I have been describing have already been investigated and published. One person, or at the utmost two, might therefore conduct it.

If the persons composing the second section, instead of delivering the numbers they calculate to the computers of the third section, were to deliver them to the engine, the whole of the remaining operations would be executed by machinery, and it would only be necessary to employ people to copy down as fast as they were able the figures presented to them by the engine. If, however, the contrivances for printing were brought to perfection and employed, even this labour would be unnecessary, and a few superintendents would manage the machine and receive the calculated pages set up in type. Thus the number of calculators employed, instead of amounting to ninety-six, would be reduced to twelve. This number might, however, be considerably diminished, because when an engine is used the intervals between the differences calculated by the second section may be greatly enlarged. In the tables of logarithms M. Prony caused the differences to be calculated at intervals of two hundred, in order to save the labour of the third section: but as that would now devolve on machinery, which would scarcely move the slower for its additional burthen, the intervals might properly be enlarged to three or four times that quantity. This would cause a considerable diminution in the labour of the second section. If to this diminution of mental labour we add that which arises from the whole work of the compositor being executed by the machine, and the total suppression of that most annoying of all literary labour, the correction of the errors of the press,* I think I am justified in presuming that if engines were made purposely for this object, and were afterwards useless, the tables could be produced at a much cheaper rate; and of their superior accuracy there could be no doubt. Such engines would, however, be far from useless: containing within themselves the power of generating, to an almost unlimited extent, tables whose accuracy would be unrivalled, at an expense comparatively moderate, they would become active agents in reducing the abstract inquiries of geometry to a form and an arrangement adapted to the ordinary purposes of human society.

I should be unwilling to terminate this Letter without noticing

---

* I have been informed that the publishers of a valuable collection of mathematical tables, now reprinting, pay to the gentleman employed in correcting the press, at the rate of three guineas a sheet, a sum by no means too large for the faithful execution of such a laborious duty.

another class of tables of the greatest importance, almost the whole of which are capable of being calculated by the method of differences. I refer to all astronomical tables for determining the positions of the sun or planets: it is scarcely necessary to observe that the constituent parts of these are of the form $a \sin \theta$, where a is a constant quantity, and θ is what is usually called the argument. Viewed in this light they differ but little from a table of sines, and like it may be computed by the method of differences.

I am aware that the statements contained in this Letter may perhaps be viewed as something more than Utopian, and that the philosophers of Laputa may be called up to dispute my claim to originality. Should such be the case, I hope the resemblance will be found to adhere to the nature of the subject rather than to the manner in which it has been treated. Conscious, from my own experience, of the diffi-culty of convincing those who are but little skilled in mathematical knowledge, of the possibility of making a machine which shall perform calculations, I was naturally anxious, in introducing it to the public, to appeal to the testimony of one so distinguished in the records of British science. Of the extent to which the machinery whose nature I have described may be carried, opinions will necessarily fluctuate, until experiment shall have finally decided their relative value: but of that engine which already exists I think I shall be supported, both by yourself and by several scientific friends who have examined it, in stating that it performs with rapidity and precision all those calcu-lations for which it was designed.

Whether I shall construct a larger engine of this kind, and bring to perfection the others I have described, will in a great measure depend on the nature of the encouragement I may receive.

Induced, by a conviction of the great utility of such engines, to withdraw for some time my attention from a subject on which it has been engaged during several years, and which possesses charms of a higher order, I have now arrived at a point where success is no longer doubtful. It must, however, be attained at a very considerable expense, which would not probably be replaced, by the works it might produce, for a long period of time, and which is an undertaking I should feel unwilling to commence, as altogether foreign to my habits and pursuits.

                    I remain, my dear Sir,
                        Faithfully yours,
                            C. BABBAGE.

DEVONSHIRE STREET, PORTLAND PLACE,
    July 3rd, 1822.

ON THE THEORETICAL PRINCIPLES OF THE MACHINERY FOR CALCULATING TABLES

By CHARLES BABBAGE

From a Letter to Dr. Brewster dated November 6, 1822, reprinted in *Brewster's Journal of Science*

MY DEAR SIR,

Having, during the last two or three months, laid aside the further construction of machinery for calculating tables, I have occasionally employed myself in examining the theoretical principles on which it is founded. Several singular results having presented themselves in these inquiries, I am induced to communicate some of them to you, less from the importance of the analytical difficulties they present, than from the curious fact which they offer in the history of invention.

I had mentioned to you that, before I left London, I had completed a small engine which calculated tables by means of differences. On considering this machine, a new arrangement occurred to me, by which an engine might be constructed that should calculate tables of other species, whose analytical laws were unknown. On this suggestion, I proceeded to write down a table which might have been made, had such an engine existed; and finding that there were no known methods of expressing its nth term, I thought the analytical difficulty which was thus brought to light, was itself worthy of examination. The following are the first thirty terms of a series of this kind:

0 ...	2	11 ...	222	22 ...	924
1 ...	2		264		1010
2 ...	4		310		1096
3 ...	10		356	25 ...	1188
4 ...	16	15 ...	408		1288
5 ...	28		468		1396
	48		536		1510
	76		610		1624
	110		684	30 ...	1742
	144	20 ...	762		1862
10 ...	182		842		1984

The law of formation of which is, that the first term is 2, its first difference 2, and its second difference equal to the units figure of the second term; and generally, the second difference corresponding to any term, is always equal to the units figure of the next succeeding term. This engine, when once set, would continue to produce term after term of this series without end, and without any alteration; but we are not in possession of methods of determining its nth term, without passing through all the previous ones. If u_n represent any term, then u_n must be determined from the equation

$$\Delta^2 u_n = \text{the units figure of } u_{n+1};$$

an equation of differences of a species which I have never met with in treatises on that subject.

If we push the inquiry one step further, it is possible to express the units figure of any number in an analytical form. Thus, let S_v represent the sum of the vth powers of the tenth roots of unity, then will

$$0S_v + 9S_{v+1} + 8S_{v+2} + \dots \qquad 1S_{v+9}$$

represent the units of the number v. Now, if we put u_{n+1} instead of v in the above equation, we have

$$\Delta^2 u_n = 0S + 9S + 8S + \dots \qquad 1S$$
$$u_{n+1} \quad u_{n+1}+1 \quad u_{n+1}+2 \qquad u_{n+1}+9$$

an equation whose mode of solution is as yet quite unknown. Finding the difficulty of a direct attempt so considerable, I employed two other processes; one was a kind of induction, and the other was quite unexceptionable. From these I have deduced the following formula:

$$u_n + \overline{(a)} + 20b(10b + 2a - 1) + 2,$$

where a is the units figure of n; b is the number n, when its unit figure is cut off; and $\overline{(a)}$ represents whatever number is opposite to it in the subsidiary table below:

If $a = 0$ ... 0
       1 ... 0
       2 ... 2
       3 ... 8   EXAMPLE: Required the 27th term of the series
       4 ... 14   here, $a = 7$, and $b = 2$: hence, $10b = 2a - 1$
       5 ... 26 $= 20 + 14 - 1 = 33$.
       6 ... 46
       7 ... 74
       8 ... 108
       9 ... 142

then

(a) or the number opposite 7, is   74

$20b\ (10b+2a-1) =$ 1320

           2     $=$ 2

                     1396 $= u_{27}$, or the 27th term.

similarly if $n = 1121,\ \ a = 1,\ \ b = 112,$

then $u_{1121} = 251106.$

Another series of a similar kind, but more simple in its form, is derived from the following equation:

$$\Delta u_z = \text{units figure of } u_z.$$

If the constant or first term is equal to 2, then we may express u_z thus,

$$u_z = 20b + 2a,$$

where a is any of the numbers 1, 2, 3, 4, which, taken from z, leaves the remainder divisible by 4, and b is the quotient of that division: the series is,

1 ... 2	48	EXAMPLE: Let $z =$	13
4	56	1 being subtracted,	1
8	62		—
16	64		12 which,
5 ... 22	15 ... 68	divided by 4, gives	3, hence,
24	76	$a = 1, b = 3, u_{13} = 20.3+2 = 62.$	
28	82		
36	84		
42	88		
10 ... 44	20 ... 96		

Innumerable other series might be formed by the same engine, the differences of any order depending on the value of the figure which might occur in the units, or the tens, or the hundreds place, or in any one or more determinate places of the same, or the next, or preceding terms. Other laws might be observed by the same engine, of which the following is an example. A series of cube numbers might be formed, subject to this condition, that whenever the number 2 occurred in the tens' place, that and all the succeeding cubes should be increased by ten. In such a series, of course, the second figure would never be a 2, because the addition of ten would convert it into 3.

The Series of Cubes		*The Series Proposed*
1		1
8		8
27	*	37
64		74
125		135
216	*	236
343		363
512		532
729		749
1000	*	1030
1331		1361
1728		1758
2197	*	2237
2744		2784

the stars indicating the number at which the law takes effect.   These, and other similar series, open a wide field of analytical inquiry—a subject which I shall take some other opportunity of resuming.   I will, however, mention an unexpected circumstance, as it illustrates, in a striking manner, the connection between remote inquiries in mathematics, and as it may furnish a lesson to those who are rashly inclined to undervalue the more recondite speculations of pure analysis, from an erroneous idea of their inapplicability to practical matters.   Amongst the singular and difficult equations of finite differences to which these series led, I recognised one which I had several years since met with, in an analytical attempt to solve a problem considered by Euler and Vandermonde; it relates to the knight's move at chess.   At that time, I had advanced several steps; but the equation in question proved an obstacle I was then unable to surmount.   In its present shape, although I have not yet deduced the solution from the equation, yet, as I am in possession of the former, it is not too much to anticipate a general process applicable to this class of equations; and should that be the case, I shall be able to advance some steps further in a very curious and difficult inquiry, connected with the geometry of situation.

As an erroneous idea has been entertained relative to the nature of the machinery I have contrived, I will endeavour to state to you some of the mathematical principles on which it is founded.   The contrivances of Pascal and others have, as far as I am aware, been directed to an entirely different object.   Machinery which will perform the usual operation of common arithmetic will never, in my opinion, be of that essential utility which must arise from an engine that calculates tables; and although mine is not defective in these points, and will extract the roots of numbers, and approximate to the roots of equations, and even, as I believe, to their impossible roots, yet, had this been its only office, I should have esteemed it of comparatively but little value. As far as I have inquired, I believe the method of differences has now, for the first time, been embodied into machinery; and, in speaking of this method, I am far from meaning to confine myself to calculating tables by constant differences. The same mechanical principles which I have already proved, enable me to integrate innumerable equations of finite differences, if I may be allowed to use the term *integrate*, in a sense somewhat different from its usual acceptation. My meaning is, that the equation of differences being given, I can, by setting an engine, produce, at the end of a given time, any distant term which may be required; or, if a succession of terms are sought, commencing at a distant point, these shall be produced. Thus, although I do not determine the analytical law, I can produce the numerical result which it is the object of that law to give. Some kinds of equation of differences can be adapted to machinery with much greater facility than others; and hence it will become an object of inquiry how, when we wish to calculate that of any transcendant, we may deduce from some approximate equation the differences which may be suitable to our purpose. Thus, you see, one of the first effects of machinery adapted to numbers has been to lead us to surmount new difficulties in analysis; and should it be carried to perfection, some of the most abstract parts of mathematical science will be called into practical utility, to facilitate the formation of tables. The more I examine this theoretical part, the more I feel convinced that it will be long before the novel relations which it presents will be exhausted; and if the absence of all encouragement to proceed with the mechanism I have contrived, shall prove that I have anticipated too far the period at which it shall become necessary, I will yet venture to predict that a time will arrive when the accumulating labour which arises from the arithmetical applications of mathematical formulæ, acting as a constantly retarding force, shall ultimately impede the useful progress of the science, unless this or some equivalent method is devised for relieving it from the overwhelming incumbrance of numerical detail.

I remain, my dear Sir,
Faithfully yours,
C. BABBAGE.

DEVONSHIRE STREET, PORTLAND PLACE,
    Nov. 6th, 1822.

V

OBSERVATIONS ON THE APPLICATION OF MACHINERY TO THE COMPUTATION OF MATHEMATICAL TABLES

By CHARLES BABBAGE

From the *Memoirs of the Astronomical Society*,
December 13, 1822

SINCE I had the honour of communicating to the Astronomical Society a short account of an arithmetical engine for the calculation of tables, which has been examined by several of the members of this society, I have not added much to the practical part of the subject. I have however paid some attention to the improvements of which the machinery is susceptible, and which will, if another engine is made, be greatly improved.

The theoretical inquiries to which it has conducted me are however of a singular nature; and I shall take this opportunity of briefly explaining to the society some of the principles on which they depend, as far as the nature of the subject will permit me to do this without the introduction of too many algebraic operations, which are rarely intelligible when read to a large assembly.

Of the variety of tables which are required in the present state of science, by far the larger portion are intimately connected with that department of it which it is the peculiar object of this society to promote.

The importance of astronomical science, whether viewed as the proudest triumph of intellectual power, or considered as the most valuable present of abstract science to the comfort and happiness of mankind, equally claims for it the first assistance from any new method for condensing the processes of reasoning or abridging the labour of calculation. Astronomical tables were therefore the first objects on which I turned my attention, when attempting to improve the power of the engine, as they had formed the first motive for constructing it.

I have already stated to the society, in my former communication, that the first engine I had constructed was solely destined to compute tables having constant differences. From this circumstance it will be apparent that after a certain number of terms of a table are computed, unless, as rarely happens, it has a constant order of differences, we must stop the engine and place in it other numbers, in order to produce the next portion of the table. This operation must be repeated more or less frequently according to the nature of the table. The more numerous the order of differences, the less frequent will this operation become requisite. The chance of error in such computations arises from incorrect numbers being placed in the engine: it therefore becomes desirable to limit this chance as much as possible. In examining the analytical theory of the various differences of the sine of an arc, I noticed the property which it possesses of having any of its even orders of differences equal to the sine of the same arc increased by some multiple of its increment multiplied by a constant quantity. With the aid of this principle an engine might be formed which would require but little attendance, and I believe that it might in some cases compute a table of the form $A \sin \theta$ from the 1st value of $\theta = 0$ up to $\theta = 90°$ with only one set of figures being placed in it.

It is scarcely necessary to observe what an immense number of astronomical tables are comprised under this form, nor the great accuracy which must result from having reduced to so few a number the preliminary computations which are requisite.

In pursuing into its detail the principle to which I have alluded, which lends itself so happily to numerical application, I have traced its application to other species of tables, and am enabled to point out a course of analytical investigation which will in all probability afford ready methods for constructing tables, even of the most complicated transcendent, in a manner equally easy.

I will now advert to another circumstance, which, although not immediately connected with astronomical tables, resulted from an examination of the engine by which they can be formed.

On considering the arrangement of its parts, I observed that a different mode of connecting them would produce tables of a new species altogether different from any with which I was acquainted. I therefore computed with my pen a small table such as would have been formed by the engine had it existed in this new shape, and I was much surprised at discovering that no analytical method was yet known for determining its nth term. The following is the first series I wrote down:—

	Series	Diff.		Series	Diff.		Series	Diff.
0 ...	2	0	10 ...	222	42	20 ...	924	86
1 ...	2	2		264	46		1010	86
	4	6		310	46		1096	92
	10	6		356	52	25 ...	1188	100
	16	12	15 ...	408	60		1288	108
5 ...	28	20		468	68		1396	114
	48	28		536	74		1510	114
	76	34		610	74		1624	118
	110	34		684	78	30 ...	1742	120
	144	38	20 ...	762	80		1862	122
10 ...	182	40		842	82		1984	

The equation of finite differences from which it is produced is

$$\Delta^2 u_z = \text{units fig. of } u_{z+1}$$

which is one of a class of equations never hitherto integrated. I succeeded in transforming this equation into a more analytical form: but still it presented great difficulties; I therefore undertook the investigation in a different manner, and succeeded in discovering a formula which represented its nth term. It is the following:—

TABLE

0	2
1	2
2	4
3	10
4	16
5	28
6	48
7	76
8	110
9	144

$$u_z = \overline{(a)} + 206(106 + 2a - 1)$$

where $\overline{(a)}$ represents the number opposite a in the next subsidiary table, and a is the figure in the unit's place of z, and b is that number which arises from cutting off the last figure from z. Example: let the 17th term be required, then $z = 17$, and $a = 7$, $b = 1$; the number opposite 7 in the table is

$$\overline{(7)} = . \quad . \quad . \quad . \quad . \quad 76$$
$$106 + 2a - 1 = 10 + 14 - 1 = 23$$
$$206 = 20 \quad 206 \ (106 + 2a - 1) = 460$$
$$\overline{}$$
$$536 = u_{17}$$

I have formed other series of the same class, and have succeeded in expressing any term independent of all the rest by two distinct processes. Thus I have incidentally been able to integrate the equations I have mentioned. I will just state one other of a simple form; it is the equation

$$\Delta u_z = \text{units fig. } u_z$$

whose integral is

$$u_z = 20b + 2^a$$

where a is that one of the numbers 1, 2, 3, 4, which, taken from z, leaves the remainder divisible by 4, and b is the quotient of that division.

$$
\begin{array}{ccc}
1 & - & 2 \\
  &   & 4 \\
  &   & 8 \\
4 & - & 16 \\
  &   & 22 \\
  &   & 24 \\
  &   & 28 \\
8 & - & 36 \\
  &   & 42 \\
\end{array}
$$

One of the general questions to which these researches give rise is, supposing the law of any series to be known, to find what figure will occur in the kth place of the nth term. That the mere consideration of a mechanical engine should have suggested these inquiries, is of itself sufficiently remarkable; but it is still more singular, that amongst researches of so very abstract a nature, I should have met with and overcome a difficulty which had presented itself in the form of an equation of differences, and which had impeded my progress several years since, in attempting the solution of a problem connected with the game of chess.

VI

ON THE DIVISION OF
MENTAL LABOUR

By CHARLES BABBAGE

Chapter XIX, *Economy of Manufactures and Machinery*

(241.)* WE HAVE already mentioned what may, perhaps, appear paradoxical to some of our readers,—that the division of labour can be applied with equal success to mental as to mechanical operations, and that it ensures in both the same economy of time. A short account of its practical application, in the most extensive series of calculations ever executed, will offer an interesting illustration of this fact, whilst at the same time it will afford an occasion for shewing that the arrangements which ought to regulate the interior economy of a manufactory, are founded on principles of deeper root than may have been supposed, and are capable of being usefully employed in preparing the road to some of the sublimest investigations of the human mind.

(242.) In the midst of that excitement which accompanied the Revolution of France and the succeeding wars, the ambition of the nation, unexhausted by its fatal passion for military renown, was at the same time directed to some of the nobler and more permanent triumphs which mark the era of a people's greatness,—and which receive the applause of posterity long after their conquests have been wrested from them, or even when their existence as a nation may be told only by the page of history. Amongst their enterprises of science, the French government was desirous of producing a series of mathematical tables, to facilitate the application of the decimal system which they had so recently adopted. They directed, therefore, their mathematicians to construct such tables, on the most extensive scale. Their most distinguished philosophers, responding fully to the call of their country, invented new methods for this laborious task; and a work, completely answering the large demands of the government, was produced in a remarkably short period of time. M. Prony, to

* [The numbers within parentheses identify the paragraphs of the original volume, *Economy of Manufactures and Machinery*.]

315

whom the superintendence of this great undertaking was confided, in speaking of its commencement, observes: "*Je m'y livrai avec toute l'ardeur dont j'étois capable, et je m'occupai d'abord du plan général de l'exécution. Toutes les conditions que j'avois à remplir nécessitoient l'emploi d'un grand nombre de calculateurs; et il me vint bientôt à la pensée d'appliquer à la confection de ces* Tables la division du travail, *dont les Arts de Commerce tirent un parti si avantageux pour réunir à la perfection de main-d'œuvre l'économie de la dépense et du temps.*" The circumstance which gave rise to this singular application of the principle of *the division of labour* is so interesting, that no apology is necessary for introducing it from a small pamphlet printed at Paris a few years since, when a proposition was made by the English to the French government, that the two countries should print these tables at their joint expense.

(243.) The origin of the idea is related in the following extract:

C'est à un chapitre d'un ouvrage Anglais,* justement célèbre, (I.) qu'est probablement due l'existence de l'ouvrage dont le gouvernement Britannique veut faire jouir le monde savant:—

Voici l'anecdote: M. de Prony s'était engagé, avec les comités de gouvernement, à composer, pour *la division centésimale du cercle, des tables logarithmiques et trigonométriques, qui, non-seulement ne laissassent rien à désirer quant à l'exactitude, mais qui formassent le monument de calcul le plus vaste et le plus imposant qui eût jamais été exécuté, ou même conçu.* Les logarithmes des nombres de 1 à 200,000 formaient à ce travail un supplément nécessaire et exigé. Il fut aisé à M. de Prony de s'assurer que, même en s'associant trois ou quatre habiles co-opérateurs, la plus grande durée présumable de sa vie ne lui suffirait pas pour remplir ses engagements. Il était occupé de cette fâcheuse pensée lorsque, se trouvant devant la boutique d'un marchand de livres, il aperçut la belle édition Anglaise de Smith, donnée à Londres en 1776; il ouvrit le livre au hasard, et tomba sur le premier chapitre, qui traite de *la division du travail*, et où la fabrication des épingles est citée pour exemple. A peine avait-il parcouru les premières pages, que, par une espèce d'inspiration, il conçut l'expédient de mettre ses logarithmes en *manufacture* comme les épingles. Il faisait, en ce moment, à l'école polytechnique, des leçons sur une partie d'analyse liée à ce genre de travail, *la méthode des différences*, et ses applications à *l'interpolation*. Il alla passer quelques jours à la campagne, et revint à Paris avec le plan de *fabrication*, qui a été suivi dans l'exécution. Il rassembla deux ateliers, qui faisaient séparément les mêmes calculs, et se servaient de vérification réciproque."†

---

* *An Enquiry into the Nature and Causes of the Wealth of Nations*, by Adam Smith.

† Note sur la publication, proposée par le gouvernement Anglais des grandes tables logarithmiques et trigonométriques de M. de Prony.—De l'imprimerie de F. Didot, Dec. 1, 1820, p. 7.

(244.) The ancient methods of computing tables were altogether inapplicable to such a proceeding. M. Prony, therefore, wishing to avail himself of all the talent of his country in devising new methods, formed the first section of those who were to take part in this enterprise out of five or six of the most eminent mathematicians in France.

First Section.—The duty of this first section was to investigate, amongst the various analytical expressions which could be found for the same function, that which was most readily adapted to simple numerical calculation by many individuals employed at the same time. This section had little or nothing to do with the actual numerical work. When its labours were concluded, the formulæ on the use of which it had decided, were delivered to the second section.

Second Section.—This section consisted of seven or eight persons of considerable acquaintance with mathematics: and their duty was to convert into numbers the formulæ put into their hands by the first section,—an operation of great labour; and then to deliver out these formulæ to the members of the third section, and receive from them the finished calculations. The members of this second section had certain means of verifying the calculations without the necessity of repeating, or even of examining, the whole of the work done by the third section.

Third Section.—The members of this section, whose number varied from sixty to eighty, received certain numbers from the second section, and, using nothing more than simple addition and subtraction, they returned to that section the tables in a finished state. It is remarkable that nine-tenths of this class had no knowledge of arithmetic beyond the two first rules which they were thus called upon to exercise, and that these persons were usually found more correct in their calculations, than those who possessed a more extensive knowledge of the subject.

(245.) When it is stated that the tables thus computed occupy seventeen large folio volumes, some idea may perhaps be formed of the labour. From that part executed by the third class, which may almost be termed mechanical, requiring the least knowledge and by far the greatest exertions, the first class were entirely exempt. Such labour can always be purchased at an easy rate. The duties of the second class, although requiring considerable skill in arithmetical operations, were yet in some measure relieved by the higher interest naturally felt in those more difficult operations. The exertions of the first class are not likely to require, upon another occasion, so much skill and labour as they did upon the first attempt to introduce such a method; but when the completion of a calculating-engine shall have produced a substitute for the whole of the third section of computers, the attention of analysts will naturally be directed to simplifying its application, by a new discussion of the methods of converting analytical formulæ into numbers.

(246.) The proceeding of M. Prony, in this celebrated system of calculation, much resembles that of a skilful person about to construct a cotton or silk-mill, or any similar establishment.   Having, by his own genius, or through the aid of his friends, found that some improved machinery may be successfully applied to his pursuit, he makes drawings of his plans of the machinery, and may himself be considered as constituting the first section.   He next requires the assistance of operative engineers capable of executing the machinery he has designed, some of whom should understand the nature of the processes to be carried on; and these constitute his second section.   When a sufficient number of machines have been made, a multitude of other persons, possessed of a lower degree of skill, must be employed in using them; these form the third section: but their work, and the just performance of the machines, must be still superintended by the second class.

(247.) As the possibility of performing arithmetical calculations by machinery may appear to non-mathematical readers to be rather too large a postulate, and as it is connected with the subject of the *division of labour*, I shall here endeavour, in a few lines, to give some slight perception of the manner in which this can be done,—and thus to remove a small portion of the veil which covers that apparent mystery.

(248.) *That nearly all tables of numbers which follow any law, however complicated, may be formed, to a greater or less extent, solely by the proper arrangement of the successive addition and subtraction of numbers befitting each table*, is a general principle which can be demonstrated to those only who are well acquainted with mathematics; but the mind, even of the reader who is but very slightly acquainted with that science, will readily conceive that it is not impossible, by attending to the following example.

The subjoined table is the beginning of one in very extensive use, which has been printed and reprinted very frequently in many countries, and is called *a Table of Square Numbers*.

Any number in the table, column A, may be obtained, by multiplying the number which expresses the distance of that term from the commencement of the table by itself; thus, 25 is the fifth term from the beginning of the table, and 5 multiplied by itself, or by 5, is equal to 25. Let us now subtract each term of this table from the next succeeding term, and place the results in another column (B), which may be called first-difference column.   If we again subtract each term of this first difference from the succeeding term, we find the result is always

Terms of the Table	A Table	B First Difference	C Second Difference
1	1		
		3	
2	4		2
		5	
3	9		2
		7	
4	16		2
		9	
5	25		2
		11	
6	36		2
		13	
7	49		

the number 2, (column C;) and that the same number will always recur in that column, which may be called the second difference, will appear to any person who takes the trouble to carry on the table a few terms further. Now when once this is admitted, it is quite clear that, provided the first term (1) of the Table, the first term (3) of the first differences, and the first term (2) of the second or constant difference, are originally given, we can continue the table of square numbers to any extent, merely by addition:—for the series of first differences may be formed by repeatedly adding the constant difference (2) to (3) the first number in column B, and we then have the series of numbers, 3, 5, 7, &c.: and again, by successively adding each of these to the first number (1) of the table, we produce the square numbers.

(249.) Having thus, I hope, thrown some light upon the theoretical part of the question, I shall endeavour to shew that the mechanical execution of such an engine, as would produce this series of numbers, is not so far removed from that of ordinary machinery as might be conceived.* Let the reader imagine three clocks, placed on a table

---

* Since the publication of the Second Edition of this Work, one portion of the engine which I have been constructing for some years past has been put together. It calculates, in three columns, a table with its first and second differences. Each column can be expressed as far as five figures, so that these fifteen figures constitute about one-ninth part of the larger engine. The ease and precision with which it works, leave no room to doubt its success in the more extended form. Besides tables of squares, cubes, and portions of logarithmic tables, it possesses the power of calculating certain series whose differences are not constant; and it has already tabulated parts of series formed from the following equations:

$$\Delta^3 u_x = \text{units figure of } \Delta^u{}_x$$

$$\Delta^3 u_x = \text{nearest whole no. to } \left(\frac{1}{10,000} \Delta u_x \right)$$

side by side, each having only one hand, and each having a thousand divisions instead of twelve hours marked on the face; and every time a string is pulled, let them strike on a bell the numbers of the divisions to which their hands point. Let him further suppose that two of the clocks, for the sake of distinction called B and C, have some mechanism by which the clock C advances the hand of the clock B one division, for each stroke it makes upon its own bell: and let the clock B by a similar contrivance advance the hand of the clock A one division, for each stroke it makes on its own bell. With such an arrangement, having set the hand of the clock A to the division I., that of B to III., and that of C to II., let the reader imagine the repeating parts of the clocks to be set in motion continually in the following order: viz.—pull the string of clock A; pull the string of clock B; pull the string of clock C.

The table on the following page will then express the series of movements and their results.

If now only those divisions struck or pointed at by the clock A be attended to and written down, it will be found that they produce the series of the squares of the natural numbers. Such a series could, of course, be carried by this mechanism only so far as the numbers which can be expressed by three figures; but this may be sufficient to give some idea of the construction,—and was, in fact, the point to which the first model of the calculating-engine, now in progress, extended.

The subjoined is one amongst the series which it has calculated:

0	3,486	42,972
0	4,991	50,532
1	6,907	58,813
14	9,295	67,826
70	12,236	77,602
230	15,741	88,202
495	19,861	99,627
916	24,597	111,928
1,504	30,010	125,116
2,340	36,131	139,272

The general term of this is:

$$u_x = \frac{x \cdot x - 1 \cdot x - 2}{1 \cdot 2 \cdot 3} + \text{the whole number in } \frac{x}{10} +$$
$$+ 10\Sigma^3 \left(\text{units figure of } \frac{x \cdot x + 1}{2} \right)$$

Repetitions of Process	Movements	Clock A — Hand set to I	Clock B — Hand set to III	Clock C — Hand set to II
		TABLE	First difference	Second difference
1	Pull A.	A. strikes . . . 1	. . . . . .	. . . .
	—— B.	{The hand is advanced} {(by B.) 3 divisions .}	B. strikes . . . 3	. . . .
	—— C.	. . . .	{The hand is advanced} {(by C.) 2 divisions }	C. strikes 2
2	Pull A.	A. strikes . . . 4	. . . . . .	. . . .
	—— B.	{The hand is advanced} {(by B.) 5 divisions .}	B. strikes . . . 5	. . . .
	—— C.	. . . .	{The hand is advanced} {(by C.) 2 divisions }	C. strikes 2
3	Pull A.	A. strikes . . . 9	. . . . . .	. . . .
	—— B.	{The hand is advanced} {(by B.) 7 divisions .}	B. strikes . . . 7	. . . .
	—— C.	. . . .	{The hand is advanced} {(by C.) 2 divisions .}	C. strikes 2
4	Pull A.	A. strikes . . . 16	. . . . . .	. . . .
	—— B.	{The hand is advanced} {(by B.) 9 divisions .}	B. strikes . . . 9	. . . .
	—— C.	. . . .	{The hand is advanced} {(by C.) 2 divisions .}	C. strikes 2
5	Pull A.	A. strikes . . . 25	. . . . . .	. . . .
	—— B.	{The hand is advanced} {(by B.) 11 divisions .}	B. strikes . . . 11	. . . .
	—— C.	. . . .	{The hand is advanced} {(by C.) 2 divisions .}	C. strikes 2
6	Pull A.	A. strikes . . . 36	. . . . . .	. . . .
	—— B.	{The hand is advanced} {(by B.) 13 divisions .}	B. strikes . . . 13	. . . .
	—— C.	. . . .	{The hand is advanced} {(by C.) 2 divisions .}	C. strikes 2

VII

CALCULATING ENGINES

By CHARLES BABBAGE

Chapter XIII, *The Exposition of 1851*

IT IS not a bad definition of *man* to describe him as a *tool-making animal*. His earliest contrivances to support uncivilized life, were tools of the simplest and rudest construction. His latest achievements in the substitution of machinery, not merely for the skill of the human hand, but for the relief of the human intellect, are founded on the use of tools of a still higher order.

The successful construction of all machinery depends on the perfection of the tools employed, and whoever is a master in the art of tool-making possesses the key to the construction of all machines.

The Crystal Palace, and all its splendid contents, owe their existence to *tools* as the physical means—to intellect as the guiding power, developed equally on works of industry or on objects of taste.

The contrivance and the construction of tools, must therefore ever stand at the head of the industrial arts.

The next stage in the advancement of those arts is equally necessary to the progress of each. It is the art of drawing. Here, however, a divergence commences: the drawings of the artist are entirely different from those of the mechanician. The drawings of the latter are Geometrical projections, and are of vast importance in all mechanism. The resources of mechanical drawing have not yet been sufficiently explored: with the great advance now making in machinery, it will become necessary to assist its powers by practical yet philosophical rules for expressing still more clearly by signs and by the letters themselves the mutual relations of the parts of a machine.

As we advance towards machinery for more complicated objects, other demands arise, without satisfying which our further course is absolutely stopped. It becomes necessary to see at a glance, not only every *successive* movement of each amongst thousands of different parts, but also to scrutinize all contemporaneous actions. This gave rise to the Mechanical Notation, a language of signs, which, although invented for one subject, is of so comprehensive a nature as to be applicable to many. If the whole of the facts relating to a naval or military battle were known, the mechanical notation would assist the description of it quite as much as it would that of any complicated engine.

This brief sketch has been given partly with the view of more distinctly directing attention to an important point in which England excels all other countries—the art of *contriving and making tools*; an art which has been continually forced upon my own observation in the contrivance and construction of the Calculating Engines.

When the first idea of inventing mechanical means for the calculation of all classes of astronomical and arithmetical tables occurred to me, I contented myself with making simple drawings, and with forming a small model of a few parts. But when I understood it to be the wish of the Government that a large engine should be constructed, a very serious question presented itself for consideration:

Is the present state of the art of making machinery sufficiently advanced to enable me to execute the multiplied and highly complicated movements required for the Difference Engine?

After examining all the resources of existing workshops, I came to the conclusion that, in order to succeed, it would become necessary to advance the art of construction itself. I trusted with some confidence that those studies which had enabled me to contrive mechanism for new wants, would be equally useful for the invention of new tools, or of other methods of employing the old.

During the many years the construction of the Difference Engine was carried on, the following course was adopted. After each drawing had been made, a new inquiry was instituted to determine the mechanical means by which the several parts were to be formed. Frequently sketches, or new drawings, were made, for the purpose of constructing the tools or mechanical arrangements thus contrived. This process often elicited some simpler mode of construction, and thus the original contrivances were improved. In the mean time, many workmen of the highest skill were constantly employed in making the tools, and afterwards in using them for the construction of parts of the engine. The knowledge thus acquired by the workmen, matured in many cases by their own experience, and often perhaps improved by their own sagacity, was thus in time disseminated widely throughout other workshops. Several of the most enlightened employers and constructors of machinery, who have themselves contributed to its advance, have expressed to me their opinion that if the Calculating Engine itself had entirely failed, the money expended by Government in the attempt to make it, would be well repaid by the advancement it had caused in the art of mechanical construction.

It is somewhat singular that whilst I had anticipated the difficulties of construction, I had not foreseen a far greater difficulty, which, however, was surmounted by the invention of the Mechanical Notation.

The state of the *Difference Engine* at the time it was abandoned by the Government, was as follows: A considerable portion of it had been made; a part (about sixteen figures) was put together; and the drawings, the whole of which are now in the Museum of King's College at Somerset House, were far advanced. Upon this engine the Government expended about £17,000.

The drawings of the *Analytical Engine* have been made entirely at *my own cost*: I instituted a long series of experiments for the purpose of reducing the expense of its construction to limits which might be within the means I could myself afford to supply. I am now resigned to the necessity of abstaining from its construction, and feel indisposed even to finish the drawings of one of its many general plans. As a slight idea of the state of the drawings may be interesting to some of my readers, I shall refer to a few of the great divisions of the subject.

ARITHMETICAL ADDITION.—About a dozen plans of different mechanical movements have been drawn. The last is of the very simplest order.

CARRIAGE OF TENS.—A large number of drawings have been made of modes of carrying tens. They form two classes, in one of which the carriage takes place successively: in the other it occurs simultaneously, as will be more fully explained at the end of this chapter.

MULTIPLYING BY TENS.—This is a very important process, though not difficult to contrive. Three modes are drawn; the difficulties are chiefly those of construction, and the most recent experiments now enable me to use the simplest form.

DIGIT COUNTING APPARATUS.—It is necessary that the machine should count the digits of the numbers it multiplies and divides, and that it should combine these properly with the number of decimals used. This is by no means so easy as the former operation: two or three systems of contrivances have been drawn.

COUNTING APPARATUS.—This is an apparatus of a much more general order, for treating the indices of functions and for the determination of the repetitions and movements of the Jacquard cards, on which the Algebraic developments of functions depend. Two or three such mechanisms have been drawn.

SELECTORS.—The object of the system of contrivances thus named, is to choose in the operation of Arithmetical division the proper multiple to be subtracted; this is one of the most difficult parts of the engine, and several different plans have been drawn. The one at last adopted is, considering the object, tolerably simple. Although division is an inverse operation, it is possible to perform it entirely by mechanism without any tentative process.

REGISTERING APPARATUS.—This is necessary in division to record the quotient as it arises. It is simple, and different plans have been drawn.

ALGEBRAIC SIGNS.—The means of combining these are very simple, and have been drawn.

PASSAGE THROUGH ZERO AND INFINITY.—This is one of the most important parts of the Engine, since it may lead to a totally different action upon the formulæ employed. The mechanism is much simpler than might have been expected, and is drawn and fully explained by notations.

BARRELS AND DRUMS.—These are contrivances for grouping together certain mechanical actions often required; they are occasionally under the direction of the cards; sometimes they guide themselves, and sometimes their own guidance is interfered with by the Zero Apparatus.

GROUPINGS.—These are drawings of several of the contrivances before described, united together in various forms. Many drawings of them exist.

GENERAL PLANS.—Drawings of all the parts necessary for the Analytical Engine have been made in many forms. No less than thirty different general plans for connecting them together have been devised and partially drawn; one or two are far advanced. No. 25 was lithographed at Paris in 1840. These have been superseded by simpler or more powerful combinations, and the last and most simple has only been sketched.

A large number of Mechanical Notations exist, showing the movements of these several parts, and also explaining the processes of arithmetic and algebra to which they relate. One amongst them, for the process of division, covers nearly thirty large folio sheets.

About twenty years after I had commenced the first Difference Engine, and after the greater part of these drawings had been completed, I found that almost every contrivance in it had been superseded by new and more simple mechanism, which the construction of the Analytical Engine had rendered necessary. Under these circumstances I made drawings of an entirely new Difference Engine. The drawings, both for the calculating and the printing parts, amounting in number to twenty-four, are completed. They are accompanied by the necessary mechanical notations, and by an index of letters to the drawings; so that although there is as yet no description in words, there is effectively such a description by signs, that this new Difference Engine might be constructed from them.

Amongst the difficulties which surrounded the idea of the construction of an Engine for developing Analytical formulæ, there were some which seemed insuperable if not impossible, not merely to the common understandings of well-informed persons, but even to the more practised intellect of some of the greatest masters of that science which the machine was intended to control. It still seemed, after much discussion, at least highly doubtful whether such formulæ could ever be brought within the grasp of mechanism.

I have met in the course of my inquiries with four cases of obstacles presenting the appearance of impossibilities. As these form a very interesting chapter in the history of the human mind, and are on the one hand connected with some of the simplest elements of mechanism, and on the other with some of the highest principles of philosophy, I shall endeavour to explain them in a short, and, I hope, somewhat popular manner, to those who have a very moderate share of mathematical knowledge. Those of my readers to whom they may not be sufficiently interesting, will, I hope, excuse the interruption, and pass on to the succeeding chapters.

§ The first difficulty arose at an early stage of the Analytical Engine. The mechanism necessary to add one number to another, if the carriage of the tens be neglected, is very simple. Various modes had been devised and drawings of about a dozen contrivances for carrying the tens had been made. The same general principle pervaded all of them. Each figure wheel when receiving addition, in the act of passing from nine to ten caused a lever to be put aside. An axis with arms arranged spirally upon it then revolved, and commencing with the lowest figure replaced successively those levers which might have been put aside during the addition. This replacing action upon the levers caused unity to be added to the figure wheel next above. The numerical example below will illustrate the process.

$$\left.\begin{array}{l} 597,999 \\ 201,001 \end{array}\right\} \text{Numbers to be added.}$$

| 798,990 | Sum without any carriage. |
| 1 | Puts aside lever acting on tens. |

| 798,900 | First spiral arm adds tens and |
| 1 | puts aside the next lever. |

| 798,000 | Second spiral arm adds hundreds, and |
| 1 | puts aside the next lever. |

| 799,000 | Third spiral arm adds thousands. |

Now there is in this mechanism a certain analogy with the act of memory. The lever thrust aside by the passage of the tens, is the equivalent of the note of an event made in the memory, whilst the spiral arm, acting at an after time upon the lever put aside, in some measure resembles the endeavours made to recollect a fact.

It will be observed that in these modes of *carrying*, the action must be *successive*. Supposing a number to consist of thirty places of figures, each of which is a nine, then if any other number of thirty figures be added to it, since the addition of each figure to the corresponding one takes place at the same time, the whole addition will only occupy nine units of time. But since the number added may be unity, the carriages may possibly amount to twenty-nine, consequently the time of making the carriages may be more than three times as long as that required for addition.

The time thus occupied was, it is true, very considerably shortened in the Difference Engine: but when the Analytical Engine was to be contrived, it became essentially necessary to diminish it still further. After much time fruitlessly expended in many contrivances and drawings, a very different principle, which seemed indeed at first to be impossible, suggested itself.

It is evident that whenever a carriage is conveyed to the figure above, if that figure happen to be a nine, a new carriage must then take place, and so on as far as the nines extend. Now the principle sought to be expressed in mechanism amounted to this.

1st. That a lever should be put aside, as before, on the passage of a figure-wheel from nine to ten.

2d. That the engine should then ascertain the position of all those nines which by carriage would ultimately become zero, and give notice of new carriages; that, foreseeing those events, it should anticipate the result by making all the carriages simultaneously.

This was at last accomplished, and many different mechanical contrivances fulfilling these conditions were drawn. The former part of this mechanism bears an analogy to memory, the latter to foresight. The apparatus remembers as it were, one set of events, the transits from nine to ten: examines what nines are found in certain critical places: then, in consequence of the concurrence of these events, acts at once so as to anticipate other actions that would have happened at a more distant period, had less artificial means been used.

§ The second apparent impossibility seemed to present far greater difficulty. Fortunately it was not one of immediate *practical* importance, although as a question of philosophical inquiry it possessed the highest interest. I had frequently discussed with Mrs. Somerville and my highly gifted friend the late Professor M'Cullagh of Dublin, the question whether it was possible that we should be able to treat algebraic formulæ by means of machinery. The result of many inquiries led to the conclusion that, if not really impossible, it was almost hopeless. The first difficulty was that of representing an indefinite number in a machine of finite size. It was readily admitted that if a machine afforded means of operating on *all* numbers under twenty places of figures, then that any number, or *an indefinite* number, of less than twenty places of figures might be represented by it. But such number will not be really indefinite. It would be possible to make a machine capable of operating upon numbers of forty, sixty, or one hundred places of figures: still, however, a limit must at last be reached, and the numbers represented would not be really *indefinite*. After lengthened consideration of this subject, the solution of the difficulty was discovered; and it presented the appearance of reasoning in a circle.

Algebraical operations in their most general form cannot be carried on by machinery without the capability of expressing *indefinite* constants. On the other hand, the only way of arriving at the expression of an indefinite constant, was through the intervention of Algebra itself.

This is not a fit place to enter into the detail of the means employed, further than to observe, that it was found possible to evade the difficulty by connecting *indefinite* number with the *infinite in time* instead of with the *infinite in space*.

The solution of this difficulty being found, and the discovery of another principle having been made, namely—that *the nature of a function might be indicated by its position*—algebra, in all its most abstract forms, was placed completely within the reach of mechanism.

§ The third difficulty that presented itself was one which I had long before anticipated. It was posed to me nearly at the same time by three of the most eminent cultivators of analysis then existing, M. Jacobi, M. Bessel, and Professor M'Cullagh, who were examining the drawings of the Analytical Engine. The question they proposed was this: How would the Analytical Engine be able to treat calculations in which the use of tables of logarithms, sines, &c., or any other tabular numbers should be required?

My reply was, that as at the time logarithms were invented, it became necessary to remodel the whole of the formulæ of Trigonometry, in order to adapt it to the new instrument of calculation: so when the Analytical Engine is made, it will be desirable to transform all formulæ containing tabular numbers into others better adapted to the use of such a machine. This, I replied, is the answer I give to you as mathematicians; but I added, that for others less skilled in our science, I had another answer: namely—

That the engine might be so arranged that wherever tabular numbers of any kind occurred in a formula given it to compute, it would on arriving at any required tabular number, as for instance, if it required the logarithm of 1207, stop itself, and ring a bell to call the attendant, who would find written at a certain part of the machine "Wanted log. of 1207." The attendant would then fetch from tables previously computed by the engine, the logarithm it required, and placing it in the proper place, would lift a detent, permitting the engine to continue its work.

The next step of the engine, on receiving the tabular number (in this case the logarithm of 1207) would be to *verify* the fact of its being really that logarithm. In case no mistake had been made by the attendant, the engine would use the given tabular number, and go on with its work until some other tabular number were required, when the same process would be repeated. If, however, any mistake had been made by the attendant, and a wrong logarithm had been accidentally given to the engine, it would have discovered the mistake, and have rung a louder bell to call the attention of its guide, who on looking at the proper place, would see a plate above the logarithm he had just put in with the word "*wrong*" engraven upon it.

By such means it would be perfectly possible to make all calculations requiring tabular numbers, without the chance of error.

Although such a plan does not seem absolutely impossible, it has always excited, in those informed of it for the first time, the greatest surprise. How, it has been often asked, does it happen if the engine knows when the *wrong* logarithm is offered to it, that it does not also know the right one; and if so, what is the necessity of having recourse to the attendant to supply it? The solution of this difficulty is accomplished by the very simplest means.

§ The fourth of the apparent impossibilities to which I have referred, involves a condition of so extraordinary a nature than even the most fastidious inquirer into the powers of the Analytical Engine could scarcely require it to fulfil.

Knowing the kind of objections that my countrymen make to this invention, I proposed to myself this inquiry:

Is it possible so to construct the Analytical Engine, that after the cards representing the formulæ and numbers are put into it, and the handle is turned, the following condition shall be fulfilled?

The attendant shall stop the machine in the middle of its work, whenever he chooses, and as often as he pleases. At each stoppage he shall examine all the figure wheels, and if he can, without breaking the machine, move any of them to other figures, he shall be at liberty to do so. Thus he may from time to time, falsify as many numbers as he pleases. Yet notwithstanding this, the final calculation and all the intermediate steps shall be entirely free from error. I have succeeded in fulfilling this condition by means of a principle in itself very simple. It may add somewhat, though not very much, to the amount of mechanism required; in many parts of the engine the principle has been already carried out. I by no means think such a plan *necessary*, although wherever it can be accomplished without expense it ought to be adopted.

VIII

THE ANALYTICAL ENGINE

By MAJOR-GENERAL H. P. BABBAGE

From the *Proceedings of the British Association, 1888*;
Paper read at Bath, September 12, 1888

TEN YEARS have elapsed since a committee of the British Association reported upon the Analytical Engine of my father, Charles Babbage, and I desire now, while offering a few remarks upon that Report, to endeavour to convey some idea of the mechanical arrangements of the engine to those who may be interested in it.

2. I am well assured that a time will come when such an engine will be completed and be a powerful means of enlarging not only pure mathematical science, but other branches of knowledge, and I wish, as far as in me lies, to hasten that time, and to help towards the general appreciation of the labours of my father, so little known or understood by the multitude even of the educated.

3. He considered the Paper by Menabrea,* translated with notes by Lady Lovelace, published in volume 3 of Taylor's "Scientific Memoirs," as quite disposing of the mathematical aspect of the invention. My business now is not with that.

4. The idea of the Analytical Engine arose thus: When the fragment of the Difference Engine, now in the South Kensington Museum, was put together early in 1833, it was found that, as had been before anticipated, it possessed powers beyond those for which it was intended, and some of them could be and were demonstrated on that fragment.†

5. It was evident that by interposing a few connecting-wheels, the column of Result can be made to influence the last Difference, or other part of the machine in several ways. Following out this train of thought, he first proposed to arrange the axes of the Difference Engine circularly, so that the Result column should be near that of the last Difference, and thus easily within reach of it.‡ He called this arrangement "the engine eating its own tail."§ But this soon led to

---

* This article is reprinted in this volume, p. 225.
† See The Ninth Bridgewater Treatise, London, 1838.
‡ See p. 52.
§ See Note 1, p. 345.

the idea of controlling the machine by entirely independent means, and making it perform not only Addition, but all the processes of arithmetic at will in any order and as many times as might be required. Work on the Difference Engine was stopped on 10th April, 1833, and the first drawing of the Analytical Engine is dated in September, 1834.

6. The object may shortly be given thus: It is a machine to calculate the numerical value or values of any formula or function of which the mathematician can indicate the *method* of solution. It is to perform the ordinary rules of arithmetic in any order as previously settled by the mathematician, and any number of times and on any quantities.

7. It is to be absolutely automatic, the slave of the mathematician, carrying out his orders and relieving him from the drudgery of computing.

8. The Analytical Engine is of course to print the results, or any intermediate result arrived at. He regarded this as an indispensable requisite, without which a calculating machine might indeed be useful for some purposes, but not for those of any scientific value. The perpetual risk of error in copying and transferring lines of figures is most troublesome, and tends to make results, themselves perfectly accurate, unreliable in use.

9. It is at once seen that the necessity of the engine being automatic imposes a gigantic task on the inventor. The first means employed to meet it is the use of cards to govern the engine. These are very similar to those in use in the Jacquard loom, to which we owe the figured patterns in the beautiful fabrics we see everywhere in common use.

10. One set of cards would be used to communicate the "given numbers," or constants of a problem to the machine. I shall call these throughout this paper "Number Cards."*

11. Another set of cards would be used to direct to which particular place or column in the engine these numbers, or any intermediate numbers arising in the course of the calculation, are to be conveyed or transferred; these cards I will call "Directive Cards." There would also be other "Directive Cards" for general purposes of control when necessary.

12. A third sort, called "Operation Cards,"† would direct the actual operations to be performed, these would put the engine mechanically into a condition to perform the particular operation required—Addition, Subtraction, &c., &c.

---

* See diagram on general plan, No. 25 of August 6, 1840, in Appendix, p. 378.
† See p. 62 and general plan of August 6, 1840, in Appendix, p. 378.

13. I was once asked, "How do you set the question? Do you write it on paper and put it into the machine?" Well, given the problem, the mathematician must first of all settle the operations and the particular quantities each is to be performed on and the time for each operation.

14. Then the superintendent of the engine must make a "Number Card" for each "given number," and settle the particular column in the machine on to which each "given number" is to be first received, and assign columns for every intermediate result expected to arise in the course of the calculation.

15. He will then prepare "Directive Cards" accordingly, and these, together with the necessary "Operation Cards" being placed in the engine, the question will have been set; not exactly, as my friend suggested, written on paper, but in cardboard, and motion being supplied the engine will give the answer.*

16. Now this appears a long process for what may be a simple question, but it is to be noted that the engine is designed for analytical purposes, and it would be like using the steam hammer to crush the nut, to use the Analytical Engine to solve common sums in arithmetic; or, adopting the language of Leibnitz (see page 219): "It is not made for those who sell vegetables or little fishes, but for observatories, or the private rooms of calculators, or for others who can easily bear the expense, and need a good deal of calculation." Moreover, except the "Number Cards," all the cards, once made for any given problem, can be used for the same problem, with any other "given numbers," and it would not be necessary to prepare them a second time—they could be carefully kept for future use. Each formula would require its own set of cards, and by degrees the engine would have a library of its own.*

17. Thus the values for any number of Life Insurance Policies might be calculated one after the other by merely supplying fresh cards for the age, amount, rate of interest, &c., for each individual case.

18. The separation of "Operation Cards" from the numbers to be operated on is complete. The powers of the engine are the most extended, but each set of cards makes it special for the solution of one particular problem; each individual case of which, again, requires its own "Number Cards."

19. Taking the formula used by Mr. Merrifield $(ab+c)d$ as an

* See p. 56.

illustration,* the full detail of the cards of all sorts required, and the order in which they would come into play is this:—

The four cards for the "given numbers" a, b, c and d, strung together are placed by hand on the roller, these numbers have to be placed on the columns assigned to them in a part of the machine called "The Store," where every quantity is first received and kept ready for use as wanted.

Directive Card	Operation Card	—
1st	...	Places a on column 1 of Store
2nd	...	,, b ,, 2 ,,
3rd	...	,, c ,, 3 ,,
4th	...	,, d ,, 4 ,,
5th	...	Brings a from Store to Mill
6th	...	,, b ,, ,,
...	1	Multiplies a and $b=p$
7th	...	Takes p to column 5 of Store where it is kept for use and record
8th	...	Brings p into Mill
9th	...	Brings c into Mill
...	2	Adds p and $c=q$
10th	...	Takes q to column 6 of Store
11th	...	Brings d into Mill
12th	...	,, q ,,
...	3	Multiplies $d \times q = p_2$
13th	...	Takes p_2 to column 7 of Store
14th	...	Takes p_2 to printing or stereo-moulding apparatus

20. We have thus besides the "Given Number" Cards, three "Operation Cards" used, and fourteen "Directive Cards;" each set of cards would be strung together and placed on a roller or prism of its own; this roller would be suspended and be moved to and fro. Each backward motion would cause the prism to move one face, bringing the next card into play, just as on the loom.

21. It is obvious that the rollers must be made to work in harmony, and for this purpose the levers which make the rollers turn would themselves be controlled by suitable means, or by general "Directive Cards," and the beats of the suspended rollers be stopped in the proper intervals.

22. This brings me to the second great distinguishing feature of the engine, the principle of "Chain." This enables us to deal mechanically with any single combination which may occur out of many possible, and thus to be ready for any or every contingency which may arise.

* See Note 2.

23. Supposing that it is desired to provide for a certain possible combination such as, for example, the concurrence of ten different events, it could be effected mechanically thus*—each event would be represented by an arm turning on its axis, and having at its end a block held loosely and capable of vertical motion independently of the arm which carries it. Now suppose each of these arms to be brought, on the occurrence of the event it represents, into a position so that the blocks should all be in one vertical line, then if the block in the lowest arm was raised by a lever, it would raise all the nine blocks together, and the top one could be made to ring a bell or communicate motion, &c. &c.; but if any one of the ten events had not happened, its block would be out of the "Chain," and the lowest block would be raised in vain and the bell remain silent.

24. This is the simplest form of "Chain," there may be many modifications of it to suit various purposes.

25. In its largest extent it will appear in the Anticipating Carriage† further on, but in its simplest form it appears here as a means of producing intermittent motion at uncertain intervals, which may or may not be previously known either to the mathematician or even to the superintendent of the engine who had prepared the cards.

26. In our illustration the first operation card happens to be multiplication, and the time this would occupy necessarily depends upon the number of digits in the two quantities multiplied. Now when multiplication is directed, the "Chain" to every part of the machine not wanted will be broken, and all motion thus stopped till the multiplication is completed, and whenever the last step of that is reached, the "Chain" will be restored.

27. The Chain for this purpose may be made this way, a link is cut and the two ends are brought side by side, overlapping each other; a part of each link is then cut away exactly the same in both pieces. Now if the piece in communication with the driving power is moved, the other is not, and remains stationary; but if a block is let fall into the space cut away so as to fill the gaps in both parts of the severed link, the chain will be complete and link—block—link will pass on the motion. The block is hung in the end of an arm in which it can slide, and all that is required is to move the arm sideways while the link pieces are at rest in the proper position. This is shown in the diagram on page 344.

28. In a machine such as the Analytical Engine, consisting of so many distinct trains of motion of which only a few would be in action

* See Note 3.
† See p. 54.

at a time, it may be easily conceived how useful this application of the
principle of Chain is. It helps to realize, too, another important
principle largely adopted, viz., to break up every train of motion as far
as possible into short courses, the last step of each furnishing a mere
guide for a fresh start in the mechanism from the driving power.   Of
course the link could be broken and restored also by any of the "Directive Cards" mentioned above.   The interposition of such links would
also be used to save the cards from the wear and tear unavoidable if
they were used for the actual transmission of any force.

29. I hope that I have now given some idea of the methods used
for the general control of the engine, and shall pass on to illustrate,
as well as I am able, some of the details of the machine.   It is not
to be supposed that I have mastered all these myself; nor, had I done
so, would it be possible for me to make them intelligible in the course
of a Paper, but some idea of them I hope to convey.

30. The machine consists of many parts.   I have found it easier
myself to regard these parts as so many separate machines, driven
by the same motive power and starting and stopping each other in
every possible combination, but otherwise acting independently,
though with a settled harmony towards a desired result.   When that
idea has been reached, it seems easy to imagine them all brought close
together, grouped in the positions most convenient.

31. Many general plans have been drawn.   Plan No. 25, dated
6 August, 1840 was lithographed, others followed.   The fact is, that
what suits one part best does not suit another, and the number of
possible variations of the parts renders it difficult to settle which
combination of the whole may be the best for the general plan.

32. A part of the machine called "The Store" has already been
mentioned.   It would consist of a number of vertical columns to
receive the "given numbers" and those arising in the course of the
calculations from them.

33. In the example already mentioned seven columns would be
used; perhaps for a first machine twenty would do, with twenty-five
wheels in each; each wheel might have a disc with the digits 0 to 9
engraved on its circumference, but this is not absolutely necessary.
In fact, the whole machine might be constructed without a figure
anywhere except for the printing.

34. The "Directive Cards" would put the column selected to
receive a "given number" into gear with a set of racks through
which the "given number" (expressed on its card) would be conveyed
to the Store column, each wheel being moved as many teeth as in the
corresponding place of the number.

35. One revolution of the main shaft would be sufficient to put a "given number" on a column of the Store, or to transfer it from the store to some other part of the engine.

36. The action of the "directive card" in this case is to raise the selected column so that its wheels should be level with a corresponding set of racks, and thus brought into gear with them.   Each column of the Store would require its own "directive card."

37. At the top of each column of the Store would be a wheel on which the Algebraic symbol of the quantity could be written,* and on the top of those columns assigned to receive the intermediate results this would also be done; though at first these columns would be at zero.

38. There would also be a wheel for the sign + or −.   Drawings and notations for such sign wheels exist,† and how far the operations of Algebra may be exhibited has been discussed.

39. The Mill is the part of the machine where the quantities are operated on.   Two numbers being transferred from the Store to two columns in the Mill the two are put into gear together through racks, and the reduction of one column to zero turns the other the exact equivalent, and thus adds it to the other.   Supposing there to be with each wheel a disc with the digits 0 to 9 engraved on its edge, and a screen in front of the column with a hole or window before each disc allowing one figure only to be seen at a time; during the process of addition the digits will pass before the window just as in counting till the sum is reached: thus if 5 is added to 7, the 7 will disappear and 8, 9, 0, 1, pass before the window in succession till 2 appear.   At the moment when 9 passes to 0, a lever will be moved, thus recording the necessity of a carriage to the figure above; the carriage is made subsequently, and for the Analytical Engine a method of performing the carriages all simultaneously‡ was invented by my father which he called "Anticipating" Carriage.

40. When two numbers are added, carriages may occur in any or every place except the last; where the wheels pass as just described from 9 to 0, a carriage arises directly; in those places where the figure disc comes to rest at 9, no carriage occurs, but one of two things may happen—if there is no carriage to arrive from the next right hand place, the 9 will not be changed; but should one be due, not only the 9 must be pushed on to 0, but a carriage must be passed on to the next place on the left hand, and if there happen to be a succession of

* See p. 56.
† See Note 4.
‡ See p. 54.

9's, they will all have to be pushed on. Working with twenty-five places of figures there can be no carriage ever required in the last place, but there may be in any one, any two, any three, &c., up to twenty-four places, where a carriage may arise directly and the same up to twenty-three places where the presence of 9's may indirectly cause it.

41. Now immense as is the number of these two sets of combinations, they can be successfully dealt with mechanically by the principle of "Chain"* and indeed every single combination as it arises is presented to the eye. There is a series of blocks for each place, the lower block is made to serve for the two events. The upper one has a projecting arm, which when moved circularly engages a toothed wheel and moves it on one tooth affecting the figure disc similarly. It rests on the lower block which moves it up and down with itself always. After the addition is completed, should a carriage have become due and the warning lever have been pushed aside, it is made (by motion from the main shaft) to actuate the lower block and throw it into "Chain," when this is raised (again by motion from the main shaft) it raises the upper block which thus effects ordinary carriage; but supposing no carriage to have become due, there will be at the window either a 9 or some less number. The latter case may be dismissed at once; as a carriage arriving to it will cause no carriage to be passed on. Should there be a 9 at the window the warning-lever cannot have been pushed aside, but in every place where there is a 9 another lever, again by motion from the main shaft comes into play and pushes the lower block into "Chain"; not, of course, into "Chain" for ordinary carriage, but into another position for Chain for 9's, so that should there come a carriage from below, the Chain for 9's will be raised as far as it extends and effect the carriages necessary, be it for a single place, or for several, or for many. Should there be no carriage from below to disturb it the Chain remains passive; it has been made ready for a possible event which has not occurred. A certain time is required for the preparation of "Chain," but that done all the carriages are then effected simultaneously. A piece of mechanism for Anticipating Carriage on this plan to twenty-nine figures exists and works perfectly.

42. When a large number of places of figures is being dealt with, the saving of time is very considerable, especially when it is remembered that multiplication is usually done by successive additions.

43. Another plan for carriage has also been contrived and drawn.†
It is obvious that there is no necessity, when there are many successive

---

* See Note 5.
† See Note 6.

additions, to make the carriages immediately follow each addition. The additions may be made one after the other, and the carriages having been warned or even actually made on a separate wheel in each place as they arise, can all be made in one lot afterwards; more machinery is required, but the saving of time is very considerable. This plan has been called "Hoarding Carriage," and thoroughly worked out.

44. It is interesting to note that Hoarding Carriage is seen in the little machine of Sir Samuel Moreland invented in 1666, and probably existed in that of Pascal still earlier.

45. Now it may happen in addition that two or more numbers being added together, there may not be room at the top of the column for the left hand figure of the result. This would usually happen from an oversight in preparing or arranging the cards when space should be left; but it might so happen* that the calculation led, as mathematical problems sometimes do, through infinity. In either case a bell would be rung and the engine stopped; or if the contingency had been anticipated by the mathematician, fresh directive cards previously prepared might be brought into play.

46. One addition would be executed in each turn of the main axis, the intermittent motions required being produced by cams on the main axis. These would be flat discs with projecting parts on them acting on arms with friction rollers at the end. Each cam would be double, *i.e.* have two discs; the projections on the one corresponding with depressions on the other. Such cams are easily made and fixed and adjusted, about six or seven are sufficient for addition. The illustration shows such a double cam, together with a "Chain" for throwing into and out of gear, as explained on page 335. This may or may not be wanted.

47. Subtraction is performed by the interposition of an additional pinion, which turns the figure discs the reverse way; the figures decreasing in succession as they pass the window, and the carriage arises whenever the 0 passes and a 9 appears. The same arrangement is applied to the 0's in subtraction as to the 9's in addition, and the same principle of "Chain"; indeed, the same actual mechanism serves for both, the change being made by the movement of a single lever. See page 344.

48. In subtraction, when a larger is subtracted from a smaller number, there will be a warning made in the highest place for a carriage to a place above which does not exist, zero has been passed; the warning

---

* See p. 67.

lever will ring a bell and stop the engine, unless the contingency has been anticipated and provided for by the mathematician.*

49. Several ways have been worked out† and drawn for multiplication. A skilled computer dealing with many places of figures having to multiply would make by successive additions a table of the first ten multiples of the multiplicand; if he has done this correctly the tenth multiple will be the same as the first, only all moved one place to the left and with a 0 in the units place. Using this table he picks out in succession the multiples required and puts them in the proper places, then adding all up gets his results, dispensing altogether with the multiplication table and doing nothing beyond addition. For the machine this way has been worked out and many drawings and notations exist for it.

50. Another way by the use of barrels‡ has also been drawn. The way by succession additions and stepping is perhaps the simplest.

51. When two numbers each of any number of places from one up to twenty or thirty have to be multiplied, it becomes necessary, in order to save time, to ascertain which has the fewest significant digits; special apparatus has been designed for this, called "Digit Counting Apparatus."§ The smaller of the two is made the multiplier. Both are brought into the Mill and put on the proper columns. As the successive additions are made, the figure wheels of the multiplier are successively reduced to zero; when this happens for any one figure of the multiplier a cam on its wheel pushes out a lever which breaks the link or "chain" for addition, and completes that for stepping; so that the next revolution of the main axis causes stepping instead of addition, and that being done the links are changed back and the successive additions go on.

52. By this process the multiplier column is all reduced to zero; but if need be there can be another column alongside to which it may step by step be worked on—anyhow, it stands on record in its own column in the Store.

53. Multiplication would ordinarily be performed from the highest place downwards, and from the decimal point onwards; so that the result would be complete to the lowest place of decimals; but, of course, would contain the accumulated error due to cutting off the figures beyond in each quantity operated on. As, however, there would be a counting apparatus recording the successive additions,

* See p. 67.
† See Note 7.
‡ See Note 8.
§ See Note 9.

which at the end of the multiplication, would give the sum of all the digits of the multiplier, the maximum possible error would be known, and if thought advisable, a correction ordered to be made for it; for instance, the machine might be directed to halve the total of the digits, and add it to the last figures of the result wherever cut off.

54. Division is a more troublesome operation by far than multiplication. Mr. Merrifield in his report has called this "essentially a tentative process," and so it is as regards the computer with the pen; but as regards the Analytical Engine, I do not assent to it.

55. As, however, he considered the striking part of a clock a tentative process, while I do not, the difference might be one of definition of a word, and I only notice it as maintaining that the processes of the engine are only so far tentative as the guiding spirit of the mathematician leads him to make them.

56. Division by Table and also by barrels has been thoroughly drawn and worked out—the process by successive subtraction also. The divisor and dividend being brought into the Mill, the successive subtractions proceed and their number is recorded; when the correct figure has been reached, the subtraction is thrown out by "Chain" and stepping caused.

57. There is no trial and error whatever. Some additional machinery is required for this and more time is occupied, but the result is certain; the engine stops of necessity at the right figure of the quotient, and gives the order for stepping and so on to the end.

58. For the extraction of the square root a barrel would be used, but the ordinary process might be followed step by step without a barrel.

59. Counting machines of sorts would perform important functions in the general directive. Some would be mere records of the progress of the different steps going on, but others, of which the multiplier column may serve as a sort of example, would themselves act at appointed times as might be previously arranged.

60. This principle of "Chain" is used also to govern the engine in those cases where the mathematician himself is not able to say beforehand what may happen, and what course is to be pursued, but has to let it depend on the intermediate result of the calculation arrived at. He may wish to shape it in different ways according as one or several events may occur, and "Chain" gives him the power to do it mechanically. By this contrivance machines to play simple games of skill such as "tit tat too" have been designed.

61. Take a simpler case where the mathematician desired to deal with the largest of two numbers arrived at, not knowing beforehand

which it might be.* The numbers would have been placed on the two columns of the Store previously assigned for their reception, and cards would have been arranged to direct the two numbers to be subtracted each from the other. In the one case there would be a remainder, but in the other the carriage warning lever would have been moved in the highest place, which would be made to bring an alternative set of cards, previously prepared, into play.

62. I have not been quite able to accept Mr. Merrifield's opinion as to the "capability of the engine." He says: "Its capability thus extends to any system of operations or equations which leads to a single numerical result." Now it could furnish not only one root, but every root of an equation, if there were more than one, capable of arithmetical expression, and there are many such equations. It could follow the processes of the mathematician be they tentative or direct,† wherever he could show the way to any number of numerical results. It is only a question of cards and time. Fabrics have been woven requiring several thousand cards. I possess one‡ made by the aid of over twenty thousand cards, and there is no reason why an equal number of cards should not be used if necessary, in an Analytical Engine for the purposes of the mathematician.

63. There exist over two hundred drawings, in full detail, to scale, of the engine and its parts. These were beautifully executed by a highly skilled draughtsman and were very costly.§

64. There are over four hundred notations of different parts.‖ These are, in my father's system of mechanical notation, an outline of which I had the honour to submit to this Association at Glasgow in 1855.¶ Not many years ago I was looking over one of my own drawings with a very intelligent mechanical engineer in Clerkenwell. I wished to get motion for some particular purpose. He suggested to get it from an axis which he pointed to on the drawing. I answered, "No, that will not do; I see by the drawing that it is a 'dead centre.'" He replied, "You have some means of knowing which I have not." I certainly had, for I used this mechanical notation. The system in whole or part should be taught in our Art and Technical Schools.

65. In addition to the above things there will shortly be available, as I have said before, the reprint of the various papers published

---

* See p. 67.
† See p. 65.
‡ A portrait of Jacquard woven in silk. See p. 169 of *Passages from the Life of a Philosopher.*
§ See Note 10.
‖ See Note 11.
¶ See Note 12.

relating to these machines, and a full list of the drawings and notations will be included.

66. I believe that the present state of the design would admit of the engine being executed in metal; nor do I think, as suggested by Mr. Merrifield, quantities and proportions would have to be calculated. Of course working drawings would have to be made—it would not be wise to commence any work without such drawings—but they would mostly be simply copies from the originals with such details as workmen want added. It would also be wise to make models of particular parts. The shapes of the cams, for instance, might be tried in wood which would afterwards suit as patterns for castings.

67. Mr. Merrifield doubts "whether the drawings really represent what is meant to be rendered in metal, or whether it is simply a provisional solution to be afterwards simplified." I have no doubt that the drawings do represent what at the time was intended to be put into metal; but as certainly were intended to be superseded when anything better could be found. Very few machines indeed are invented which do not undergo modification. It is almost invariably the case that the second machine made has improvements on the first design. In such a machine as an Analytical Engine, this would be important; but no one ever stops a useful invention for fear of improvements. A gun or an ironclad, for example, has been scarcely made before superseded by something better, and so it must be with all inventions, though fortunately not at the same speed.

68. As to the possible modification of the engine (Chapter VIII. of the Report) I may say that I am myself of opinion that the general design might with practical advantage be restricted. The engine would be still very useful indeed, if made not quite so automatic; even the Mill by itself would, I believe, be extensively useful, if a printing or stereo-moulding apparatus were joined to it. Perhaps, if that existed, the wants of further parts such as the Store would be felt and supplied.

69. As to the general conclusion and recommendation (Chapter IX. of the Report) there is little for me to notice. I see no hope of any Analytical Engine, however useful it might be, bringing any profit to its constructor, and beyond the preparation of this Paper, and the publication of the volume I have mentioned as shortly to follow, there is little or no temptation to do more. Those who wish for such an engine would, I think, give it a helping hand if they could show what pecuniary benefit it would bring. The History of Babbage's Calculating Machines is sufficient to damp the ardour of a dozen enthusiasts.

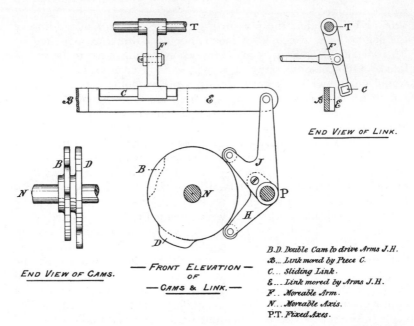

END VIEW OF LINK.

END VIEW OF CAMS.

— FRONT ELEVATION —
OF
— CAMS & LINK. —

B.D. *Double Cam to drive Arms J.H.*
B... *Link moved by Piece C.*
C... *Sliding Link.*
E... *Link moved by Arms J.H.*
F... *Moveable Arm.*
N... *Moveable Axis.*
P.T. *Fixed Axes.*

The "chain mechanism" held as "a great distinguishing feature of Babbage's engine." This device provides a means for coupling and decoupling the drive from any desired shaft as the program commands. See the description on p. 335.

NOTES BY THE EDITORS

Throughout these notes *Engines* refers to *Babbage's Calculating Engines*, edited by Maj.-Gen. H. P. Babbage and published by E. and F. N. Spon in London in 1889.

Note 1 (p. 331). See p. 268 of *Engines*. This is a Report of the Stokes Committee from the *Proceedings* of the Royal Society, January 21, 1855.

Note 2 (p. 334). See p. 326 of *Engines*. This is the Report of the Cayley Committee from the *Proceedings* of the British Association for the Advancement of Science, 1878.

Note 3 (p. 335). See Plate XII of December 4, 1836, reproduced in *Engines*.

Note 4 (p. 337). See Plate V and Drawings 138 and 139 in List, p. 291, and Notations, p. 286 of *Engines*. Plate V appears at the end of *Engines*. The drawings referred to are unpublished drawings of May, 1846 listed in *Engines*, p. 291. Three cases of drawings 2 in. × 3 in. (about 200 in all) are catalogued. "Notations" refers to a list of notations in Babbage's own symbolism from which working drawings could eventually be prepared. These are logical diagrams of the mechanism following the ideas given in Babbage's early paper, p. 346 of this volume, and his handout of 1851, p. 357. The word "Notations" in these Notes refers to this kind of material.

Note 5 (p. 338). See Plate XII and IV in the List of Drawings, p. 288 of *Engines*. Cf. Note 4.

Note 6 (p. 338). See Drawings 3, Fig. 1; 57; 180** and 183* in List, pp. 288–92 of *Engines*. Drawings 3 and 57 are dated in the 1830's and 180** and 183* are dated in 1864.

Note 7 (p. 340). See p. 275 of *Engines*. This refers to entries of 1835 and 1836 listed in the catalogue of Notations. Cf. Note 4.

Note 8 (p. 340). See List of Drawings, p. 290 of *Engines*. Drawings 87 and 88 are dated 1840.

Note 9 (p. 340). See List of Drawings, p. 289 of *Engines*. Drawings 36 and 52 are dated 1836 and 1837.

Note 10 (p. 342). See p. 288 of *Engines*. Catalogue of Drawings.

Note 11 (p. 342). See p. 271 of *Engines*. Catalogue of Notations.

Note 12 (p. 342). See pp. 242–57 of *Engines*. Babbage's 1851 paper is reprinted here on p. 357. The rest of the material includes two papers by Maj.-Gen. H. P. Babbage.

IX

ON A METHOD OF EXPRESSING BY SIGNS THE ACTION OF MACHINERY

By CHARLES BABBAGE

From the *Philosophical Transactions* of the Royal
Society, Vol. 2, 1826

IN THE construction of an engine, on which I have now been for some time occupied, for the purpose of calculating tables, and impressing the results on plates of copper, I experienced great delay and inconvenience from the difficulty of ascertaining from the drawings the state of motion or rest of any individual part at any given instant of time; and if it became necessary to inquire into the state of several parts at the same moment, the labour was much increased.

In the description of machinery by means of drawings, it is only possible to represent an engine in one particular state of the action. If indeed it is very simple in its operations a succession of drawings may be made of it in each state of its progress, which will represent its whole course; but this rarely happens, and is attended with the inconvenience and expense of numerous drawings. The difficulty of retaining in the mind all the contemporaneous and successive movements of a complicated machine, and the still greater difficulty of properly timing movements which had already been provided for, induced me to seek for some method by which I might at a glance of the eye select any particular part, and find at any given time its state of motion or rest, its relation to the motions of any other part of the machine, and if necessary trace back the sources of its movement through all its successive stages to the original moving power. I soon felt that the forms of ordinary language were far too diffuse to admit of any expectation of removing the difficulty, and being convinced from experience of the vast power which analysis derives from the great condensation of meaning in the language it employs, I was not long in deciding that the most favourable path to pursue was to have recourse to the language of signs. It then became necessary to contrive a notation which ought, if possible, to be at once simple and expressive, easily understood at the commencement, and capable of being readily retained in the memory from the proper adaptation of the signs to the circumstances they were intended to represent. The first thing to be done was obviously to make an accurate enumeration of all the moving parts and to appropriate a name to each; the multitude of different contrivances in various machinery, precluded all idea of substituting signs for these parts. They were therefore written down in succession, only observing to preserve such an order that those which jointly concur for accomplishing the effect of any separate part of the machine might be found situated near to each other: thus in a clock, those parts which belong to the striking part ought to be placed together, whilst those by which the repeating part operates ought, although kept distinct, yet to be as a whole, adjacent to the former part.

Each of these names is attached to a faint line which runs longitudinally down the page, and which may for the sake of reference be called its *indicating line*.

The next object was to connect the notation with the drawings of the machine, in order that the two might mutually illustrate and explain each other.

It is convenient in the three representations of a machine, to employ the same letters for each part; in order to connect these with the notations, the letters which in the several drawings refer to the same parts, are placed upon the indicating lines immediately under the names of the things. If circumstances should prevent us from adhering to this rule, it would be desirable to mark those things represented in the plan by the ordinary letters of the alphabet, those pointed out for one of the other projections by the letters of an accented alphabet, and the parts delineated on the third projection by a doubly accented alphabet. In engines of so complicated a nature as to require sections at various parts, as well as the three projections, this system is equally applicable, and its advantage consists in this—that the number of accents on the letter indicates at once the number of the drawing on which it appears, and when it is intended to refer to several at the same time, the requisite letters may be employed and placed in the order in which the drawings will best illustrate the part under examination.

The next circumstance which can be indicated by the system of mechanical notation which I propose, more readily than by drawings, is the number of teeth on each wheel or sector, or the number of pins or studs on any revolving barrel. A line immediately succeeding that which contains the references to the drawings is devoted to this purpose, and on each vertical line indicating any particular part of the machine, is written the number of teeth belonging to it. As there is generally a great variety of parts of machinery which do not consist of teeth; of course every vertical line will not have a number attached to it.

The three lines immediately succeeding this, are devoted to the indication of the velocities of the several parts of the machine. The first must have, on the indicating line of all those parts which have a rectilinear motion, numbers expressing the velocity with which those parts move, and if this velocity is variable, two numbers should be written, one expressing the greatest, the other the least velocity of the part. The second line must have numbers expressing the angular velocity of all those parts which revolve; the time of revolution of some one of them being taken as the unit of the measure of angular velocity.

It sometimes happens that two wheels have the same angular velocity when they move; but from the structure of the machine, one of them rests one-half of the time during which the other is in action. In this case, although their angular velocities are equal, their comparative velocities are as 1 to 2; for the second wheel makes two revolutions, whilst the other only makes one. A line is devoted to the numbers which thus arise, and it is entitled, Comparative Angular Velocity.

The next object to be considered is the course through which the moving power is transmitted, and the particular modes by which each part derives its movement from that immediately preceding it in the order of action. The sign which I have chosen to indicate this transmission of motion (an arrow), is one very generally employed to denote the direction of motion in mechanical drawings; it will therefore readily suggest the direction in which the movement is transmitted. There are however various ways by which motion is communicated; and it becomes a matter of some importance to consider whether, without interfering with the sign just selected, some modification might not be introduced into its minor parts, which, leaving it unaltered in the general form, should yet indicate the peculiar nature of the means by which the movement is accomplished.

On enumerating those modes in which motion is usually communicated, it appeared that they may be reduced to the following:

One piece may receive its motion from another by being permanently attached to it as a pin on a wheel, or a wheel and pinion on the same axis. } This may be indicated by an arrow with a bar at the end.

One piece may be driven by another in such a manner that when the driver moves, the other also always moves; as happens when a wheel is driven by a pinion.

An arrow without any bar.

———————→

One thing may be attached to another by stiff friction.

An arrow formed of a line interrupted by dots.

— · — · — · —→

One piece may be driven by another, and yet not always move when the latter moves: as is the case when a stud lifts a bolt once in the course of its revolution.

By an arrow, the first half of which is a full line, and the second half a dotted one.

——————· · · ·➤

One wheel may be connected with another by a ratchet, as the great wheel of a clock is attached to the fusee.

By a dotted arrow with a ratchet tooth at its end.

· ⋀ · · · · ·➤

Each of the vertical lines, representing any part of the machine, must now be connected with that representing the part from which it receives its movement, by an arrow of such a kind as the preceding table indicates; and if any part derives motion from two or more sources, it must be connected by the proper arrows with each origin of its movement.   It will in some cases contribute to the better understanding of the machine, if those parts which derive movement from two or more sources have their names connected by a bracket, with two or more vertical lines, which may be employed to indicate the different motions separately.   Thus if a shaft has a circular as well as a longitudinal motion, the two lines attached to its name should be characterised by a distinguishing mark, such as (vert. motion) and (circ. motion).   Whenever any two or more motions take place at the same time, this is essential; and when they do not, it is convenient for the purpose of distinguishing them.

All machines require, after their parts are finished and put together, certain alterations which are called adjustments.   Some of these are permanent, and, when once fixed by the maker, require no further care.   Others depend on the nature of the work they are intended to perform, as in the instance of a corn mill; the distance between the stones is altered according to the fineness of the flour to be ground: these may be called usual adjustments; whilst there are others depending on the winding up of a weight, or spring, which may be called periodic adjustments.   As it is very desirable to know all the adjustments of a machine, a space is reserved, below that in which the connections of the moving parts are exhibited, where these may be indicated; if there are many adjustments, this space may be subdivided into three, and appropriated to each of the three species just enumerated, the permanent, the usual, and the periodic; if their number is small, it is better merely to distinguish them by a modification of their signs.   It is sometimes impossible to perform such adjustment, except in a particular succession; and it is always convenient to adhere to one particular order.   Numbers attached to their respective lines denote the order in which the parts are to be adjusted; and as it will sometimes happen that two or more adjustments must be made at the same time, in that case the same numbers must be written on the lines belonging to all those parts which it is necessary to adjust simultaneously.

If it is convenient to distinguish between the species of adjustments without separating them by lines, this may be accomplished by putting a line above or below the figures, or inclosing them in a circle, or by some similar mode.   I have attached the letter P to those which are periodic.

It would add to the knowledge thus conveyed, if the sign indicating adjustment also gave us some information respecting its nature; there are, however, so many different species, that it is perhaps better in the first instance to confine ourselves to a few of the most common, and to leave to those who may have occasion to employ this kind of notation, the contrivance of signs, fitted for their more immediate purpose.

One of the most common adjustments is that of determining the distance between two parts, as between the point of suspension and the centre of oscillation of a pendulum.   This might be indicated by a

small line crossing the vertical line attached to the part.   ⊣⊢ │ │   If

the distance between two parts, which are represented by different lines, is to be altered, their lines may be connected by an horizontal line.

Thus the adjustment of A in the above figure depends on its distance from D.   This adjustment often determines the length of a stroke, and sometimes the eccentricity of an eccentric, which depends on the linear distance between its centre and its centre of motion.

The next most frequently occurring adjustment is that which is sometimes necessary in fixing two wheels, or a wheel and an arm on the same axis.   This relation of angular position, may be indicated by a circle and two radii placed at the requisite angle, or that angle may be stated in figures and enclosed within the circle; this circle ought however to be connected by a line with the other part, with which the angle is to be formed, thus,

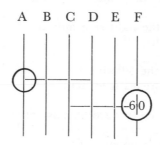

which means that an adjustment is to be made by fixing A on its axis, making a right angle with D and that F must also make with C an angle of sixty degrees.   In speaking of these angles it should always be observed that they refer to the angles made by the parts on one of the planes of projection.

When it is thought requisite to enter into this minute detail of adjustments, it will be necessary, in order to avoid confusion, to put the lines indicating the order of adjustment, above and distinct from these signs.

The last and most essential circumstance to be represented, is the succession of the movements which take place in the working of the machine.   Almost all machinery after a certain number of successive operations, recommences the same course which it had just completed, and the work which it performs usually consists of a multitude of repetitions of the same course of particular motions.

It is one of the great objects of the notation I am now explaining, to point out a method by which, at any instant of time in this course or cycle of operations of any machine, we may know the state of motion or rest of every particular part; to present a picture by which we may, on inspection, see not only the motion at that moment of time, but the whole history of its movements, as well as that of all the contemporaneous changes from the beginning of the cycle.

In order to accomplish this, each of the vertical indicating lines representing any part of the machine, has, adjacent to it, other lines drawn in the same direction: these accompanying lines denote the state of motion or rest of the part to which they refer, according to the following rules:

1. Unbroken lines indicate motion.

2. Lines on the right side indicate that the motion is from right to left.

3. Lines on the left side indicate that the direction of the motion is from left to right.

4. If the movements are such as not to admit of this distinction, then, when lines are draw adjacent to an indicating line, and on opposite sides of it, they signify motions in opposite directions.

5. Parallel straight lines denote uniform motion.

6. Curved lines denote a variable velocity. It is convenient as far as possible to make the ordinates of the curve proportional to the different velocities.

7. If the motion may be greater or less within certain limits; then if the motion begin at a fixed moment of time, and it is uncertain when it will terminate, the line denoting motion must extend from one limit to the other, and must be connected by a small cross line at its commencement with the indicating line. If the beginning of its motion is uncertain, but its end determined, then the cross line must be at its termination. If the commencement and the termination of any motion are both uncertain, the line representing motion must be connected with the indicating line in the middle by a cross line.

8. Dotted lines imply rest. It is convenient sometimes to denote a state of rest by the absence of any line whatever.

9. If the thing indicated be a click, bolt, or valve, its dotted line should be on the right side if it is out of action, unbolted, or open, and on the left side if the reverse is the case.

10. If a bolt may rest in three positions: 1st, bolted on the right side; 2nd, unbolted; 3rd, bolted on the left side. When it is unbolted, and in the middle station, use two lines whilst in the act of unbolting, and two lines of dots, one on each side of, and close to the indicating line, whilst it rests in this position. When it is bolted on the right side, a line or a line of dots at a greater distance on the right hand from the indicating line will represent it. And if it is bolted on the left, a similar mode of denoting it must be used on that side. Any explanation may, if required, be put in words at the end of the notation, as will be observed in that of the hydraulic ram, Plate X.

I have now explained means of denoting by signs almost all those circumstances which usually occur in the motion of machinery: if other modifications of movement should present themselves, it will not be difficult for any one who has rendered himself familiar with the symbols employed in this paper, to contrive others adapted to the new combinations which may present themselves.

The two machines which I have selected as illustrations of the application of this method, are, the common eight-day clock, and the hydraulic ram. The former was made choice of from its construction being very generally known, and I was induced to choose the latter from the apparent difficulty of applying this method to its operations.

The advantages which appear to result from the employment of this *mechanical notation*, are to render the description of machinery considerably shorter than it can be when expressed in words. The signs, if they have been properly chosen, and if they should be generally adopted, will form as it were an universal language; and to those who become skilful in their use, they will supply the means of writing down at sight even the most complicated machine, and of understanding the order and succession of the movements of any engine of which they possess the drawings and the mechanical notation. In contriving machinery, in which it is necessary that numerous wheels and levers, deriving their motion from distant parts of the engine, should concur at some instant of time, or in some precise order, for the proper performance of a particular operation, it furnishes most important assistance; and I have myself experienced the advantages of its application to my own calculating engine, when all other methods appeared nearly hopeless.

DESCRIPTION OF THE PLATES*

Plates VII. and VIII. are different representations of an eight-day clock, for the purpose of comparing it with the notation.

Plate IX. represents the mechanical notation of the same clock.

Under the names of each part follow the letters which distinguish them in the plates.

The next line contains numbers which mark the number of teeth in each wheel, pinion, or sector.

The following line is intended to contain numbers expressing the linear velocity of the different parts: in the eight-day clock this line is vacant, because almost all the motions are circular.

The next line indicates the angular velocity of each part; and in order to render the velocity of the striking parts comparable with those of the time part, I have supposed one revolution of the striking fusee to be made in one minute: I have also taken one revolution of the scapement wheel, or one minute of time as the unit of angular velocity.

The space entitled comparative angular velocity, expresses the number of revolutions one wheel makes during one revolution of some other; it differs from the real angular velocity, because one wheel may be at rest during part of the time. Thus the clock strikes 78 strokes during twelve hours, or one revolution of the hour hand: if this be called unity, the pin wheel moves through 78 pins or $9\frac{3}{4}$ revolutions in the same time; its comparative angular velocity is therefore $9\frac{3}{4}$.

The space in which the origin of motion is given, will not require any explanation after reading the description of the signs employed in this paper.

The adjustments are numbered in the order in which they are to be made. No. 1, is attached to the crutch: the first adjustment is to set the clock in beat. No. 2, is to adjust the length of the pendulum to beat seconds. No. 3, occurs in three different places at the hour and minute hands, and at the snail on the hour wheel. It is necessary that when the hour hand is at a given figure, three o'clock for instance, that the minute hand should be set to twelve o'clock; it is also necessary that the snail should be in such a position that the clock may strike three. These adjustments must be made at the same time. No. 4, is for the adjustment of the seconds hand to 60 seconds. No. 5, is

* See p. 380.

double, and is for the adjustment of the minute and hour hand to the next whole minute to that which is indicated by the watch by which the clock is set.   No. 6, is for the pendulum, which must be held aside at its extreme arc until the instant at which the watch reaches the time set on the face of the clock; it must then be set free.

The remaining part of the notation indicates the action of every part at all times; but as the whole cycle of twelve hours would occupy too much space, a portion only is given about the hour of four: from this the machine may be sufficiently understood.   As an instance of its use, let us inquire what movements are taking place at seven seconds after four o'clock.   On looking down on the left-hand side to the time just mentioned, we observe between the end of the sixth, and end of the seventh second, that the pendulum and crutch begin to move from the right to the left, increasing their velocity to a maximum, and then diminishing it; that the whole train of wheels of the time part are at rest during the greater part of that second, and all move simultaneously a little before its termination.   The greater part of the train of the striking part is moving uniformly; but two parts, the cross piece, and the other moving the hammer, being at the commencement of this second in a state of motion from right to left, suddenly have that motion reversed for a short time: this is at the moment of striking: two other pieces, the hawk's bill, and the gathering pallet, appear to act at the same moment.

If the course of movement of any one part is required throughout the whole cycle of the machine's action, we have only to follow its indicating line.   If it is required to find what motions take place at the same time, we have only to look along the horizontal line marked by the time specified.

Let us now inquire into the source of motion of the minute hand. On looking down to the space in which the origin of motion is given, we observe an arrow point, which conveys us to the

Cannon pinion, with which it is connected permanently.

The cannon pinion is driven by the centre or hour wheel, with which it is connected by stiff friction.

The hour wheel is driven by its pinion, to which it is permanently attached.

The hour wheel pinion is driven by the great wheel, into which it works.

The great wheel is driven by the fusee, with which it is connected by a ratchet.

The fusee is driven by the spring barrel or main spring, which is the origin of all the movements.

When that part of the notation which relates to the successive movements of the machine is of considerable extent, it is convenient to write on a separate piece of paper the names of every part, at the same distances from each other as the indicating lines, and exactly as they are placed at the top. By sliding this paper down the page to any part which is under consideration, the trouble of continual reference to the top of the drawing will be avoided.

Plate X. represents the hydraulic ram: its mechanical notation is added below it.

A, is the supplying pipe.

B, is the great valve.

C, the valve into the air vessel.

D, the air vessel.

E, the ascending water.

F, the small air valve: its office is to supply a small quantity of air at each stroke; it opens when the valve C is just closed, and a re-gurgitation takes place in the supplying pipe just previous to the opening of the great valve. Without this contrivance, the pressure on the air in the air vessel would cause it to be soon absorbed by the water, and the engine would cease to act.

In this notation two indicating lines A, A, are allowed to the supplying water, because it takes three different courses during the action of the machine. The first of these marks the time of its motion when it enters the air vessel, and the second indicates its course when passing through the great valve, and also its course when, owing to the elasticity of the materials, its motion is for an instant reversed, at which moment air is taken in at the air valve F.

The action of the machine is as follows: the supplying water rushing along the great pipe passes out at the great valve; it acquires velocity until the pressure of the effluent water against the under part of the great valve causes it to close suddenly. At this moment the whole momentum of the water is directed against the sides of the machine, and the air valve being the weakest part gives way, and admits a small quantity of water; the air spring soon resists sufficiently to close the air valve: at this moment the elasticity of the apparatus reacting on the water in the great pipe, drives it back for an instant, during which the pressure of the atmosphere opens the air valve, and a small quantity of air enters; this finds its way to the air chambers, which easily discharges it through the ascending pipe if too much air has entered.

X

LAWS OF MECHANICAL NOTATION

By CHARLES BABBAGE

This paper was given away by Mr. Babbage
during and after the Great Exhibition of 1851

Chapter I—On Lettering Drawings

ALL MACHINERY consists of—

Framing		Parts, or Pieces
Fixed	Moveable	Moveable as axes, springs, &c.

Every *Piece* possesses one or more *Working Points*. These are divided into two classes, those by which the Piece acts on others, and those by which it receives action from them: these are called *Driving* and *Driven Points*. A *Working Point* may fulfil both these offices, as, for example, the same teeth which are driven by one wheel may in another part of their course drive other wheels.

The following alphabets of large letters are used in Drawings:—

Etruscan		Roman		Writing	
A	*A*	A	*A*	*𝒜*	*𝒜*
B	*B*	B	*B*	*ℬ*	*ℬ*
C	*C*	C	*C*	*𝒞*	*𝒞*

The following alphabets of small letters are used:—

a	*b*	*c*	*d*	.	.	.
a	b	c	d	.	.	.

It is most convenient, and generally sufficient, to use only the letters *a, c, e, i, m, n, o, r, s, u, v, w, x, z* of both these latter alphabets.

Rule 1.—Every separate portion of *Frame-work* must be indicated by a large *upright* letter.

Rule 2.—Every *Working Point* of Frame-work must be indicated by a *small* printed letter.

Rule 3.—*Frame-work* which is itself moveable must be represented by a *large upright* letter, with the sign of motion in its proper place below it (*see Signs of Motion*), as

G   **H**
─   ◡

Rule 4.—In lettering Drawings, commence with the axes. These must be lettered with *large inclined* letters of either of the three alphabets. Whenever the wheels or arms of any two or more adjacent axes cross each other on the plan, avoid denoting those axes by letters of the *same* alphabet.

Rule 5.—In lettering *Pieces*, as wheels, arms, &c., belonging to any axis, whether they are fixed to it or moveable upon it, always use *inclined capitals* of the *same* alphabet as that of the letter representing the Axis.

Rule 6.—Beginning with the lowest *Piece* upon an Axis, assign to it any *capital* letter of the *same* alphabet. To the Piece next above, assign any other *capital* letter which occurs *later* in the *same* alphabet. Continue this process for each *Piece*.

Thus, although the succession of the letters of the *same* alphabet need not be continuous, yet their occurrence in *alphabetic* order will never be violated.

Rule 7.—In lettering *Pieces* upon axes perpendicular to the elevation, or to the end views, looking from the left side, the earliest letters of the alphabet must be placed on the Pieces most remote from the eye.

Rule 8.—No axis which has a *Piece* crossing any other *Piece* belonging to an adjacent axis, must have the same identity as that axis.

If there are many *Pieces* on the same axis, it may be necessary to commence with one of the earlier letters of the alphabet.

Rule 9.—In placing letters representing any *Piece* on which portions of other *Pieces* are projected, it is always desirable to select such a situation that no doubt can be entertained as to which of those *Pieces* the letter is intended to indicate. This can often be accomplished by placing the letter upon some portion of its own *Piece* which extends beyond the projected parts of the other *Piece*.

Rule 10.—When *Pieces* are very small, or when they are crossed by many other lines, it is convenient to place the letter representing them outside the *Piece* itself, and to connect it with the *Piece* it indicates by an arrow. This arrow should be a short fine line terminated by a head, abutting on, or perhaps projecting into, the *Piece* represented by the letter.

Rule 11.—When upon any Drawing, a letter having a dot beneath it occurs, it marks the existence of a *Piece* below.

Rule 12.—In case another *Piece*, exactly similar to one already represented and lettered, exists below it, it cannot be expressed by any visible line. It may, however, be indicated by placing its proper letter outside, and connecting that letter with a *dotted* arrow abutting on the upper *Piece*.

Rule 13.—The permanent connexion of two pieces of matter, or the permanent gearing of two wheels, is indicated by a short line crossing, at right angles, the point of contact. The sign | indicates, in a certain sense, fixed connexion. This sign will be found very useful for indicating the boundaries of various pieces of framing.

It is to be observed that letters of the simplest and least ornamented style ought to be preferred: such are more quickly apprehended by the eye, and more easily recalled by the memory.

Of the Indices of Letters

Rule 14.—Various indices and signs may be affixed to letters. Their position and use are indicated in the subjoined letter:—

[Sign of Form.]

(Identity.)   **H**   (Cir. Pos^n.)

(New Alph^t.)   (Linear Pos^n.)

[Sign of Motion.]

Rule 15.—The index on the left-hand upper corner is used to mark the identity of two or more parts of a *Piece* which are permanently united; each being denoted by a letter with the *same* index.

Rule 16.—It is used also to connect any *Piece* itself with its various working points. Thus all the small letters which indicate the working points, must have the *same* index of identity as the letter expressing the *Piece* itself.

Rule 17.—Every *Working Point* must be marked by the *same small letter* as the *Working Point* of the *Piece* upon which it acts.

Rule 18.—The bearings in which axes work, as well as the working surface of the axes themselves, and also the working surfaces of slides, are *Working Points*, and must be lettered as such.

Of the Index of Linear Position

The successive order in which the various *Pieces* upon one axis succeed each other, is indicated by the alphabetic succession.

It may, however, in some cases be convenient to distinguish between the relative heights of the various arms or wheels which constitute one *Piece*.

This may be easily accomplished by means of the index of *linear position*.

Every Piece may be represented as a whole, by one letter, with its proper index of identity. If, however, it is necessary to distinguish the different arms or parts of which it is composed, so as to indicate their relative position above the plane of projection, this may be accomplished by means of the indices of linear position.

Rule 19.—If ^3P represent the whole of any *Piece*, $^3P_1$, $^3P_2$, $^3P_3$, $^3P_4$, &c., will represent in succession the several arms or parts of which ^3P is composed: $^3P_1$ indicating that which is most distant from the eye.

Of the Index of Circular Position

It may occasionally be desirable to indicate the order of succession in angular position of the various arms belonging to the same *Piece*, when projected on a plane. The index on the right-hand upper corner is applied to this purpose.

Rule 20. ^6R representing any *Piece*,

$^6R^1$ will represent any arm as the origin,

$^6R^2$ the next arm in angular position in the direction "*screw*," that is, from left to right,

$^6R^3$ the next, &c.

Thus,

$$^6R^1_1, \ ^6R^2_2, \ ^6R^3_3, \ldots ^6R^n_n$$

would represent n arms placed spirally round an axis at various heights above it.

Of the Index of New Alphabet

In case the three alphabets given above are found insufficient, the index on the left lower side is reserved to mark new alphabets. In the most complicated drawing I have scarcely ever had occasion to use it. It might in some cases be desirable to have a fourth alphabet, differing in form from those already given.

The following lithographic Plate contains the signs of form and those of motion.

These signs of form have been the subject of much thought and discussion. A good test of their fitness arose under the following circumstances:—Three signs had been selected for the representation of various link motions, such as those of the parallel motions connected with the beam of a steam-engine.

Twelve of the motions of which links are susceptible are represented in the list; but, after a time, I observed that there were four other combinations which had not been represented, because they did not admit of motion.

On examining the combinations of these signs, it was found that, although not moveable, they represented real mechanical combinations.

The first twelve were formed according to the following laws:

1st. The circles at the ends of each line represent *axes* which are hollow if the axis is hollow, and are dark if the axis is solid.

2nd. If the axis is a *fixed* axis, then its circle has a vertical line passing through its centre.

It will be observed that links marked 23 are all moveable about their left-hand fixed centre, whilst those marked 25 are all moveable about their right-hand fixed centre.

Those links marked No. 24, which have no bar, are moveable centres like some of the rods of the parallel motion of a steam-engine.

There are, however, four other possible combinations of these signs.

It therefore becomes an interesting inquiry to ascertain whether these represent any known mechanical contrivances.

On interpreting them literally, it appears that the first is a bar having a solid stud fixed at each end, whilst the last is a bar having two holes in it, by which it may be screwed to any other piece of matter. The other pair represent a bar having a stud at one end, and a hole for a screw or bolt at the other.

There are two other chapters necessary to complete this subject.

Chapter II.—On the Notation of Periods

The object of this is to give a minute account of the time at every motion and of every action throughout the cycle of the movement of the machine to be described.

Chapter III.—On the Trains

The special object of this chapter is, to give an account of the directions of the various courses through which ,he active forces of the machine are developed. But the times of every action can be combined with it, and, to a certain extent, the forms of every moving part.

Some further notice of the mechanical notation will be found in the Introduction to this work.

MR. BABBAGE *will feel obliged by any criticism, or additions to these Rules of Drawing, and to the Mechanical Alphabet, and requests they may be addressed to him by post, at No.* 1, *Dorset Street, Manchester Square.*

JULY, 1851.

PART III

APPENDIX OF
MISCELLANEOUS PAPERS

CONTENTS

I

ON THE AGE OF STRATA, AS INFERRED FROM THE RINGS OF TREES EMBEDDED IN THEM

By CHARLES BABBAGE

Note M from *The Ninth Bridgewater Treatise*

THE INDELIBLE records of past events which are preserved within the solid substance of our globe, may be in some measure understood without the aid of that refined analysis on which a complete acquaintance with them depends. The remains of vegetation, and of animal life, embedded in their coeval rocks, attest the existence of far distant times; and as science and the arts advance, we shall be enabled to read the minuter details of their living history. The object of the present note is to suggest to the reader a line of inquiry, by which we may still trace some small portion of the history of the past in the fossil woods which occur in so many of our strata.

It is well known that dicotyledonous trees increase in size by the deposition of an additional layer annually between the wood and the bark, and that a transverse section of such trees presents a series of nearly concentric though irregular rings, the number of which indicates the age of the tree. The relative thickness of these rings depends on the more or less flourishing state of the plant during the years in which they were formed. Each ring may, in some trees, be observed to be subdivided into others, thus indicating successive periods of the same year during which its vegetation was advanced or checked. These rings are disturbed in certain parts by irregularities resulting from branches; and the year in which each branch first sprung from the parent stock may be ascertained by proper sections.

It has been found by experiment, that even the motion imparted to a tree by the winds has an influence on its growth. Two young trees of equal size and vigour were selected and planted in similar circumstances, except that one was restrained from having any motion in the direction of the meridian, by two strong ropes fixed to it, and connecting it to the ground, at some distance towards the north and

367

south. The other tree was by similar means prevented from having any motion in the direction of east and west. After several years, both trees were cut down, and the sections of their stems were found to be oval; but the longer axis of the oval of each was in the direction in which it had been capable of being moved by the winds.

These prominent effects are obvious to our senses; but every shower that falls, every change of temperature that occurs, and every wind that blows, leaves on the vegetable world the traces of its passage; slight, indeed, and imperceptible, perhaps, to us, but not the less permanently recorded in the depths of those woody fabrics. All these indications of the growth of the living tree are preserved in the fossil trunk, and with them also frequently the history of its partial decay.

Let us now enquire into the use we may make of these details relative to individual trees, when examining forests submerged by seas, embedded in peat mosses, or transformed, as in some of the older strata, into stone. Let us imagine, that we possessed sections of the trunks of a considerable number of trees, such as those occurring in the bed called the *Dirt-bed*,* in the island of Portland. If we were to select a number of trees of about the same size, we should probably find many of them to have been contemporaries. This fact would be rendered probable if we observed, as we doubtless should do, on examining the annual rings, that some of them conspicuous for their size occurred at the same distances of years in several trees. If, for example, we found on several trees a remarkably large annual ring, followed at the distance of seven years by a remarkably thin ring, and this again, after two years, followed by another large ring, we should reasonably infer that seven years after a season highly favourable to the growth of these trees, there had occurred a season peculiarly unfavourable to them: that after two more years another very favourable season had happened, and that all the trees so observed had existed at the same period of time. The nature of the season, whether hot or cold, wet or dry, might be conjectured with some degree of probability, from the class of tree under consideration. This kind of evidence, though slight at first, receives additional and great confirmation by the discovery of every new ring which supports it; and, by an extensive concurrence of such observations, the succession of seasons might be in some measure ascertained at remote geological periods.

On examining the shape of the sections of such trees, we might

---

* The reader will find an account of these fossil trees, and of the strata in which they occur, in several papers by Mr. Webster, Dr. Buckland, Mr. De la Beche, and Dr. Fitton, in the Transactions of the Geological Society of London, vol. iv. Series 2.

perceive some general tendency towards a uniform inequality in their diameters; and we should perhaps find that the longer axes of the sections most frequently pointed in one direction. If we knew from the species of tree that it possessed no natural tendency to such an inequality, then we might infer that, during the growth of these trees, they were bent most frequently in one direction; and hence derive an indication of the prevailing winds at that time. In order to determine from which of the two opposite quarters these winds came, we might observe the centres of these sections; and we should *generally* find that the rings on one side were closer and more compressed than those on the opposite side. From this we might infer the most exposed side, or that from which the wind most frequently blew. Doubtless there would be many exceptions arising from local circumstances— some trees might have been sheltered from the direct course of the wind, and have only been acted upon by an eddy. Some might have been protected by adjacent large trees, sufficiently near to shelter them from the ruder gales, but not close enough to obstruct the light and air by which they were nourished. Such a tree might have a series of large and rather uniform rings, during the period of its protection by its neighbour; and these might be followed by a series of stinted and irregular ones, occasioned by the destruction of its protector. The same storm might have mutilated some trees, and half uprooted others: these latter might strive to support themselves for years, making but little addition, by stinted layers, to the thickness of their stems; and then, having thrown out new roots, they might regain their former rate of growth, until a new tempest again shook them from their places. Similar effects might result from floods and the action of rivers on the trees adjacent to their banks. But the effect of all these local and peculiar circumstances would disappear, if a sufficient number of sections could be procured from fossil trees, spread over a considerable extent of country.

The annual rings might however furnish other intimations of the successive existence of these trees.

On examining some rings remarkable for their size and position, let us suppose that we find, in one section, two remarkably large rings, separated from another large ring, by one very stinted ring, and this followed, after three ordinary rings, by two very small and two very large ones. Such a group might be indicated by the letters—

$$o\,L\,L\,s\,o\,o\,o\,s\,L\,L\,o\,o$$

where *o* denotes an ordinary year or ring, *L* a large one, and *s* a small or stinted ring.   If such a group occurred in the sections of several different trees, it might fairly be attributed to general causes.

Let us now suppose such a group to be found near the centre of one tree, and towards the external edge or bark of another; we should certainly conclude, that the tree near whose bark it occurred was the more ancient tree; that it had been advanced in age when *that* group of seasons occurred which had left their mark near the pith of the more recent tree, which was young at the time those seasons happened.   If, on counting the rings of this younger tree, we found that there were, counting inward from the bark to this remarkable group, three hundred and fifty rings, we should justly conclude that, three hundred and fifty years before the death of this tree, which we will call A, the other, which we will call B, and whose section we possess, had then been an old tree.   If we now search towards the centre of the second tree B, for another remarkable group of rings; and if we also find a similar group near the bark of a third tree, which we will call C; and if, on counting the distance of the second group from the first in B, we find an interval of 420 rings, then we draw the inference that the tree A, 350 years before its destruction, was influenced in its growth by a succession of ten remarkable seasons, which also had their effect on a neighbouring tree B, which was at that time of a considerable age.   We conclude further, that the tree B was influenced in its youth, or 420 years before the group of the ten seasons, by another remarkable succession of seasons, which also acted on a third tree, C, then old. Thus we connect the time of the death of the tree A with the series of seasons which affected the tree C in its old age, at a period 770 years antecedent.   If we could discover other trees having other cycles of seasons, capable of identification, we might trace back the history of the ancient forest, and possibly find in it some indications for conjecturing the time occupied in forming the stratum in which it is embedded.

The application of these principles to ascertaining the age of submerged forests, or to that of peat mosses, may possibly connect them ultimately with the chronology of man.   Already we have an instance of a wooden hut with a stone hearth before it, and burnt wood on it, and a gate leading to a pile of wood, discovered at a depth of fifteen feet below the surface of a bog in Ireland: and it was found that this hut had probably been built when the bog had only reached half its present thickness, since there were still fifteen feet of turf below it.

The realization of the views here thrown out would require the united exertions of many individuals patiently exerted through a series of years.   The first step must be to study fully the relations of the annual rings in every part of an individual tree.   The effect of a favourable or unfavourable season on a section near the root must be compared with the influence of the same circumstance on its growth towards the top of the tree. Vertical sections also must be examined in order to register the annual additions to its height, and to compare them with its increase of thickness. Every branch must be traced to its origin, and its sections be registered. The means of identifying the influence of different seasons in various sections of the same individual tree and its branches being thus attained, the conclusions arrived at must be applied to several trees under similar circumstances, and such modifications must be applied to them as the case may require; and before any general conclusions can be reached respecting a tract of country once occupied by a forest, it will be necessary to have a considerable number of sections of trees scattered over various parts of it.

LIST OF MR. BABBAGE'S
PRINTED PAPERS

From *Passages from the Life of a Philosopher*

1. The Preface; jointly with Sir John Herschel. } *Memoirs of the Analytical Society.* 4to. Cambridge, 1813.

2. On Continued Products.

3. An Essay towards the Calculus of Functions.—*Phil. Trans.* 1815.

4. An Essay towards the Calculus of Functions, Part. 2.—*Phil. Trans.* 1816. p. 179.

5. Demonstrations of some of Dr. Matthew Stewart's General Theorems, to which is added an Account of some New Properties of the Circle.—*Roy. Inst. Jour.* 1816. Vol. i. p. 6.

6. Observations on the Analogy which subsists between the Calculus of Functions and other branches of Analysis.—*Phil. Trans.* 1817. p. 179.

7. Solution of some Problems by means of the Calculus of Functions. —*Roy. Inst. Jour.* 1817. p. 371.

8. Note respecting Elimination.—*Roy. Inst. Jour.* 1817. p. 355.

9. An Account of Euler's Method of Solving a Problem relating to the Knight's Move at Chess.—*Roy. Inst. Jour.* 1817. p. 72.

10. On some new Methods of Investigating the Sums of several Classes of Infinite Series.—*Phil. Trans.* 1819. p. 245.

11. Demonstration of a Theorem relating to Prime Numbers.— *Edin. Phil. Jour.* 1819. p. 46.

12. An Examination of some Questions connected with Games of Chance.—*Trans. of Roy. Soc. of Edin.* 1820. Vol. ix. p. 153.

13. Observations on the Notation employed in the Calculus of Functions.—*Trans. of Cam. Phil. Soc.* 1820. Vol. i. p. 63.

14. On the Application of Analysis, &c. to the Discovery of Local Theorems and Porisms.—*Trans. of Roy. Soc. of Edin.* Vol. ix. p. 337. 1820.

15. *Translation of the Differential and Integral Calculus of La Croix,* 1 vol. 1816. } These two works were executed in conjunction with the Rev. G. Peacock (Dean of Ely) and Sir John Herschel, Bart.

16. *Examples to the Differential and Integral Calculus.* 2 vols. 8vo. 1820.

17. *Examples of the Solution of Functional Equations.* Extracted from the preceding. 8vo. 1820.

18. Note respecting the Application of Machinery to the Calculation of Mathematical Tables.—*Memoirs of the Astron. Soc.* June, 1822. Vol. i. p. 309.

19. A Letter to Sir H. Davy, P.R.S., on the Application of Machinery to the purpose of calculating and printing Mathematical Tables. 4to. July, 1822.

20. On the Theoretical Principles of the Machinery for calculating Tables.—*Brewster's Edin. Jour. of Science.* Vol. viii. p. 122. 1822.

21. Observations on the application of Machinery to the Computations of Mathematical Tables, Dec. 1822.—*Memoirs of Astron. Soc.* 1824. Vol. i. p. 311.

22. On the Determination of the General Term of a new Class of Infinite Series.—*Trans. Cam. Phil. Soc.* 1824. Vol. ii. p. 218.

23. Observations on the Measurement of Heights by the Barometer. —*Brewster's Edin. Jour. of Science*, 1824. p. 85.

24. On a New Zenith Micrometer.—*Mem. Astro. Soc.* March, 1825.

25. Account of the repetition of M. Arago's Experiments on the Magnetism manifested by various substances during Rotation. By C. Babbage, Esq. and Sir John Herschel.—*Phil. Trans.* 1825. p. 467.

26. On the Diving Bell.—*Ency. Metrop.* 4to. 1826.

27. On Electric and Magnetic Rotation.—*Phil. Trans.* 1826. Vol. ii. p. 494.

28. On a method of expressing by Signs the Action of Machinery.— *Phil. Trans.* 1826. Vol. ii. p. 250.

29. On the Influence of Signs in Mathematical Reasoning.—*Trans. Cam. Phil. Soc.* 1826. Vol. ii. p. 218.

30. *A Comparative View of the different Institutions for the Assurance of Life.* 1 vol. 8vo. 1826. German Translation. Weimar, 1827.

31. On Notation.—*Edinburgh Encyclopedia.* 4to.

32. On Porisms.—*Edinburgh Encyclopedia.* 4to.

33. *A Table of the Logarithms of the Natural Numbers, from* 1 *to* 108,000. Stereotyped. 1 vol. 8vo. 1826.

34. Three editions on coloured paper, with the Preface and Instructions translated into German and Hungarian, by Mr. Chas. Nagy, have been published at Pesth and Vienna. 1834.

35. Notice respecting some Errors common to many Tables of Logarithms.—*Mem. Astron. Soc.* 4to. 1827. Vol. iii. p. 65.

Evidence on Savings-Banks, before a Committee of the House of Commons, 1827.

36. Essay on the general Principles which regulate the Application of Machinery.—*Ency. Metrop.* 4to. 1829.

37. Letter to T. P. Courtenay on the Proportion of Births of the two Sexes amongst Legitimate and Illegitimate Children.—*Brewster's Edin. Jour. of Science.* Vol. ii. p. 85. 1829. This letter was translated into French and published by M. Villermé, Member of the Institute of France.

38. Account of the great Congress of Philosophers at Berlin, on 18 Sept. 1828.—Communicated by a Correspondent [C. B.]. *Edin. Journ. of Science by David Brewster.* Vol. x. p. 225. 1829.

39. Note on the Description of Mammalia.—*Edin. Jour. of Science,* 1829. Vol. i. p. 187. *Ferussac Bull,* vol. xxv. p. 296.

40. *Reflections on the Decline of Science in England, and on some of its Causes.* 4to. and 8vo. 1830.

41. Sketch of the Philosophical Characters of Dr. Wollaston and Sir H. Davy. Extracted from the *Decline of Science.* 1830.

42. On the Proportion of Letters occurring in Various Languages, in a letter to M. Quételet.—*Correspondence Mathématique et Physique.* Tom. vi. p. 136.

43. *Specimen of Logarithmic Tables,* printed with different coloured inks and on variously-coloured papers, in twenty-one volumes 8vo. London. 1831.

The object of this Work, of which *one single copy only* was printed, is to ascertain by experiment the tints of the paper and colours of the inks least fatiguing to the eye.

One hundred and fifty-one variously-coloured papers were chosen, and the same two pages of my stereotype Table of Logarithms were printed upon them in inks of the following colours: light blue, dark blue, light green, dark green, olive, yellow, light red, dark red, purple, and black.

Each of these twenty volumes contains papers of the same colour, numbered in the same order, and there are two volumes printed with each kind of ink.

The twenty-first volume contains metallic printing of the same specimen in gold, silver, and copper, upon vellum and on variously-coloured papers.

For the same purpose, about thirty-five copies of the complete table of logarithms were printed on thick drawing paper of various tints.

An account of this work may be found in the *Edin. Journ. of Science (Brewster's),* 1832. Vol. vi. p. 144.

44. *Economy of Manufactures and Machinery.* 8vo. 1832.

There are many editions and also American reprints, and several Translations of this Work into German, French, Italian, Spanish, &c.

45. Letter to Sir David Brewster, on the Advantage of a Collection of the Constants of Nature and Art.—*Brewster's Edin. Jour. of Science.* 1832. Vol. vi. p. 334. Reprinted by order of the British Association for the Promotion of Science. Cambridge, 1833. See also pp. 484, 490, Report of the Third Meeting of the British Association. Reprinted in *Compte Rendu des Travaux du Congrès Général de Statistique,* Bruxelles, Sept. 1853.

46. Barometrical Observations, made at the Fall of the Staubbach, by Sir John Herschel, Bart., and C. Babbage, Esq.—*Brewster's Edin. Jour. of Science.* Vol. vi. p. 224. 1832.

47. Abstract of a Paper, entitled Observations on the Temple of Serapis, at Pozzuoli, near Naples; with an attempt to explain the causes of the frequent elevation and depression of large portions of the earth's surface in remote periods, and to prove that those causes continue in action at the present time. Read at Geological Society, 12 March, 1834. See *Abstract of Proceedings of Geol. Soc.* Vol. ii. p. 72.
This was the first *printed* publication of Mr. Babbage's Geological Theory of the Isothermal Surfaces of the Earth.

48. The Paper itself was published in the *Proceedings of the Geological Soc.* 1846.

49. Reprint of the same, with Supplemental Conjectures on the Physical State of the Surface of the Moon. 1847.

50. Letter from Mr. Abraham Sharpe to Mr. J. Crosthwait, Hoxton, 2 Feb. 1721–22. Deciphered by Mr. Babbage. See *Life of Flamsteed,* by Mr. F. Baily. Appendix, pp. 348, 390. 1835.

51. *The Ninth Bridgewater Treatise.* 8vo. May, 1837; Second Edition, Jan. 1838.

52. On some Impressions in Sandstone.—*Proceedings of Geological Society.* Vol. ii. p. 439. Ditto, *Phil. Mag.* Ser. 3. Vol. x. p. 474. 1837.

52*. Short account of a method by which Engraving on Wood may be rendered more useful for the Illustration and Description of Machinery.—*Report of Meeting of British Association at Newcastle.* 1838. p. 154.

53. *Letter to the Members of the British Association.* 8vo. 1839.

54. General Plan, No. 25, of Mr. Babbage's Great Calculating or Analytical Engine, lithographed at Paris. 24 by 36 inches. 1840.

55. *Statement of the circumstances respecting Mr. Babbage's Calculating Engines.* 8vo. 1843.

56. Note on the Boracic Acid Works in Tuscany.—*Murray's Handbook of Central Italy.* First Edition, p. 178. 1843.

57. On the Principles of Tools for Turning and Planing Metals, by Charles Babbage. Printed in the Appendix of Vol. ii. *Holtzapffel Turning and Mechanical Manipulation.* 1846.

58. On the Planet Neptune.—*The Times,* 15th March, 1847.

59. *Thoughts on the Principles of Taxation, with reference to a Property Tax and its Exceptions.* 8vo. 1848. Second Edition, 1851. Third Edition, 1852.
An Italian translation of the first edition, with notes, was published at Turin, in 1851.

60. Note respecting the pink projections from the Sun's disc observed during the total solar eclipse in 1851.—*Proceedings of the Astron. Soc.*, vol. xii., No. 7.

61. *Laws of Mechanical Notation*, with Lithographic Plate. Privately printed for distribution. 4to. July, 1851.

62. Note respecting Lighthouses (Occulting Lights). 8vo. Nov., 1851.
Communicated to the Trinity House, 30 Nov. 1851.
Reprinted in the Appendix to the Report on Lighthouses presented to the Senate of the United States, Feb. 1852.
Reprinted in the *Mechanics' Magazine*, and in various other periodicals and newspapers. 1852–3.
It was reprinted in various parts of the Report of Commissioners appointed to examine into the state of Lighthouses. Parliamentary Paper. 1861.

63. *The Exposition of 1851; or, Views of the Industry, the Science, and the Government of England.* 6s. 6d. Second Edition, 1851.

64. On the Statistics of Light-houses. *Compte Rendu des Travaux du Congrès Général*, Bruxelles, Sept. 1853.

65. A short description of Mr. Babbage's Ophthalmoscope is contained in the Report on the Ophthalmoscope by T. Wharton Jones, F.R.S.—*British and Foreign Medical Review.* Oct. 1854. Vol. xiv. p. 551.

66. On Secret or Cipher Writing. Mr. T.'s Cipher Deciphered by C.—*Jour. Soc. Arts*, July, 1854, p. 707.

67. On Mr. T.'s Second Inscrutable Cipher Deciphered by C.—*Jour. Soc. Arts.* p. 777, Aug. 1854.

68. On Submarine Navigation.—*Illustrated News*, 23rd June, 1855.

69. Letter to the Editor of *The Times*, on Occulting Lights for Lighthouses and Night Signals. Flashing Lights at Sebastopol. 16th July, 1855.

70. On a Method of Laying Guns in a Battery without exposing the men to the shot of the enemy. *The Times*, 8 Aug., 1855.

71. Sur la Machine Suédoise de M. Scheutz pour Calculer les Tables Mathématiques. 4to. *Comptes Rendus et l'Académie des Sciences.* Paris, Oct. 8, 1855.

72. On the Action of Ocean-currents in the Formation of the Strata of the Earth.—*Quarterly Journal Geological Society*, Nov. 1856.

73. Observations by Charles Babbage, on the Mechanical Notation of Scheutz's Difference Engine, prepared and drawn up by his Son, Major Henry Prevost Babbage, addressed to the Institution of Civil Engineers. *Minutes of Proceedings*, vol. xv. 1856.

74. Statistics of the Clearing-House. Reprinted from *Trans. of Statistical Soc.* 8vo. 1856.

75. *Observations on Peerage for Life.* July, 1833. Reprinted, 1856.

76. Observations addressed to the President and Fellows of the Royal Society on the Award of their Medals for 1856. 8vo.

77. Table of the Relative Frequency of Occurrence of the Causes of Breaking Plate-glass Windows.—*Mech. Mag.* 24th Jan. 1857.

78. On remains of Human Art, mixed with the Bones of Extinct Races of Animals. *Proceedings of Roy. Soc.* 26th May, 1859.

79. *Passages from the Life of a Philosopher.* 8vo. 1864.

80. [In the press.] *History of the Analytical Engine.* 4to. It will contain Chapters V., VI., VII., and VIII., of the present Volume. Reprint of The Translation of General Menabrea's Sketch of the Analytical Engine invented by Charles Babbage. From the *Bibliothèque Universelle de Genève*, No. 82, Oct. 1842. Translated by the late Countess of Lovelace, with extensive Notes by the Translator.

GENERAL PLAN OF ENGINE No. 1

Plan 25, dated August 6, 1840

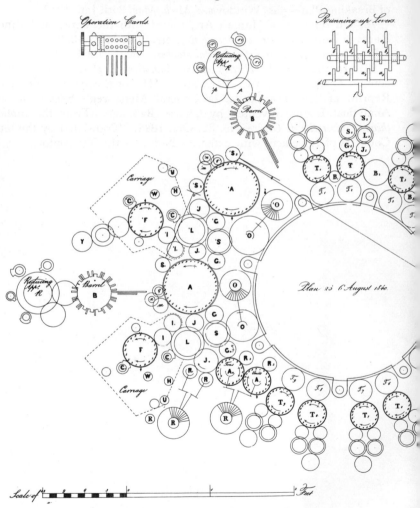

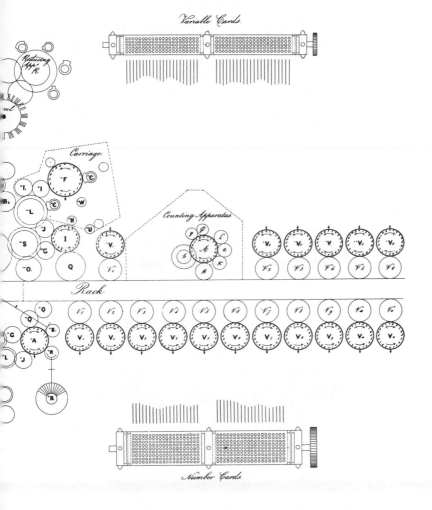

Variable Cards.

Carriage

Counting Apparatus

Rack

Number Cards.

PLATES VII-X

From the *Philosophical Transactions* of the
Royal Society, Vol. 2, 1826

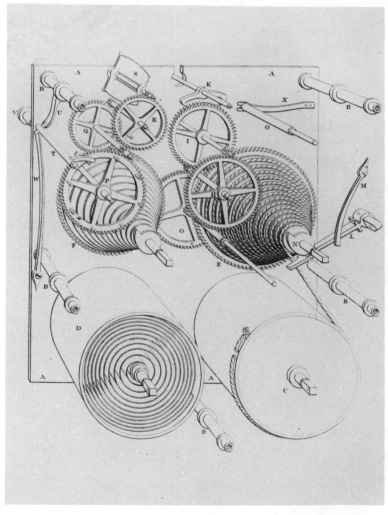

Plate VII.   See p. 354.

Plate VIII

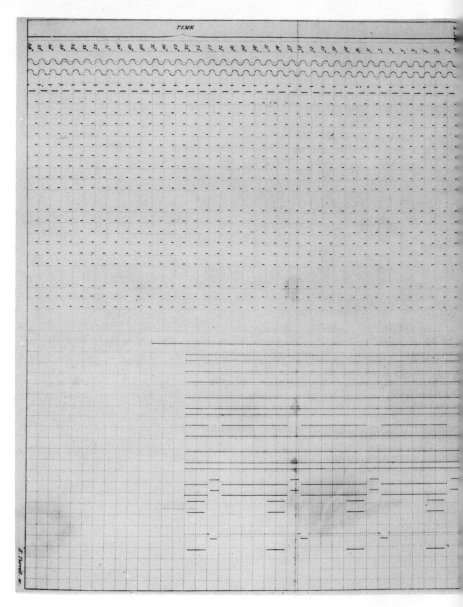

Plate IX

382

EIGHT DAY CLOCK.

		NAMES
TIME PART	Escapement	Gravity
		Pendulum
		Crutch
		Pallet a)
		Pallet b)
		Escapement or 3rd Wheel
		Pinion on ditto
		2nd Wheel
		Pinion on do
		Centre or hour Wheel
		Pinion on hour Wheel
		Great Wheel
		Fusee
		Barrel for main Spring
	Motion	Cannon pinion
		Pinion of report
		Small pinion on do
		12 hour Wheel
		Pin on Pinion of Report
		Snail on 12 hour Wheel
GOING OFF Part		Hour hand
		Minute hand
		Seconds hand
STRIKING PART and REPEATING PART		Arbor of warning Piece
		Fanner
		Pinion on axis of do
		Cross Wheel
		Pin on Cross Wheel
		Pinion on do
		Tumbler Wheel
		Pinion of do
		Pin Wheel
		Pins on do 8
		Pinion to do
		Fusee Wheel
		Striking Fusee
		Striking spring
		Arbor moving hammer for bell
		Cross-piece
		Rack
		Rack-tail
		Rack-tail Spring
		Hawks bill
		Gathering pallet
		Warning piece

Phil. Trans. MDCCCXLVI. Plate IX. p. 384

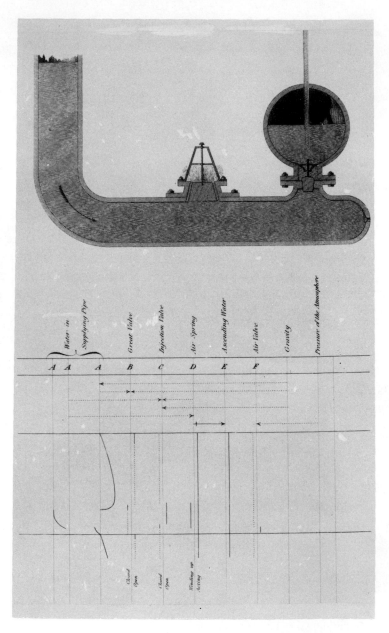

Plate X

V

TABLE OF CONTENTS

From *Passages from the Life of a Philosopher*

published in 1864

CHAPTER VI

Statement relative to the Difference Engine, drawn up by the late Sir H. Nicolas from the Author's Papers

CHAPTER VII

Difference Engine No. 2

Difference Engine No. 2—The Earl of Rosse, President of the Royal Society, proposed to the Government a Plan by which the Difference Engine No. 2 might have been executed—It was addressed to the Earl of Derby, and rejected by his Chancellor of the Exchequer.

CHAPTER VIII

Of the Analytical Engine

Built Workshops for constructing the Analytical Engine—Difficulties about carrying the Tens—Unexpectedly solved—Application of the Jacquard Principle —Treatment of Tables—Probable Time required for Arithmetical Operations— Conditions it must fulfil—Unlimited in Number of Figures, or in extent of Analytical Operations—The Author invited to Turin in 1840—Meetings for Discussion—Plana, Menabrea, MacCullagh, Mossotti—Difficulty proposed by the latter—Observations on the Errata of Astronomical Tables—Suggestions for a Reform of Analytical Signs.

CHAPTER IX

Of the Mechanical Notation

Art of Lettering Drawings—Of expressing the Time and Duration of Action of every Part—A New Demonstrative Science—Royal Medals of 1826.

CHAPTER X

The Exhibition of 1862

Mr. Gravatt suggests to King's College the exhibition of the Difference Engine No. 1, and offers to superintend its Transmission and Return—Place allotted to it most unfit—Not Exhibited in 1851—Its Loan refused to New York—Refused to the Dublin Exhibition in 1847—Not sent to the great French Exhibition in 1855— Its Exhibition in 1862 entirely due to Mr. Gravatt—Space for its Drawings refused—The Payment of Six Shillings a Day for a competent person to explain it refused by the Commissioners—Copy of Swedish Difference Engine made by English Workmen not exhibited—Loan of various other Calculating Machines offered—Anecdote of Count Strzelecki's—The Royal Commissioners' elaborate taste for Children's Toys—A plan for making such Exhibitions profitable— Extravagance of the Commissioners to their favourite—Contrast between his Treatment and that of Industrious Workmen—The Inventor of the Difference Engine publicly insulted by his Countrymen in the Exhibition of 1862.

CHAPTER XI

The late Prince Consort

Count Mensdorf mentions to the Duke of Wellington his wish to see the Difference Engine—An appointment made—Prince Albert expresses his intention of accompanying his uncle—Time of appointment altered—Their visit, accompanied by the Duke of Wellington—Portrait of Jacquard—Anecdote of Wilkie—Afghanistan arms—Extract from the Author's work on the Exhibition of 1862.

CHAPTER XVII

Experience amongst Workmen

Visit to Bradford—Clubs—Co-operative Shops—The Author of the "Economy of Manufactures" welcomed by the Workmen—Visit to the Temple of Eolus— The Philosopher moralises—Commiserates the unsuccessful Statesman—Points to the Poet a Theme for his Verse—Immortalises both.

CHAPTER XVIII

Picking Locks and Deciphering

Interview with Vidocq—Remarkable Power of altering his Height—A Bungler in picking Locks—Mr. Hobb's Lock and the Duke of Wellington—Strong belief that certain Ciphers are inscrutable—Davies Gilbert's Cipher—The Author's Cipher both deciphered—Classified Dictionaries of the English Language— Anagrams—Squaring Words—Bishop not easily squared—Lesser Dignitaries easier to work upon.

CHAPTER XIX

Experience in St. Giles's

Deep-snow—Beggar in Belgravia wanted work—He said he was a Watchmaker— Gave his address—It was false—Met him months after—The same story—The same untruth—Children hired for the purpose of Begging—Cellar in St. Giles's— Inquired for a Poor Woman and Child—Landlady told me of a Man almost starving in her back kitchen—He turned out to be an accomplished Swindler— Pot-boys—Caught him at last—Took him to Bow Street.

CHAPTER XX

Theatrical Experience

The Philosopher in a Tableau at the Feet of Beauty—Tableau encored— Philosopher at the Opera of "Don Juan"—Visits the Water-works above and the dark expanse below the Stage—Seized by two Devils on their way up to fetch Juan—Cheated the Devils by springing off to a beam at an infinite distance, just as his head appeared to the Audience through the trap-door—The Philosopher writes a Ballet—Its rehearsal—Its high moral tone—Its rejection on the ground of the probable combustion of the Opera-house.

CHAPTER XXI

Electioneering Experience

The late Lord Lyndhurst candidate for the University of Cambridge—The Philosopher refuses to vote for him—The reason why—Example of unrivalled virtue—In 1829 Mr. Cavendish was a Candidate for that University—The Author was Chairman of his London Committee—Motives for putting men on Committees—Of the pairing Sub-Committee—Motives for Voting—Means of influencing Voters—Voters brought from Berlin and from India—Elections after the Reform Bill, 1832—The Author again requested to be Chairman of Mr. Cavendish's Committee—Reserves three days in case of a Contest for Bridgenorth—It occurs, but is arranged—Bridgenorth being secure, the Author gets up a Contest for Shropshire—Patriotic Fund sends 500*l.* to assist the Contest —It lasts three days—Reflections on Squibs—Borough of Finsbury—Adventure in an Omnibus—A judicious Loan—Subsequent invitation to stand for Stroud— Declined—Reflections on improper influence on Voters.

CHAPTER XXII

Scene from a New After-Piece
"Politics and Poetry"; or, "The Decline of Science".

CHAPTER XXIII

Experience at Courts

Pension to Dr. Dalton—Inhabitants of Manchester subscribe for a Statue by Chantrey—The Author proposed that he should appear at a Levee—Various difficulties suggested and removed—The Chancellor approves and offers to present him—Mentions it to King William IV.—Difficulties occur—Dalton as a Quaker could not wear a Sword—Answer, he may go in his Robes as Doctor of Laws of Oxford—As a Quaker he could not wear Scarlet Robes—Answer, Dalton is afflicted with Colour-blindness—Crimson to him is dirt-colour—Dr. Dalton breakfasts with the Author—First Rehearsal—Second Rehearsal at Mr. Wood's—At the Levee—The Church in danger—Courtiers jealous of the Quaker—Conversation at Court sometimes interesting, occasionally profitable.

CHAPTER XXIV

Experience at Courts

The Author invited to a Meeting at Turin of the Philosophers of Italy, 1840—The King, Charles Albert—Reflections on Shyness—Question of Dress—Electric Telegraph—Theory of Storms—Remark of an Italian Friend in the evening at the Opera—Various Instruments taken to the Palace, and shown to the young Princes—The Queen being absent—The reason why—The young Princes did great credit to their Governor—The General highly gratified—The Philosopher proposes another difficult question—It is referred to the King himself—An audience is granted to ask the King's permission to present the woven Silk Engraving of Jacquard to Her Majesty—Singular but Comic Scene—The final Capture of the Butterflies—Visit to Raconigi—The Vintage.

CHAPTER XXV

Railways

Opening of Manchester and Liverpool Railway—Death of Mr. Huskisson—Plate-glass Manufactory—Mode of separating Engine from Train—Broad-gauge Question—Experimental Carriage—Measure the Force of Traction, the Vertical, Lateral, and End Shake of Carriage, also its Velocity by Chronometer—Fortunate Escape from meeting on the same Line Brunel on another Engine—Sailed across the Hanwell Viaduct in a Waggon without Steam—Meeting of British Association at Newcastle—George Stephenson—Dr. Lardner—Suggestions for greater Safety on Railroads—George Stephenson's Opinion of the Gauges—Railways at National Exhibitions.

CHAPTER XXVI

Street Nuisances

Various Classes injured—Instruments of Torture—Encouragers; Servants, Beer-Shops, Children, Ladies of elastic virtue—Effects on the Musical Profession—Retaliation—Police themselves disturbed—Invalids distracted—Horses run away—Children run over—A Cab-stand placed in the Author's street attracts Organs—Mobs shouting out his Name—Threats to Burn his House—Disturbed in the middle of the night when very ill—An average number of Persons are always ill—Hence always disturbed—Abusive Placards—Great Difficulty of getting Convictions—Got a Case for the Queen's Bench—Found it useless—A Dead Sell—Another Illustration—Musicians give False Name and Address—Get Warrant for Apprehension—They keep out of the way—Offenders not yet found and arrested by the Police—Legitimate Use of Highways—An Old Lawyer's Letter to *The Times*—Proposed Remedies; Forbid entirely—Authorize Police to seize the Instrument and take it to the Station—An Association for Prevention of Street Music proposed.

CHAPTER XXVII

Wit

Poor Dogs—Puns Double and Triple—History of the Silver Lady—Disappointed by the Milliner—The Philosopher performs her functions—Lady Morgan's Criticism—Allsop's Beer—Sydney Smith—Toss up a Bishop—Lady M . . . and the Gipsy in Spain—Epigram on the Planet Neptune—Epigram on Henry Drummond's attack upon Catholics in the House of Commons—On Catholic Miracles.

CHAPTER XXVIII

Hints for Travellers

New Inventions—Stomach Pump—Built a Carriage—Description of Thames Tunnel—Barton's Iridescent Buttons—Chinese Orders of Nobility—Manufactory of Gold Chains at Venice—Pulsations and Respirations of Animals—Punching a Hole in Glass without cracking it—Specimen of an Enormous Smash—Proteus Anguineus—Travellers' Hotel at Sheffield—Wentworth House.

CHAPTER XXIX

Miracles

Difference Engine set so as to follow a given law for a vast period—Thus to change to another law of equally vast or of greater duration, and so on—Parallel between the successive creations of animal life—The Author visited Dublin at the first Meeting of the British Association—Is the Guest of Trinity College—Innocently wears a Waistcoat of the wrong colour—Is informed of the sad fact—Rushes to a Tailor to rectify it—Finds nothing but party-colours—Nearly loses his Breakfast, and is thought to be an amazing Dandy—The Dean thinks better of the Philosopher, and accompanied him to Killarney—The Philosopher preaches a Sermon to the Divine by the side of the Lake.

CHAPTER XXX

Religion

The à priori proof of the existence of a Deity—Proof from Revelation—Dr Johnson's definition of Inspiration—Various Meanings assigned to the word "Revelation"—Illustration of transmitted Testimony—The third source of proof of the existence of a Deity—By an examination of His Works—Effect of hearing the Athanasian Creed read for the first time.

CHAPTER XXXI

A Vision

CHAPTER XXXII

Various Reminiscences

On preventing the forgery of bank-notes—An émeute—Letters of credit—The speaker—Ancient music.

CHAPTER XXXIII

The Author's Contributions to Human Knowledge

Scientific Societies—Analytical Society—Astronomical Society—Grand Duke of Tuscany, Leopold II.—Scientific Meeting at Florence—Also at Berlin—At Edinburgh—At Cambridge—Origin of the Statistical Society—Statistical Congress at Brussels—Calculus of Functions—Division of Labour—Verification part of Cost—Principles of Taxation—Extension to Elections—The two Pumps—Monopoly—Miracles.

CHAPTER XXXIV

The Author's Further Contributions to Human Knowledge

Glaciers—Uniform Postage—Weight of the Bristol Bags—Parcel Post—Plan for transmitting Letters along Aërial Wires—Cost of Verification is part of Price—Sir Rowland Hill—Submarine Navigation—Difference Engine—Analytical Engine—Cause of Magnetic and Electric Rotations—Mechanical Notation—Occulting Lights—Semi-occultation may determine Disturbances—Distinction of Lighthouses numerically—Application from the United States—Proposed Voyage —Loss of the Ship and Mr. Reed—Congress of Naval Officers at Brussels in 1853 —My Portable Occulting Light exhibited—Night Signals—Sun Signals—Solar Occulting Lights—Afterwards used at Sebastopol—Numerical Signals applicable to all Dictionaries—Zenith Light Signals—Telegraph for Ships on Shore—Greenwich Time Signals—Theory of Isothermal Surfaces to account for the Geological Facts of the successive Uprising and Depression of various parts of the Earth's Surface—Games of Skill—Tit-tat-to—Exhibitions—Problem of the Three Magnetic Bodies.

CHAPTER XXXV

Results of Science

Board of Longitude—Professorship of Mathematics at the East India College—Professorship of Mathematics at Edinburgh—Secretaryship of the Royal Society—Master of the Mint—Ditto—Ditto—Registrar-General of Births, Deaths, and Marriages—Ditto—Commissioner of Railways—Ditto—Ditto Abolished.

CHAPTER XXXVI

Agreeable Recollections

Appendix

INDEX

A CATALOGUE OF SELECTED DOVER BOOKS
IN ALL FIELDS OF INTEREST

A CATALOGUE OF SELECTED DOVER
BOOKS IN ALL FIELDS OF INTEREST

CONDITIONED REFLEXES, Ivan P. Pavlov. Full translation of most complete statement of Pavlov's work; cerebral damage, conditioned reflex, experiments with dogs, sleep, similar topics of great importance. 430pp. 5⅜ x 8½. 60614-7 Pa. $4.50

NOTES ON NURSING: WHAT IT IS, AND WHAT IT IS NOT, Florence Nightingale. Outspoken writings by founder of modern nursing. When first published (1860) it played an important role in much needed revolution in nursing. Still stimulating. 140pp. 5⅜ x 8½. 22340-X Pa. $3.00

HARTER'S PICTURE ARCHIVE FOR COLLAGE AND ILLUSTRATION, Jim Harter. Over 300 authentic, rare 19th-century engravings selected by noted collagist for artists, designers, decoupeurs, etc. Machines, people, animals, etc., printed one side of page. 25 scene plates for backgrounds. 6 collages by Harter, Satty, Singer, Evans. Introduction. 192pp. 8⅞ x 11¾. 23659-5 Pa. $5.00

MANUAL OF TRADITIONAL WOOD CARVING, edited by Paul N. Hasluck. Possibly the best book in English on the craft of wood carving. Practical instructions, along with 1,146 working drawings and photographic illustrations. Formerly titled *Cassell's Wood Carving*. 576pp. 6½ x 9¼. 23489-4 Pa. $7.95

THE PRINCIPLES AND PRACTICE OF HAND OR SIMPLE TURNING, John Jacob Holtzapffel. Full coverage of basic lathe techniques—history and development, special apparatus, softwood. turning, hardwood turning, metal turning. Many projects—billiard ball, works formed within a sphere, egg cups, ash trays, vases, jardiniers, others—included. 1881 edition. 800 illustrations. 592pp. 6⅛ x 9¼. 23365-0 Clothbd. $15.00

THE JOY OF HANDWEAVING, Osma Tod. Only book you need for hand weaving. Fundamentals, threads, weaves, plus numerous projects for small board-loom, two-harness, tapestry, laid-in, four-harness weaving and more. Over 160 illustrations. 2nd revised edition. 352pp. 6½ x 9¼. 23458-4 Pa. $6.00

THE BOOK OF WOOD CARVING, Charles Marshall Sayers. Still finest book for beginning student in. wood sculpture. Noted teacher, craftsman discusses fundamentals, technique; gives 34 designs, over 34 projects for panels, bookends, mirrors, etc. "Absolutely first-rate"—E. J. Tangerman. 33 photos. 118pp. 7¾ x 10⅝. 23654-4 Pa. $3.50

AMERICAN BIRD ENGRAVINGS, Alexander Wilson et al. All 76 plates. from Wilson's *American Ornithology* (1808-14), most important ornithological work before Audubon, plus 27 plates from the supplement (1825-33) by Charles Bonaparte. Over 250 birds portrayed. 8 plates also reproduced in full color. 111pp. 9⅜ x 12½.   23195-X Pa. $6.00

CRUICKSHANK'S PHOTOGRAPHS OF BIRDS OF AMERICA, Allan D. Cruickshank. Great ornithologist, photographer presents 177 closeups, groupings, panoramas, flightings, etc., of about 150 different birds. Expanded *Wings in the Wilderness*. Introduction by Helen G. Cruickshank. 191pp. 8¼ x 11.   23497-5 Pa. $6.00

AMERICAN WILDLIFE AND PLANTS, A. C. Martin, et al. Describes food habits of more than 1000 species of mammals, birds, fish. Special treatment of important food plants. Over 300 illustrations. 500pp. 5⅜ x 8½.
20793-5 Pa. $4.95

THE PEOPLE CALLED SHAKERS, Edward D. Andrews. Lifetime of research, definitive study of Shakers: origins, beliefs, practices, dances, social organization, furniture and crafts, impact on 19th-century USA, present heritage. Indispensable to student of American history, collector. 33 illustrations. 351pp. 5⅜ x 8½.   21081-2 Pa. $4.50

OLD NEW YORK IN EARLY PHOTOGRAPHS, Mary Black. New York City as it was in 1853-1901, through 196 wonderful photographs from N.-Y. Historical Society. Great Blizzard, Lincoln's funeral procession, great buildings. 228pp. 9 x 12.   22907-6 Pa. $8.95

MR. LINCOLN'S CAMERA MAN: MATHEW BRADY, Roy Meredith. Over 300 Brady photos reproduced directly from original negatives, photos. Jackson, Webster, Grant, Lee, Carnegie, Barnum; Lincoln; Battle Smoke, Death of Rebel Sniper, Atlanta Just After Capture. Lively commentary. 368pp. 8⅜ x 11¼.   23021-X Pa. $8.95

TRAVELS OF WILLIAM BARTRAM, William Bartram. From 1773-8, Bartram explored Northern Florida, Georgia, Carolinas, and reported on wild life, plants, Indians, early settlers. Basic account for period, entertaining reading. Edited by Mark Van Doren. 13 illustrations. 141pp. 5⅜ x 8½.   20013-2 Pa. $5.00

THE GENTLEMAN AND CABINET MAKER'S DIRECTOR, Thomas Chippendale. Full reprint, 1762 style book, most influential of all time; chairs, tables, sofas, mirrors, cabinets, etc. 200 plates, plus 24 photographs of surviving pieces. 249pp. 9⅞ x 12¾.   21601-2 Pa. $7.95

AMERICAN CARRIAGES, SLEIGHS, SULKIES AND CARTS, edited by Don H. Berkebile. 168 Victorian illustrations from catalogues, trade journals, fully captioned. Useful for artists. Author is Assoc. Curator, Div. of Transportation of Smithsonian Institution. 168pp. 8½ x 9½.
23328-6 Pa. $5.00

A MAYA GRAMMAR, Alfred M. Tozzer. Practical, useful English-language grammar by the Harvard anthropologist who was one of the three greatest American scholars in the area of Maya culture. Phonetics, grammatical processes, syntax, more. 301pp. 5⅜ x 8½. 23465-7 Pa. $4.00

THE JOURNAL OF HENRY D. THOREAU, edited by Bradford Torrey, F. H. Allen. Complete reprinting of 14 volumes, 1837-61, over two million words; the sourcebooks for *Walden*, etc. Definitive. All original sketches, plus 75 photographs. Introduction by Walter Harding. Total of 1804pp. 8½ x 12¼. 20312-3, 20313-1 Clothbd., Two-vol. set $70.00

CLASSIC GHOST STORIES, Charles Dickens and others. 18 wonderful stories you've wanted to reread: "The Monkey's Paw," "The House and the Brain," "The Upper Berth," "The Signalman," "Dracula's Guest," "The Tapestried Chamber," etc. Dickens, Scott, Mary Shelley, Stoker, etc. 330pp. 5⅜ x 8½. 20735-8 Pa. $4.50

SEVEN SCIENCE FICTION NOVELS, H. G. Wells. Full novels. *First Men in the Moon, Island of Dr. Moreau, War of the Worlds, Food of the Gods, Invisible Man, Time Machine, In the Days of the Comet*. A basic science-fiction library. 1015pp. 5⅜ x 8½. (Available in U.S. only) 20264-X Clothbd. $8.95

ARMADALE, Wilkie Collins. Third great mystery novel by the author of *The Woman in White* and *The Moonstone*. Ingeniously plotted narrative shows an exceptional command of character, incident and mood. Original magazine version with 40 illustrations. 597pp. 5⅜ x 8½. 23429-0 Pa. $6.00

MASTERS OF MYSTERY, H. Douglas Thomson. The first book in English (1931) devoted to history and aesthetics of detective story. Poe, Doyle, LeFanu, Dickens, many others, up to 1930. New introduction and notes by E. F. Bleiler. 288pp. 5⅜ x 8½. (Available in U.S. only) 23606-4 Pa. $4.00

FLATLAND, E. A. Abbott. Science-fiction classic explores life of 2-D being in 3-D world. Read also as introduction to thought about hyperspace. Introduction by Banesh Hoffmann. 16 illustrations. 103pp. 5⅜ x 8½. 20001-9 Pa. $2.00

THREE SUPERNATURAL NOVELS OF THE VICTORIAN PERIOD, edited, with an introduction, by E. F. Bleiler. Reprinted complete and unabridged, three great classics of the supernatural: *The Haunted Hotel* by Wilkie Collins, *The Haunted House at Latchford* by Mrs. J. H. Riddell, and *The Lost Stradivarius* by J. Meade Falkner. 325pp. 5⅜ x 8½. 22571-2 Pa. $4.00

AYESHA: THE RETURN OF "SHE," H. Rider Haggard. Virtuoso sequel featuring the great mythic creation, Ayesha, in an adventure that is fully as good as the first book, *She*. Original magazine version, with 47 original illustrations by Maurice Greiffenhagen. 189pp. 6½ x 9¼. 23649-8 Pa. $3.50

GEOMETRY, RELATIVITY AND THE FOURTH DIMENSION, Rudolf Rucker. Exposition of fourth dimension, means of visualization, concepts of relativity as Flatland characters continue adventures. Popular, easily followed yet accurate, profound. 141 illustrations. 133pp. 5⅜ x 8½.
23400-2 Pa. $2.75

THE ORIGIN OF LIFE, A. I. Oparin. Modern classic in biochemistry, the first rigorous examination of possible evolution of life from nitrocarbon compounds. Non-technical, easily followed. Total of 295pp. 5⅜ x 8½.
60213-3 Pa. $4.00

PLANETS, STARS AND GALAXIES, A. E. Fanning. Comprehensive introductory survey: the sun, solar system, stars, galaxies, universe, cosmology; quasars, radio stars, etc. 24pp. of photographs. 189pp. 5⅜ x 8½. (Available in U.S. only)
21680-2 Pa. $3.75

THE THIRTEEN BOOKS OF EUCLID'S ELEMENTS, translated with introduction and commentary by Sir Thomas L. Heath. Definitive edition. Textual and linguistic notes, mathematical analysis, 2500 years of critical commentary. Do not confuse with abridged school editions. Total of 1414pp. 5⅜ x 8½.   60088-2, 60089-0, 60090-4 Pa., Three-vol. set $18.50

Prices subject to change without notice.

Available at your book dealer or write for free catalogue to Dept. GI, Dover Publications, Inc., 31 East Second Street, Mineola, N.Y. 11501. Dover publishes more than 175 books each year on science, elementary and advanced mathematics, biology, music, art, literary history, social sciences and other areas.